15th Russian Youth Conference on Physics and Astronomy

(PhysicA.SPb)

Journal of Physics: Conference Series Volume 461

St. Petersburg, Russia
23-24 October 2012

Editors:

Grigorii S. Sokolovskii   **Nikita S. Averkiev**

ISBN: 978-1-62993-098-5
ISSN: 1742-6588

Printed from e-media with permission by:

Curran Associates, Inc.
57 Morehouse Lane
Red Hook, NY 12571

Some format issues inherent in the e-media version may also appear in this print version.

Printed by Curran Associates, Inc. (2013)

For permission requests, please contact the Institute of Physics at the address below.

Institute of Physics
Dirac House, Temple Back
Bristol BS1 6BE UK

Phone: 44 1 17 929 7481
Fax: 44 1 17 920 0979

techtracking@iop.org

Additional copies of this publication are available from:

Curran Associates, Inc.
57 Morehouse Lane
Red Hook, NY 12571 USA
Phone: 845-758-0400
Fax: 845-758-2634
Email: curran@proceedings.com
Web: www.proceedings.com

TABLE OF CONTENTS

PHYSICS OF SOLID STATE AND SEMICONDUCTORS

ASTRONOMY, ASTROPHYSICS AND PHYSICS OF PLASMA

NANOSTRUCTURED AND THIN-FILM MATERIALS

SEMICONDUCTOR LASERS, LIGHT-EMITTING DIODES AND PHOTORECEIVERS

IMPURITIES AND DEFECTS IN SEMICONDUCTORS

HYDRO- AND AERO-DYNAMICS

PHYSICS AND TECHNOLOGY OF ENERGY CONVERSION

PHYSICS AND QUANTUM STRUCTURES

PHYSICS OF ELEMENTARY PARTICLES

NUMERICAL METHODS IN PHYSICS

POSTDEADLINE

Journal of Physics: Conference Series **461** (2013) 011001 doi:10.1088/1742-6596/461/1/011001

Preface

XV Russian Youth Conference on Physics and Astronomy PhysicA.SPb

The fifteenth Russian Youth Conference on Physics and Astronomy PhysicA.SPb was held 23–24 October 2012 in Saint-Petersburg, Russia. The Conference continues the tradition of Saint-Petersburg Seminars on Physics and Astronomy originating from the mid-90s. The main feature of PhysicA.SPb since then, remains the combination of both scientific and educational quality of the contributions delivered to the young audience. This feature makes it possible to combine the whole spectrum of modern Physics and Astronomy within one conference.

PhysicA.SPb 2012 has brought together more than 150 students, young scientists and their professors from many universities and research institutes across Russia, as well as from Belarus, Ukraine, Finland, France and the United Kingdom. Oral and poster presentations were combined into a few well-defined sections among which one should name Astronomy and Astrophysics, Physics of semiconductors, Physics of solid state, Physics and technology of the alternative energetics, Nanostructured and thin-film materials, THz and UHF materials and devices, and Physics of the quantum-sized structures.

This issue of the *Journal of Physics: Conference Series* presents the extended contributions from participants of PhysicA.SPb 2012 that were peer-reviewed by expert referees through processes administered by the Presiders of the Organising and Programme Committees to the best professional and scientific standards.

Grigorii S. Sokolovskii and Nikita S. Averkiev
Editors

Journal of Physics: Conference Series **461** (2013) 011002 doi:10.1088/1742-6596/461/1/011002

Peer review statement

All papers published in this volume of *Journal of Physics: Conference Series* have been peer reviewed through processes administered by the proceedings Editors. Reviews were conducted by expert referees to the professional and scientific standards expected of a proceedings journal published by IOP Publishing.

Journal of Physics: Conference Series **461** (2013) 012001 doi:10.1088/1742-6596/461/1/012001

Nonradiative resonance energy transfer between quantum dots

A S Stepashkina[1], **D M Samosvat, O P Chikalova-Luzina, G G Zegrya**

Ioffe Physic Technical Institute
26 Polytekhnicheskaya, St Petersburg 194021, Russian Federation

E-mail: stepashkina.anna@yandex.ru

Abstract. In the work we examined the mechanism of nonradiactive resonant energy transfer between quantum dots (QD), the probability of this process was calculated. The valence band has difficult structure due to the additional matrix element connected with another polarization of heavy holes. Dependences of transfer probability on distance between quantum dots and barrier heights for electrons were studied.

1. Introduction

The modern medicine passes to new more modern diagnostics methods with usage of quantum dots for detecting of DNA helix hybridization/splitting that can help us to study a genetic code and to warn genetic diseases, to create special DNA chips, to investigate the delivery of drugs to bodies, to detect the location of cancerous growth, etc. In this case it is necessary to construct the theory of this process [1,2]. One of mechanisms of realization of this process is nonradiactive resonant energy transfer. First this process was considered by Forster for interaction between protein molecule [3].

In the work we examined the mechanism of nonradiactive resonant energy transfer between quantum dots (QD), the probability of this process was calculated. The task was considered in Kane model where the nonparabolic zones are considered. The valence band has difficult structure due to the additional matrix element connected with another polarization of heavy holes. Dependences of transfer probability on distance between quantum dots and barrier heights for electrons were studied.

The accounting of the barrier finiteness leads to transitions with participation of electron and hole conditions with different main quantum numbers.

There is a strong dependence of energy transfer probability on distance between QD: $W_{if} \propto \frac{1}{r^6}$. This dependence turns out at consideration of the dipolar allowed transitions. Due to such behavior practical realization of nonradiactive resonant energy transfer is possible: detecting of linkage or not linkage of DNA helixes with the attached QD both in vitro, and in vivo

2. Nonradiactive resonant energy transfer

2.1. Model

[1] To whom any correspondence should be addressed.

Journal of Physics: Conference Series **461** (2013) 012001 doi:10.1088/1742-6596/461/1/012001

Let's consider the system of two QD located from each other at final distance r in the dielectric environment ε. QD are made of the same material. QD have final potential barriers to electrons and holes. In an initial condition the electron in the energy donor QD is in a conduction band (condition 1), and in the acceptor QD the electron in a valence band (condition 3). As a result of the Coulomb interaction ($V = \frac{e^2}{|\mathbf{r} + \mathbf{r}_d - \mathbf{r}_a|}$), where $\mathbf{r}_d$ and $\mathbf{r}_a$ - QD radius), there is the nonradiactive resonant energy transfer: the birth an electron - hole pair in the acceptor due to the transferred excitement (transition of electrons from conditions 1, 3 in conditions 2, 4 respectively – fig.1). In fact, this process can be carried to the Auger process. This process takes place only under condition of energy conjunction of the donor and the acceptor accurate within half-width.

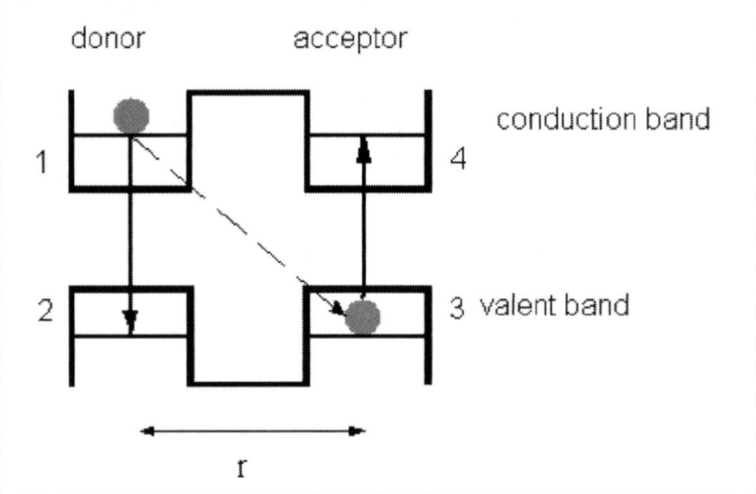

Figure 1. The model of the nonradiactive resonant energy transfer

For resonant energy transfer studying it is necessary to find wave functions of charge carriers. We will calculte wave functions within Kane's three-band model. Energy is counted from a bottom of the conduction zone. backs - orbital interaction is neglected,therefore $\Delta_{SO} = 0$.

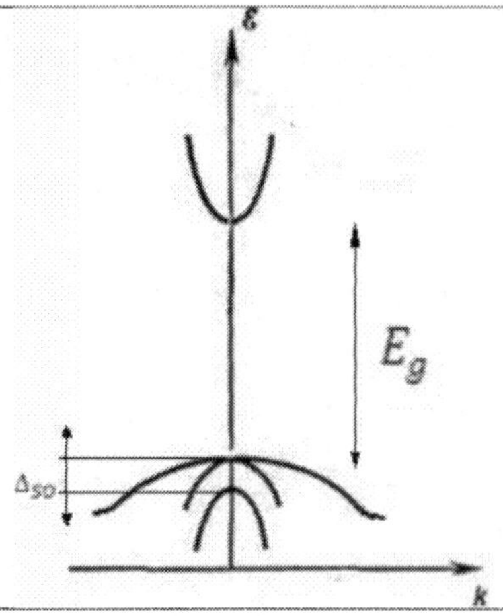

Figure 2. Kane's model

Wave functions can be written down in the following look:

$$\psi = \psi_s |s\rangle + \boldsymbol{\psi}|\mathbf{p}\rangle, \tag{1}$$

where $|s\rangle$ and $|\mathbf{p}\rangle$ are Bloch wave functions. Functions of s-type describe a condition in a conduction band, and p-type describe a condition of a valence band. Functions ψ_s and $\boldsymbol{\psi}$ are envelope wave functions. Kane's equations for envelope functions have the following appearance:

$$\begin{cases} (E_g - E)\psi_s - i\hbar\gamma\nabla\boldsymbol{\psi} = 0 \\ -E\boldsymbol{\psi} - i\hbar\gamma\psi_s - \dfrac{\hbar^2}{2m}(\gamma_1 + 4\gamma_2)\nabla(\nabla\boldsymbol{\psi}) - \dfrac{\hbar^2}{2m}(\gamma_1 - 2\gamma_2)\nabla\times(\nabla\times\boldsymbol{\psi}) = 0, \end{cases} \tag{2}$$

where γ- Kane's matrix element, γ_1 and γ_2 – Lattinzher's generalized parameters, m- free electron mass.

As we have spherical QD, so the equation (3) can be written is spherical coordinates, in this case variables are divided and thewave function can be presented in a radial part and spherical function:

$$\psi_s = R(r)Y_{lm}(\theta,\varphi) \tag{3}$$

The following equation for the radial part $R(r)$ takes place:

$$\frac{1}{r^2}\frac{\partial}{\partial r}\left(r^2\frac{\partial R}{\partial r}\right) + \left(k^2 - \frac{l(l+1)}{r^2}\right)R = 0 \tag{4}$$

For envelope functions of electrons in QD we received the following expressions (wave functions of electrons and light holes):

$$\begin{aligned} &\psi_s = Aj_l(k_e r)Y_{lm}(\theta,\varphi) \\ &\boldsymbol{\psi} = -\frac{i\hbar\gamma}{E}Ak_e\left(\left(\frac{l+1}{2l+1}\right)^{1/2} j_{l+1}(k_e r)\mathbf{Y}_{lm}^{l+1}(\theta,\varphi) + \left(\frac{l}{2l+1}\right)^{1/2} j_{l-1}(k_e r)\mathbf{Y}_{lm}^{l-1}(\theta,\varphi)\right) \end{aligned} \tag{5}$$

Here $\mathbf{Y}_{lm}^{l\pm1}(\theta,\varphi)$ is vector spherical harmonicas, l and m – values of the full angular moment and its projection to an axis z, $j_l(kr)$ – Bessel's spherical function, k_e – a wave vector of electrons, A – a normalizing constant.

By analogy we received wave functions under a barrier. For heavy holes we receive:

$$\begin{aligned} &\psi_s = 0 \\ &\boldsymbol{\psi}_{h2} = Bj_l(k_h r)\mathbf{Y}_{lm}^{l}(\theta,\varphi) \\ &\boldsymbol{\psi}_{h1} = C\left(\left(\frac{l}{2l+1}\right)^{1/2} j_{l+1}(k_h r)\mathbf{Y}_{lm}^{l+1}(\theta,\varphi) + \left(\frac{l+1}{2l+1}\right)^{1/2} j_{l-1}(k_h r)\mathbf{Y}_{lm}^{l-1}(\theta,\varphi)\right) \end{aligned} \tag{6}$$

where k_h – wave vector of holes. B and C – normalizing constants.

Here are boundary conditions fo electrons and light holes:

$$\begin{cases} \psi_s^< = \psi_s^> \\ \dfrac{1}{E_g - E}\dfrac{\partial\psi_s^<(R)}{\partial r} = \dfrac{1}{E_g - E - V}\dfrac{\partial\psi_s^>(R)}{\partial r} \end{cases} \tag{7}$$

for heavy holes

$$\left\{\psi_s \equiv 0, \quad \boldsymbol{\psi}^<(R) = \boldsymbol{\psi}^>(R), \quad \frac{d\psi_\perp^<(R)}{dr} = \frac{d\psi_\perp^>(R)}{dr}\right. \tag{8}$$

Indexes «<» and «>» means that parametrs undertake respectively at the left and the right side of heterobarrier.

2.2. Matrix element

For finding of probability of nonradiactive resonant energy transfer we will use Fermi's "gold" rule within a first order perturbation theory:

$$W_{if} = \frac{2\pi}{\hbar}|M_{if}|^2\delta(E_i - E_f) \tag{9}$$

Our task is reduced to find a matrix element of transition from an initial condition (in the donor there is an electron - hole pair, the acceptor is empty) to the final (the donor is empty, in an acceptor there is an electron - hole pair):

$$M_{if} = \langle\psi_4(r_d)\psi_2(r_2)|V|\psi_1(r_d)\psi_3(r_a)\rangle = \frac{e^2}{\varepsilon}\iint d^3r_d d^3r_a \frac{\psi_3\psi_4^*\psi_1\psi_2^*}{|\mathbf{r}+\mathbf{r_d}-\mathbf{r_a}|} \tag{10}$$

where r_d and r_a are coordinates of the electron - hole pairs in the donor and the acceptor respectively, ψ_1 and ψ_2 wave functions of the initial and final conditions of the electron in the donor, ψ_3 and ψ_4 wave functions of the initial and final conditions of the electron in the acceptor. In our work we consider two types of the hole polarization, two matrix elements are possible, they answer to different polarization: M_{if1} and M_{if2} . Let's consider a matrix element with polarization for heavy holes. For the calculation of a matrix element we will use Fourier representation for Coulomb interaction:

$$\frac{e^2}{|\mathbf{r}+\mathbf{r_d}-\mathbf{r_a}|} = \frac{4\pi e^2}{\varepsilon}\int\frac{d^3q}{(2\pi)^3q^2}\exp(i\mathbf{q}(\mathbf{r}+\mathbf{r_d}-\mathbf{r_a})) \tag{11}$$

Matrix element has the following view

$$M_{if1} = \frac{4\pi e^2}{(2\pi)^3}\int\frac{d^3q}{q^2}\exp(i\mathbf{qr})\int d^3r_d\psi_1\psi_2^*\exp(i\mathbf{qr_d})\left(\int d^3r_a\psi_3\psi_4^*\exp(i\mathbf{qr_d})\right)^* \tag{12}$$

Substituting of wave functions for electrons and heavy holes in this expression, we receive the following type of a matrix element for heavy holes:

$$\begin{aligned}
M_{if1} &= \frac{2e^2}{\varepsilon}\left(\frac{P}{E_g}\right)^2 A_1A_2A_3A_4\frac{1}{r^3} \\
&\int_0^{R_d} r_d^2dr_d\left(\left(\frac{l_{hd}}{2l_{hd}+1}\right)^{1/2} C_{l_{hd}+1,m_{hd},1,0}^{l_{hd},m_{hd}} j_{l_{hd}+1}(k_{hd}r_d)j_{l_{ed}}(k_{ed}r_d)\delta_{l_{ed},l_{hd}+1}\delta_{m_{ed}m_{hd}} + \right. \\
&\left.+\left(\frac{l_{hd}+1}{2l_{hd}+1}\right)^{1/2} C_{l_{hd}-1,m_{hd},1,0}^{l_{hd},m_{hd}} j_{l_{hd}-1}(k_{hd}r_d)j_{l_{ed}}(k_{ed}r_d)\delta_{l_{ed},l_{hd}-1}\delta_{m_{ed}m_{hd}}\right)* \\
&\int_0^{R_a} r_a^2dr_a\left(\left(\frac{l_{ha}}{2l_{ha}+1}\right)^{1/2} C_{l_{ha}+1,m_{ha},1,0}^{l_{ha},m_{ha}} j_{l_{ha}+1}(k_{ha}r_a)j_{l_{ea}}(k_{ea}r_a)\delta_{l_{ea},l_{ha}+1}\delta_{m_{ea}m_{ha}} + \right. \\
&\left.+\left(\frac{l_{ha}+1}{2l_{ha}+1}\right)^{1/2} C_{l_{ha}-1,m_{ha},1,0}^{l_{ha},m_{ha}} j_{l_{ha}-1}(k_{ha}r_a)j_{l_{ea}}(k_{ea}r_a)\delta_{l_{ea},l_{ha}-1}\delta_{m_{ea}m_{ha}}\right)
\end{aligned} \tag{13}$$

The similar expression we received for other polarization of heavy holes is too enormous and doesn't shown

Journal of Physics: Conference Series **461** (2013) 012001 doi:10.1088/1742-6596/461/1/012001

2.3. Discussion of results

Let's consider dependence of probability of nonradiactive resonant energy transfer on the following parameters: distances between QD, height of a potential barrier. Carried out above calculation showed that the probability of nonradiactive resonant energy transfer depends on distance between quantum dots as: $W_{if} \sim \frac{1}{r^6}$ for the dipolar allowed transitions (fig. 3).

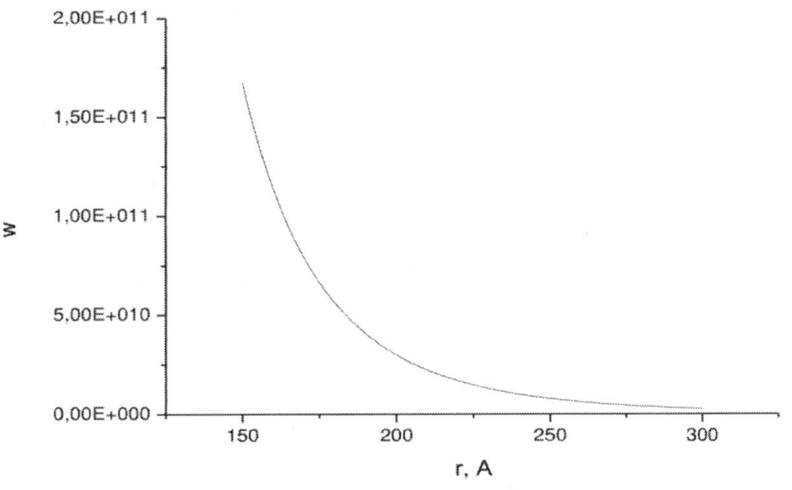

Figure 3. Dependence of probability of nonradiactive resonant energy transfer on distance between quantum dots. Y axis denote energy transfer probability in s^{-1}

Dependence of probability of a recombination on height of a potential barrier. We consider two different cases.

- the electron and the hole are in the basic condition;
- the electron is in the basic condition, and the hole is in the raised.

In case the electron and the hole are in the main condition, dependence of recombination probability grows with increase in height of a barrier. It occurs at the expense of the following effect: if the barrier is higher, the electrons in QD localized stronger therefore the overlap integral becomes stronger and the probability of the process increases (fig. 4).

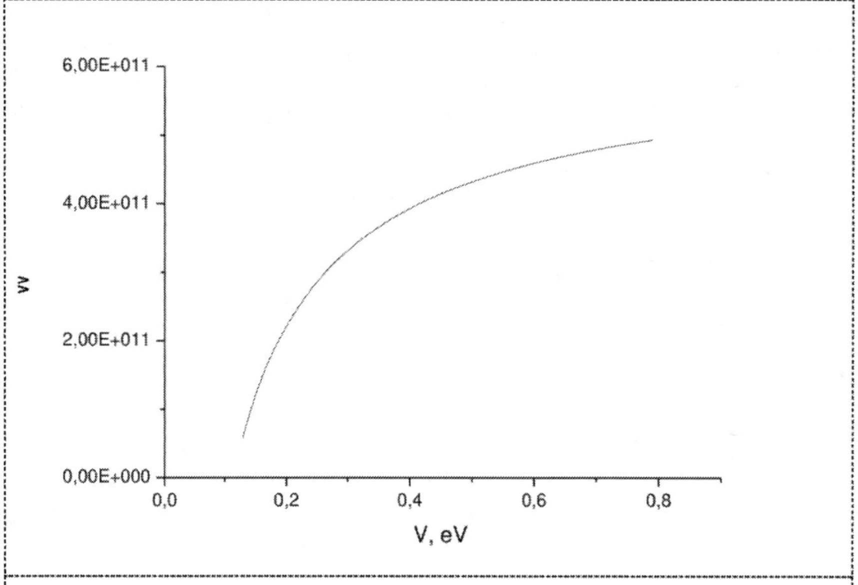

Figure 4. Dependence of probability of nonradiactive resonant energy transfer from heights of a potential barrier to electrons in case the electron and the hole are in the main conditions. Y axis denote energy transfer probability in s^{-1}

In other case if the electron is in the main condition, the hole is in the excited state at some limited height of the potential barrier, the following result turns out. On the graphics the maximum is observed: after some value of barrier height bending-around wave functions with different main quantum numbers become orthogonal (fig. 5).

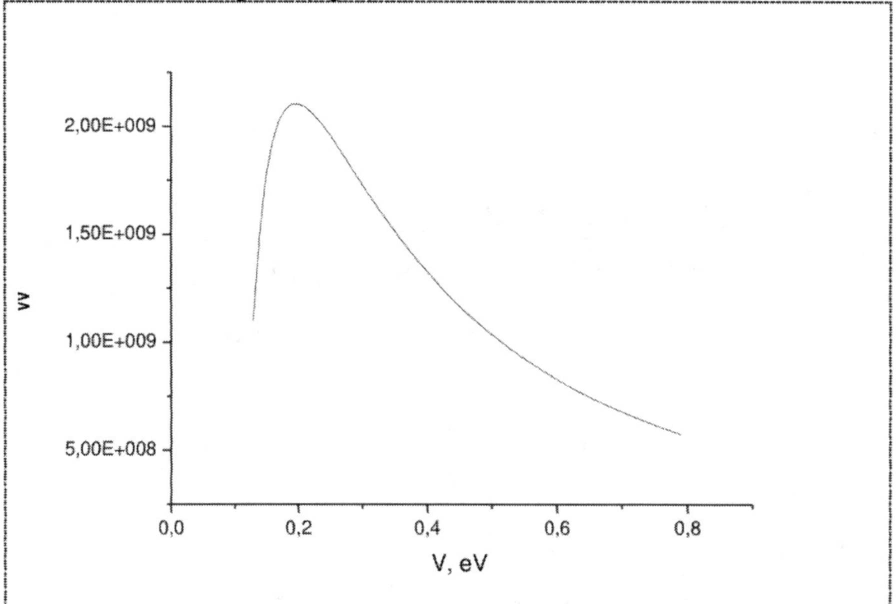

Figure 5. Dependence of probability of nonradiactive resonant energy transfer from heights of a potential barrier to electrons in case the electron is in the main condition, and the hole is in the raised. Y axis denote energy transfer probability in s^{-1}

5. Conclusion

During the carried-out work it was shown that nonradiactive resonant energy transfer from one QD to another is possible. Process has many common features with Forster's theory constructed for interaction between protein molecules. The process was considered in the Kane model where the nonparabolic dispersion law of electrons is considered. The probability of energy transfer was received. Due to the nonparabolic matrix element additional sum appears, it increases probability of the transfer process.

The dependence of recombination probability on distance between QD gives . Also in connection with complcated structure of the valent band there is the additional matrix element connected with other polarization of heavy holes. The accounting of a barrier extremity leads to possibility of transitions with participation of electrons and holes with different main quantum numbers.

The phenomenon of nonradiactive resonant energy transfer finds practical application in a modern science. This theory describes experiments with the colloidal QD surrounded with a layer of a semi-conductor material which creates a potential barrier. Such structures can be used as sensors for indication of biological objects as by in vitro, and in vivo.

References

[1] R Wargnier, Chan, Shuming Nie 1998, *Science.* **281**, , 2016-2018

[2] Bairamov B H , Voitenko V A , Toropov V V, Shi P, Dutta M 2007 Selective Functiolization of Semiconductor Quantum Dots with Short Peptides and Intergins of Cancer Cell for Biophotonic Application in Physics, Chemistry, and Application of Nanostructures, *Nanomeeting-2007*, World scientific, 511-515

[3] Th Forster. 1948, *Ann. Phys.*, 55-57

Journal of Physics: Conference Series **461** (2013) 012002 doi:10.1088/1742-6596/461/1/012002

Nonlinear absorption of femtosecond light pulses in bulk crystals under interband multiphoton resonance conditions

E. Yu. Perlin[1,2], K. A. Eliseev[1], E. G. Idrisov[1,2] and Ya. T. Khalilov[2]

[1]St. Petersburg State University of Information Technologies, Mechanics and Optics, Informational Optical Technologies Center, St. Petersburg, 199034 Russia

[2]Department of Physics and Mechanics, St. Petersburg State Polytechnic University, St. Petersburg, 195251 Russia

E-mail: idrisoft@gmail.com

Abstract. A theory of the nonsteady-state absorption of supershort light pulses in bulk materials and heterostructures with quantum wells and wires when multiphoton resonance at interband transitions occurs is developed.The absorbed energy dependence on the detunings of the multiphoton resonance and on the pulse width is investigated. The interaction between two consecutive short pulses (pump-probe spectroscopy within the femtosecond regime) with bulk crystals and low-dimensional structures under the conditions of two-photon resonance at interband transitions and transitions between size-quantized sublevels is studied.

1. Introduction

Femtosecond light pulses are widely used in studying interaction of powerful laser radiation with transparent insulators and semiconductor nanostructures [1].The carrier time τ_p (known as the relaxation time of momentum of carrier) is usually determined by scattering on longitudinal optical lattice vibrations and ordinarily is hundreds of femtoseconds in bulk crystals. In heterostructures τ_p is also determined by intrasubband transitions due to interaction of electrons with optical phonons. When the pulse width $\sqrt{\sigma} < \tau_p$, conventional methods of calculating the optical transition probabilities, relying on the concept of the number of transitions per unit time, are not adequate. Thus, the formulas obtained for the case of quasi-steady-state electromagnetic fields [2] cannot be used for a consistent interpretation of experimental data concerning the nonlinear response of materials to femtosecond light pulses [3], and other methods are required.

2. Transitions between discrete electron levels

Let a linearly polarized light with frequency ω propagate a lenght L in a medium along x direction. The absorbed power per unit area is then determined by [3]

$$J = -\omega L F_0(t) P_s(t) \tag{1}$$

where $F_0(t)$ is the electric field amplitude of the wave, while $P_s(t)$ is the medium polarization of the reactive nature. We assume that the field does not change significantly as it passes through the lenght L the medium. We take the light pulse in a Gaussian shape:

$$F_0(t) = f_0 \exp\left[-(t-t_0)^2/\sigma\right] \tag{2}$$

Futher to simplify the calculations, we considered the nonlinear polarization for the case of three-level system with discrete levels (for example, quantum dots and impurities) when multiphoton resonance corresponds to a transition between levels 0 and 1, i.e., $E_1 - E_0 = n\hbar\omega$ (Fig.1a). For even values of n, the allowed transitions are $0 \longrightarrow 2$ and $1 \longrightarrow 2$, while, for odd n, the transitions $0 \longrightarrow 1$ and $1 \longrightarrow 2$ are allowed.

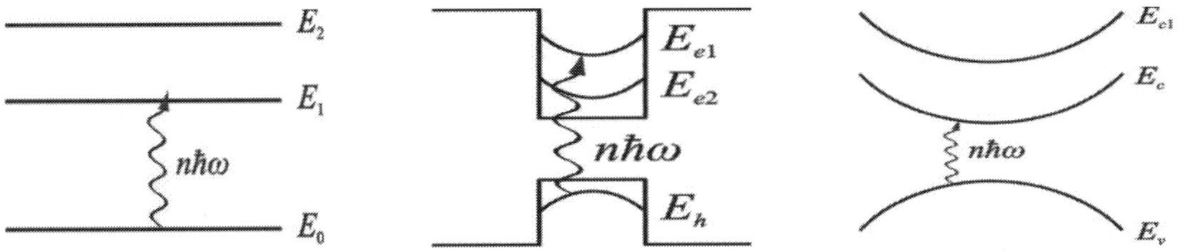

a–dicrete system *b–heterostructure* *c–bulk crystal*

Fig.1. Models of the band materials

In what follows, we also consider absorption in a three-band model of a one-, two- or three-dimensional systems (Fig.1b-c). Using Eq.(1), for the energy absorbed in unit volume the following approximate formula is obtained

$$J_n \approx -\omega \left|M_{if}^{(n)}\right|^2 n_0 \frac{2^{\frac{7n}{2}-\frac{3}{2}}\pi^{\frac{n}{2}}\sqrt{5}e^{2n}}{\varepsilon_\infty^{n/2} c^n \hbar^{2n-1}\sigma^{-1+n/2}} W^n \exp\left(-\frac{\omega_R^2}{n\sqrt{\pi}}\sigma\right)\lambda_n \tag{3}$$

where e is the electron charge, c is the speed of light, ε_∞ is the high-frequency permittivity of the medium, n_0 is the number of three-level systems (quantum dots or impurity atoms) in unit volume, $\omega_R = n\omega - \omega_{fi}$ is the detuning of then n-photon resonance under assumption that $\omega_R \ll \omega$, W is the total energy passing through a unit area during the laser pulse, $M_{if}^{(n)}$ are the n-order compound matrix elements of the electron coordinate operator between the initial i and final f states [4] and

$$\lambda_n = \sum_{p=1}^{n} \exp\left(-\Theta_p^{(n)}\gamma\sqrt{\sigma}\right) \tag{4}$$

The following approximations for the quantities entering Eq.(4) can be made

$$\Theta_1^{(n)} \approx \frac{1}{8}(n+9), \Theta_2^{(n)} \approx \frac{1}{3\sqrt{2\pi}}\left(1+3\sqrt{7n}\right), \Theta_3^{(n)} \approx \frac{6\sqrt{2\pi}}{13}\left(1+\frac{2}{\sqrt{\pi n}}\right),$$

$$\Theta_4^{(n)} \approx \frac{1}{3\sqrt{\pi}}\left(1-\frac{23\sqrt{3\pi}}{2n}\right), \Theta_5^5 \approx \frac{1}{8}\sqrt{\frac{37\pi}{3}}.$$

Let us note, that the preexponential factor in dependence (3) is proportional to $\sigma^{1-n/2}$. In addition, the σ dependence is contained in the exponent and in the factors λ_n in the right-hand side of Eq. (3). While, in the case of long pulses ($\gamma\sqrt{\sigma} \gg 1$) the absorbed energy is proportional to $\sigma^{1-n/2}$ only. *Thus, for systems with discrete levels, including quantum dots, impurity centers, etc., the nonlinear multiphoton resonance response to ultrashort and longer laser pulses, may be essentially different.*

3. Nonlinear multiphoton resonance response of heterostructures with a continuous spectrum

In the case of interband (intersubband) transitions, the electron system of a crystal can be approximately represented as a collection of noninteracting three-level systems each of which is characterized by its own wave vector $\boldsymbol{k}$ (three-, two-, or one-dimensional) and the corresponding energies $E_i(\boldsymbol{k})$.

Let us denote the dimensionality of the structure by D ($D = 3$ in the case of bulk crystals, $D = 2$ for quantum wells, and $D = 1$ for quantum wires). Then, the energy of the ultrashort pulse absorbed by the D-dimensional system under n-photon resonance conditions is determined by the formula

$$J_n^{(D)} = 2\Xi_D \int_0^\infty dk k^{D-1} J_n(k), \tag{5}$$

where $\Xi_1 = Ln_1/(2\pi), \Xi_2 = Sn_2/(2\pi), \Xi_3 = \Omega/(4\pi)$, L is the length of quantum wire, S is the area of quantum well, n_1 is the number of quantum wires per unit area normal to their direction, and n_2 is the number of quantum wells per unit length in the growth direction. Integration over k space is carried out by the saddle-point method. For the energy absorbed in unit volume from (5) we get

$$J_n^{(D)} = \eta_D \frac{2^{(7n-4)/2}\pi^{(2n-1)/4}5^{1/2}n^{1/2}\omega\mu^{1/2}e^{2n}}{\varepsilon_\infty^{n/2}c^n\hbar^{(4n-1)/2}\sigma^{(n-1)/2}} W^n\lambda_n \left|M^{(n)}\right|^2. \tag{6}$$

where

$$\eta_1 = n_1/\Delta_n^{1/2}, \eta_2 = (2\mu/\hbar)^{1/2}n_2, \eta_3 = 2\mu\Delta_n^{1/2}/(\pi\hbar)$$

and μ is the reduced mass of electron and hole and $\hbar\Delta_n = n\hbar\omega - \tilde{E}$, $\omega_R = \Delta_n - \frac{\hbar k^2}{2\mu}$. In the case of three-dimensional crystal,$\tilde{E} = E_g$, where E_g is the bandgap width. In the case of quantum wells or quantum wires, $\tilde{E}$ stands for the energy gap between initial and final subbands of quantum confinement at $\boldsymbol{k} = 0$.

It should be noted that the dependence of the absorbed energy on the detuning of resonance from the fundamental band edge, Δ_n, leads to the same result as in the case of quasi-steady-state field:

$$J_n^{(3)} \propto \sqrt{\Delta_n} \tag{7}$$

This Eq. corresponds to the density of electron states for bulk crystal ($D = 3$). Similarly, it is found that the dependence of the absorbed energy on duration of field action also is the same as in the case of quasi-stationary.

Here, the situation differs from the case of a system with discrete levels. It is clear to understand if we take into account the following reasons. When we consider the continuous electron spectrum, all the spectral components that form the ultrashort light pulse, participate in the resonant multiphoton transitions on interband transitions, each Fourier component in the proper point of the Brillouin zone. For this reason, the situation becomes similar to the case of steady-state absorption of monochromatic light.

The dependences of the absorbed energy on the pulse width in the case of low-dimensional structures ($D = 1, 2$) have the same character as in bulk crystals. The explanation of this fact is the same as in the case of bulk materials.

The dependences of the absorbed energy on the detuning from resonance Δ_n in the case of two- and one-dimensional structures again coincides with their quasi-steady-state analogs. For example, in two-photon absorption process $J_2^{(1)} \propto \Delta_2^{-1/2}$ and $J_2^{(2)} \propto const$ for $D = 1, 2$, respectively.

Journal of Physics: Conference Series **461** (2013) 012002 doi:10.1088/1742-6596/461/1/012002

4. Pump-Probe spectroscopy within the femtosecond two-photon resonance regime

In recent years, in experimental studies of multiphoton absorption by wide-band semiconductors and insulators the method of the femtosecond "pump-probe" spectroscopy is used widely [5]. Pump-probe spectroscopy within the femtosecond regime uses two consecutive laser pulses. The first pulse (known as pump) exites the sample, the second pulse (known as probe) probes the change in the optical properties, due to action of the first powerful pump pulse. Let the system interact with two consecutive ultrashort pulses with frequencies ω_1 and ω_2, respectively. The time delay between the pulses is of the same order as the pulse durations, that is $\tau \leq 100$ fs. The lights are linearly polarized and they propagate in optically thin medium, i.e. single-photon transitions in this medium are absent. Two-photon resonance corresponds to a transition between levels 0 and 1, i.e., $E_1 - E_0 = \hbar(\omega_1 + \omega_2)$. In the dipole approximation the interaction operator is

$$\hat{V}_t = \sum_{i=1}^{2} \hat{d} F_{\omega_i}(t) e^{-\imath(\omega_i t - k_{\omega_i} z)} + H.c. \tag{8}$$

where k_{ω_i} are the wave vectors of light pulses. So, using the formulas (1) we can write the final expression for absorbed energy from probe pulse in unit volume

$$J^{(0)}_{2,pp}(\omega_R, \tau, \sigma_1, \sigma_2, \gamma) \approx \sum_{i=1}^{2} M_i f_i, \tag{9}$$

where

$$M_i = n_0 \omega_2 \frac{\sigma_1 \sigma_2}{\sigma_1 + \sigma_2} \sqrt{\pi} \frac{e^4 (x_{20} x_{21})^2}{\hbar^3} \left(\frac{32\pi W_1 W_2}{c^2 \varepsilon_\infty \sqrt{\sigma_1 \sigma_2}} \right) \left(\frac{1}{\omega_{20} - \omega_i} \right) \left(\frac{1}{\omega_{20} - \omega_2} + \frac{1}{\omega_{20} - \omega_1} \right),$$

$$f_i = \frac{71}{40} \exp\left[-\frac{2(\tau + a_i \sigma_1 \gamma/4)^2}{\sigma_1 + \sigma_2} \right] \exp\left[\frac{2(a_i \sigma_1 \gamma)^2}{\sigma_1 + \sigma_2} - b_i \gamma \sqrt{\sigma} \right] \exp\left[-\frac{\sigma \omega_R^2}{2} \right]$$

and $\omega_{ij} = (E_j - E_i)/\hbar$ are the frequencies of the transitions between the jth and ith states in the electron system, ω_R is the detuning of two-photon resonance, $\sqrt{\sigma_i}$ are the pulse widths. $a_1 = 47/16, b_1 = 5/46, a_2 = 251/86, b_2 = 21/328$. These coefficients are obtained by numerical evaluation.

Using Eq.(9) we get the second pulse energy absorbed by D-dimensional system

$$J^{(D)}_{2,pp} = \Delta_2^{(D-2)/2} \frac{\alpha^{(D)} \mu^{D/2} e^4}{\hbar^{3+D/2}} \frac{\left(x^{(D)}_{20} x^{(D)}_{21} \right)^2 W_1 W_2}{c^2 \varepsilon^{(D)}_\infty} \frac{\omega_2}{\sqrt{\sigma_1 + \sigma_2}} \Omega_{ij} S_{ji}, \tag{10}$$

where we introduced the following notations

$$S_{22} = f_1, S_{11} = f_2, S_{12} = S_{21} = (f_1 + f_2)/2 \tag{11}$$

and f_i is the same as in formula (9) and

$$(\omega_{20} - \omega_i)^{-1} (\omega_{20} - \omega_j)^{-1} = \Omega_{ij}. \tag{12}$$

In Eq.(10) the summation over dummy indices $i, j = 1, 2$ is implied. $\alpha^{(D)}$ are the numerical coefficients for the D-dimension structures: $\alpha^{(1)} = 568/5, \alpha^{(2)} = 284\sqrt{2\pi}/5, \alpha^{(1)} = 568\pi/5$. For one- and two- dimensional systems the right hand side of formula (10) must be multiplied by the number of quantum wires per unit area normal to their direction, n_1 or by the number of quantum wells per unit length in the growth direction, n_2, respectively. It should be noted that the terms $a_i\sigma_1\gamma/4$ in both exponents in Eqs.(9) and (10) are closely related to polarization induced by the high-powered pump pulse.

The expressions obtained above show that the dependence of the absorbed energy on detuning of two-photon resonance from probe pulse has Gaussian form, i. e. $J_{2,pp}^{(0)} \propto \exp(-\sigma\omega_R^2/2)$.

The absorbed energy dependence on the delay between two pulses also has Gaussian shape.

5. Conclusion

We analyzed the nonlinear response of bulk materials and heterostructures with quantum wells, wires, and dots to supershort light pulses with widths less than the intraband (intrasubband) electron or hole relaxation time. The dependence of the absorbed energy on the detunings of multiphoton resonances is obtained for nanostructures of different dimensions. Similarly, the dependence of the absorbed light energy on the width of a supershort pulse with fixed energy per pulse is investigated. It is shown that these dependences for zero-dimensional objects (quantum dots, impurity centers) substantially differ from those that occur in the case of relatively long pulses. This circumstance can be imporatant in interpretation of experimental data on the nonlinear response of heterostructures with quantum wells to supershort light pulses. The dependence of the absorbed energy on the detuning of the two-photon resonance and on the delay between the pump and probe pulses is also studied. It is shown that in this case the main role in absorption of supershort pulses plays the polarization induced by pump pulse.

References

[1] M. Lenzner, J. Kruger, S. Sartania, Z. Cheng, C. H. Spielmann, G. Mourou, W. Kautek, and F. Krausz 1998 *Phys. Rev. Lett.* **80** 4076
[2] L.V. Keldysh 1964 *Zh. Eksp. Teor. Phys.* **47** 1945-1957
[3] E. Yu. Perlin 1994 *Zh. Eksp. Teor. Phys.* **105(1)**186
[4] E. Yu. Perlin, K. A. Eliseev, E. G. Idrisov and Ya. T. Khalilov 2012 *Opt. Spektrosk.* **112** 920-927
[5] Cerullo G, De Silvestri, Nisoli M, Sartania S, Svelto O 2000 *J. of Selected Topics in Quantum Electronics* **6** **(6)** 948-958

Journal of Physics: Conference Series **461** (2013) 012003 doi:10.1088/1742-6596/461/1/012003

Thermodynamics of binary solutions of the eutectic type with several intermediate phases of constant composition

G A Panov[1] **and M A Zakharov**[2]

[1] Department of General and Experimental Physics, Novgorod State University, 173003, Novgorod the Great, Russia

[2] Department of Solid State Physics and Microelectronics, Novgorod State University, 173003, Novgorod the Great, Russia

E-mail: mef.gena@rambler.ru

Abstract. An universal method is proposed for calculating the phase diagrams of binary solutions with several intermediate phases of constant composition. The proposed method is based on the nonlinear transformations of the concentration axes and makes it possible to correctly reduce the calculation of the phase diagrams with several intermediate phases to a sequence of phase diagrams of individual subsystems. The thermodynamics of binary solutions of the eutectic type with several intermediate phases of constant composition is described using the proposed method within the framework of the generalized lattice model. The results of the calculations performed are compared with phase diagrams of real binary solutions.

1. Introduction

Calculation of phase diagrams is one of the most interesting problem of physics of multicomponent systems [1, 2]. The present work is devoted to the development of a new direction in the theory of solutions with chemical bonding. The purpose of this study was to devise a consistent method for calculating the phase diagrams of binary solutions with several intermediate phases of constant composition. The proposed method is based on nonlinear transformations of concentration variables of solutions. These transformations allow one to correctly reduce the calculation of the aforementioned phase diagrams to a sequence of phase diagrams of individual subsystems.

2. Method

Let us consider the phase diagram of an A–B binary solution with a negligible solubility of the components in the solid state in the presence of two intermediate phases of constant composition (i.e., a congruently melting chemical compounds $A_{m_1}B_{n_1}$ and $A_{m_2}B_{n_2}$). Without a loss in the generality of the subsequent analysis, we examine a binary system of the eutectic type. The phase diagram of this system is schematically presented in Figure 1. It is easy to see that the left, central and right parts of the initial phase diagram are similar to a conventional phase diagram with a simple eutectic. The fundamental difference between all parts of the phase diagram in Figure 1 and the phase diagram with the simple eutectic is that the mole fractions x of the B

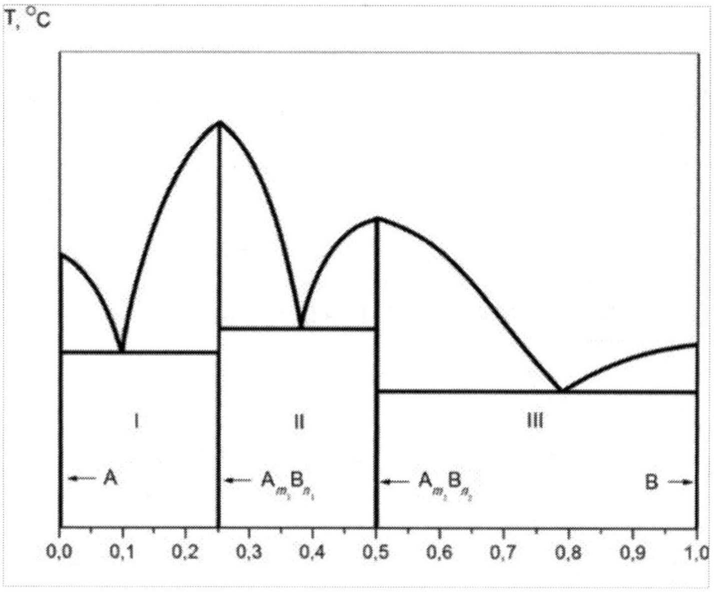

Figure 1. Phase diagram of the eutectic type with two intermediate phases of constant composition in the absence of the mutual solubility of the components in the solid state.

component (reckoned along the concentration axis) vary in the range $0 \leq x \leq n_1/(m_1+n_1)$ for the left subsystem A–$A_{m_1}B_{n_1}$, in the range $n_1/(m_1+n_1) \leq x \leq n_2/(m_2+n_2)$ for the central subsystem $A_{m_1}B_{n_1}$–$A_{m_2}B_{n_2}$ and in the range $n_2/(m_2+n_2) \leq x \leq 1$ for the right subsystem $A_{m_2}B_{n_2}$–B. As a consequence, the lengths of the segments that correspond in the concentration axis to the left, central and right parts of the diagram are not equal to unity, whereas the length of the concentration axis in any phase diagram should be equal to unity. These differences do not permit one to directly use the well-developed methods for calculating phase diagrams in the case of simulation of the components of phase diagrams for systems with intermediate phases. Nevertheless, this problem can be solved with the corresponding change in the concentration variables according to the procedure described below.

At the beginning we shall consider the central part of Figure 1. Here, the chemical compound $A_{m_2}B_{n_2}$ considered an independent molecular component of a new binary system $A_{m_1}B_{n_1}$–$A_{m_2}B_{n_2}$. Then, the mole fractions of the molecular components can be written as follows:

$$y = \frac{N_{A_{m_2}B_{n_2}}}{N_{A_{m_1}B_{n_1}+A_{m_2}B_{n_2}}}, \qquad 1-y = \frac{N_{A_{m_1}B_{n_1}}}{N_{A_{m_1}B_{n_1}+A_{m_2}B_{n_2}}}, \tag{1}$$

where $N_{A_{m_1}B_{n_1}}$ and $N_{A_{m_2}B_{n_2}}$ are the total numbers of particles of the molecular components of the $A_{m_1}B_{n_1}$–$A_{m_2}B_{n_2}$ binary system, y and $1-y$ are the mole fractions of the molecular components $A_{m_2}B_{n_2}$ and $A_{m_1}B_{n_1}$, respectively. From the above expression, we have

$$N_{A_{m_2}B_{n_2}} = \frac{y}{1-y} N_{A_{m_2}B_{n_2}}. \tag{2}$$

On the other hand, the old independent concentration variable x of the A–B initial solution is defined as follows:

$$x = \frac{\bar{N}_A}{\bar{N}_A + \bar{N}_B}, \tag{3}$$

where

$$\bar{N}_A = m_1 N_{A_{m_1}B_{n_1}} + m_2 N_{A_{m_2}B_{n_2}}, \qquad \bar{N}_B = n_1 N_{A_{m_1}B_{n_1}} + n_2 N_{A_{m_2}B_{n_2}}. \tag{4}$$

From equations (1)–(4), we obtain

$$x(y) = \frac{n_1 + y(n_2 - n_1)}{m_1 + n_1 + y(m_2 + n_2 - m_1 - n_1)}, \qquad y(x) = \frac{x(m_1 + n_1) - n_1}{n_2 - n_1 - x(m_2 + n_2 - m_1 - n_1)}. \tag{5}$$

Expressions (5) describe a sought one-to-one correspondence between the old (x) and new (y) concentration variables for the central system $A_{m_1}B_{n_1}$–$A_{m_2}B_{n_2}$. In additional, we obtain the correct limiting values $y(n_1/(m_1+n_1)) = 0$ and $y(n_2/(m_2+n_2)) = 1$ and the desired range of the new variable $0 \le y \le 1$.

The rest binary systems (A–$A_{m_1}B_{n_1}$ and $A_{m_2}B_{n_2}$–B) are considered in a similar way. In particular these cases were studied in detail in paper [3]. For this reason, we do not write out the explicit expressions for the nonlinear transformations of the concentration axes, which can be found in the paper mentioned above.

It should be noted that transformations (5) were derived without recourse to any specific model of a solution, and, in this sense, they are universal. It is evident that relations (5) can be used for calculating not only eutectic phase diagrams but also any phase diagrams with several intermediate phases of constant composition. Below, we will illustrate how these relationships can be used to calculate the phase diagrams with several intermediate phases of constant composition in terms of the generalized lattice model [4, 5].

3. Phase diagram calculation

The thermodynamics of binary solutions of the eutectic type with several intermediate phases of constant composition was described using the proposed method within the framework of the generalized lattice mode [4, 5] (see also [6]).

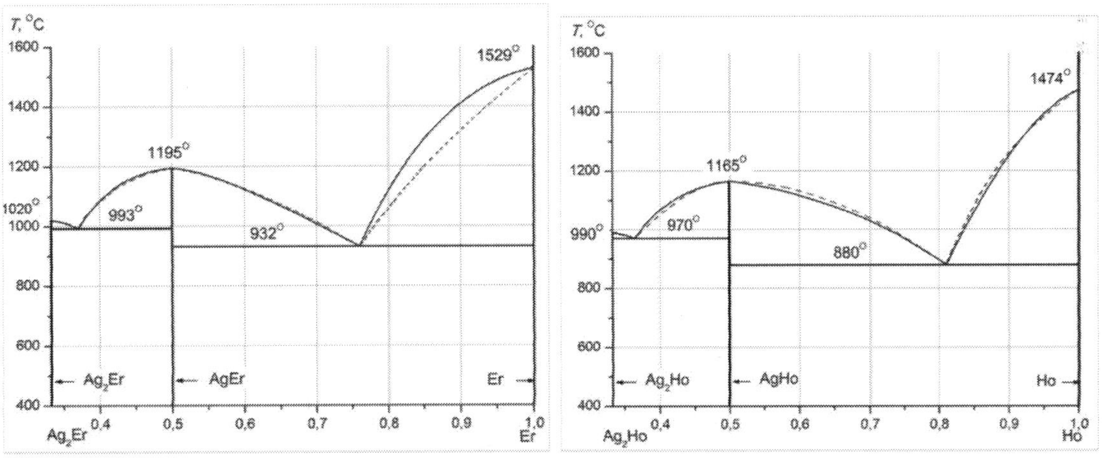

Figure 2. Phase diagrams of the Ag–Er and Ag–Ho systems. Dashed lines are the experimental data taken from [7–10] . Solid lines indicate the results obtained from theoretical calculations.

The corresponding phase diargam are shown in Figure 2. It can be seen from Figure 2 that the phase diagrams are consistent with the experimental data available in the literature [7–10].

In conclusion, we briefly note the prospects for further development of the proposed theory. First, in the model, the possible dissociation of the compounds was ignored. However, it is well known that the dissociation of chemical compounds has a substantial effect on the liquidus curves in the vicinity of the dystectic point. Second, it is interesting to investigate systems with intermediate phases of variable composition in the case when the components are mutually soluble in the solid state.

References

[1] Lukas H L, Fries S G and Sundman B 2007 *Computational Thermodynamics, CALPHAD Method* (Cambridge: Cambridge University Press)
[2] Spencer P J 2008 *CALPHAD* **32** 1
[3] Zakharov M A 2007 *Phys. Solid State* **49** 2312
[4] Zakharov A Yu and Terekhov S V 1985 *Fiz. Met. Metalloved.* **95** 261
[5] Korzun E L and Terekhov S V 1987 *Zh. Fiz. Khim.* **61** 1186
[6] Zakharov A Yu, Zakharov M A and Lebedev V 2005 *Int. J. Quant. Chem.* **104** 133
[7] Gschneidner-Jr K A and Calderwood F W 1985 *Bull. Alloy Phase Diagrams* **6** 17
[8] Gebhardt E and von Erdberg M 1966 *J. Less-Common Met.* **11** 141
[9] McMasters O D , Gschneidner-Jr K A and Venteicher R F 1970 *Acta Crystallogr.* **26** 1224
[10] Gschneidner-Jr K A and Calderwood F W 1985 *Bull. Alloy Phase Diagrams* **6** 140

Semiconductor nanostructure properties. Molecular Dynamic Simulations.

N I Podolska[1,2,3] **and A I Zhmakin**[1,2]

SPb Branch of Joint Supercomputer Center of the Russian academy of sciences
Politekhnicheskaya str. 26, 194021, Saint-Petersburg, Russia

Ioffe Physical-Technical Institute of the Russian academy of sciences,
Politekhnicheskaya str. 26, 194021 Saint-Petersburg, Russia

Saint-Petersburg Academic University of the Russian academy of sciences
Khlopina str. 8/3, 194021, Saint-Petersburg, Russia

E-mail: n.i.podolska@gmail.com

Abstract. The need for research is based on the fact that development of non-planar semiconductor nanosystems and nanomaterials with controlled properties is an important scientific and industrial problem. So, final scientific and technological problem is the creation of adequate modern methods and software for growth and properties simulation and optimization of various III-V (GaAs, InAs, InP, InGaAs etc.) nanostructures (e.g. nanowires) with controlled surface morphology, crystal structure, optical, transport properties etc. Accordingly, now we are developing a specialized computer code for atomistic simulation of structural (distribution of atoms and impurities, elastic and force constants, strain distribution etc.) and thermodynamic (mixing energy, interaction energy, surface energy etc.) properties of the nanostructures. Some simulation results are shown too.

1. Introduction

Nanowires based on semiconductor materials have recently drawn much attention due to their importance for nanoscale electronics and photonics [1]-[6]. These nanowires are usually fabricated via the Vapor-Liquid-Solis (VLS) growth mechanism [7] developed long ago for micrometer scale "whiskers" [8,9]. Using modern epitaxy techniques such as MOCVD [5,7,10] or MBE [11,12], nanowires or nanowhiskers (NWs) with diameters of several tens nanometers and length up to tens microns can be obtained. The fundamental principle underlying the VLS mechanism is the catalytic effect of a liquid metal droplet (Au, for example) assisting the NW growth from a supersaturated liquid alloy [8]. The integration of NWs components into working devices requires a high degree of control that is still lacking, and many fundamental questions concerning NW nucleation and growth remain to be answered (for instance, chemical potential in the droplet, NW and droplet surface energies, solid-liquid or/and liquid-vapor interface energies etc).

The Vapor-Liquid-Solid mechanism of NWs nucleation and growth are usually investigated using continuous medium theories of crystal growth [13]-[17]. It should be noted, that many parameters (for example, chemical potentials, surface energies, interface energies etc) have not been determined experimentally or using some continuous medium theories and their effect on nanostructures growth is

Journal of Physics: Conference Series **461** (2013) 012004 doi:10.1088/1742-6596/461/1/012004

not clear. Consequently, numerical simulations should be able to probe the NW nucleation and growth processes in their full atomistic detail.

2. Valence Force Field for Nanostructures

Since the high number of atoms (about thousands to hundreds of thousands) are involved in the nucleation and growth of NWs, *ab initio* models become prohibitively expensive. Therefore, most of the atomistic simulations (such as molecular dynamics and Monte Carlo) are based on an empirical interatomic potential, which can be fitted to experimental and/or *ab initio* data. Both molecular dynamics and Monte Carlo are important and very powerful methods for a study of atomic processes in low-dimensional (less than ten micrometers) systems. These methods allow study of kinetics of atomic processes in complex physical and chemical systems [18-40].

Strain has significant effect on the electronic and optical properties of lattice mismatched semiconductor heterostructures such as quantum dots or nanowires. In inhomogeneous systems, the strain can be conveniently computed with continuum elasticity theory, which is believed to hold down to nanometer-size systems [24]. These strains can then be used directly as input for electronic structure calculations within the effective mass or **kp** approximations [26, 25], which are also continuous medium theories. The strains do not, however, provide enough information for atomistic electronic structure methods such as tight-binding [27, 28] or pseudo-potentials [29, 30], which need atomic positions as input. Indeed, the atomic positions can hardly be reconstructed from the strains, whose analytic structure is very complex around, e.g., surfaces and interfaces, and which do not characterize the internal strains [31] within the unit cell (i.e., the displacement of one sublattice with respect to an other). Therefore, the atomic positions used as input for such methods are usually computed with semi-empirical force fields such as Keating's Valence Force Field (VFF) [32], Tersoff [33], or Stillinger-Weber potentials [34].

One of the powerful methods for modelling structural and thermodynamic properties of III-V semiconductor solid solutions (including both bulk and nanostructures) is Valence Force Field [19-24] and Keating potential as an empirical interatomic potential (See Figure 1, for example). Since the Keating's VFF model only accounts for elastic relaxation (no kinetic of dislocations), the comparison between the atomic positions obtained by the simulation and the positions obtained by an experimental method like TEM can suggest the presence or absence of dislocations in a system [22-24]. The strain distribution of differently shaped quantum dots, quantum dots and disks in nanowires, heterostrucutres in nanowires and others can also be simulated by using VFF too.

The popularity of VFF models for studying the thermodynamic properties of semiconductor alloys is due to a few reasons. First, even in the approximate form, the VFF approach has been proved to provide quantitatively accurate estimates for the mixing energies of conventional III-V compounds [25,26] and III-N compounds [19-24, 38-40]. Second, the VFF models invoke a limited number of parameters, most of which can be found directly from the experimental data on binary constituents of the alloy. Third, even with large atomic clusters, the VFF simulations can be now routinely performed on personal computers. Forth, Keating's VFF model provides a good balance between accuracy and efficiency.

3. Valence Force Field Simulations

Sometimes ago, we have derived a statistical model of ternary $A^{III}B^{V}$ alloys by using a new methodology, based on numerical calculations of their configuration partition function [38-40]. This statistical model allows calculating chemical potentials, mixing energy, mixing enthalpy, mixing entropy, phase diagrams etc. To find the mean mixing energy and the density of states in an alloy, the VFF simulations have been carried out with the Keating potential. The use of multiple computation runs (N_{exp}) is found to improve the representative statistics of the VFF computations, which is necessary for an accurate prediction of the mixing energy dispersion. The computational runs were performed for the same solid composition, differing in the random group-III atomic distributions in the cation sublattice and/or group-V atomic distributions in the anion sublattice.

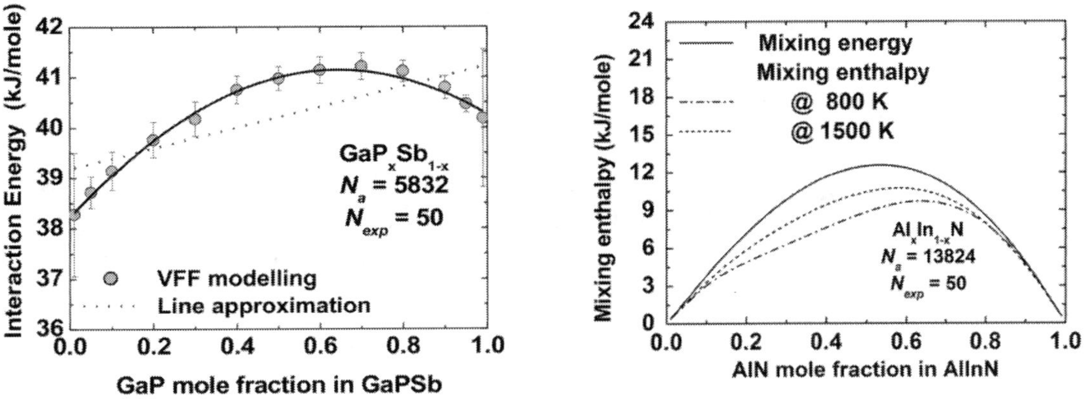

Figure 1. (a) Interaction energy as a function of GaP mole fraction in GaPSb. (b) Mixing energy and mixing enthalpy as functions of AlN mole fraction in AlInN. Number of atoms N_a is 5832 and 13824 and number of computation runs N_{exp} is 50.

Thus, this developed earlier optimized VFF method and this new methodology can be applied also to nanostructures, and not only to bulk materials. To do this, the boundary conditions for surface atoms of a modeling structure should be changed from periodic boundary conditions on free surface condition because it is no longer necessary to reduce the contribution of the surface atoms. In other aspects the simulation method is the same. The initial atomic positions are chosen to match those of a virtual crystal with the lattice constant dependent on the alloy composition via the Vegard law (the so-called "strained" modelling crystal). Both anions and cations are then allowed to relax lowering the distortion energy of the domain determined from the interatomic potential. The relaxation is allowed until the total distortion energy was converged with a relative accuracy of 10^{-5}. After minimizing the distortion energy we get the so-called "relaxed" modelling crystal.

4. Some Simulation Results

As an example, some surface energy numerical calculations will be shown here. Surface energy E_{surf} quantifies the disruption of intermolecular bonds that occurs when a surface is created. In the physics of solids, surfaces must be intrinsically less energetically favourable than the bulk of a material (the molecules on the surface have more energy compared with the molecules in the bulk of the material), otherwise there would be a driving force for surfaces to be created, removing the bulk of the material. The surface energy may be calculated as the excess energy at the surface of a material compared to the bulk $E_{surf}=(E_{PBC}-E_{FBC})/S_{free}$, where E_{PBC} is the energy of the modelling crystals with the periodic boundary conditions (PBC), E_{FBC} is the energy of the modelling crystals with the free surface boundary conditions (FBC), and S_{free} is the free surface area (here is the surface area of the cubic of side L). Table 1 shows the values of free surface area of cubic crystals with a different number of atoms. It must be remembered that *ab initio* calculations and simulations can be held only for crystals with atoms of less than 1000 whereas a III-V cubic crystal with a side of less than 10 nm composed of about 32000 atoms.

Figure 2 shows that $(E_{PBC}-E_{FBC})$ and free surface area are functions of number of atoms in the cubic modelling crystals, i.e. the surface energy will be a function of the free surface area. This means that the surface energy calculated for the modelling crystals with various sizes of crystals (number of atoms and/or free surface area) can vary greatly.

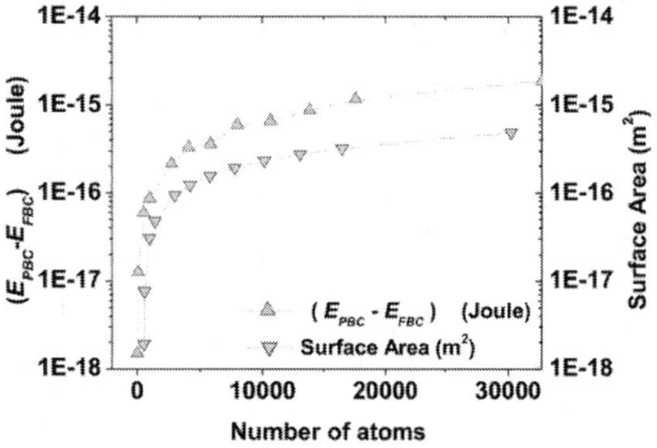

Figure 2. Excess energy at the surface of a material (cubic GaAs crystals, in this case) compared to the bulk and surface area as functions of number of atoms in modelling crystals.

Na	Cube side, A	S_{free}, m²
8	5.65	$1.92\cdot10^{-18}$
64	11.31	$7.67\cdot10^{-18}$
512	22.61	$3.07\cdot10^{-17}$
1000	28.27	$4.8\cdot10^{-17}$
2744	39.57	$9.4\cdot10^{-17}$
...	...	...
32768	90.45	$4.91\cdot10^{-16}$

Table 1. Number of atoms (*Na*) in the cubic modelling crystals, length of the cube side, and surface area of the cubes.

We can see in Figure 3 surface energy as a function of number of atoms and/or free surface area, computed for zinc-blend InAs (111) by Valence Force Field (see Section 3). It can be seen that the surface energy of the "strained" crystals (see Figure 3a) does not vary with the free surface area of the crystals and is 0.47 ± 0.01 J/m². In the case of the "relaxed" crystals (Figure 3b) surface energy is a function of the free surface area. A similar behaviour of the surface energy is also observed for other materials, such as GaAs (111) (see Figure 4).

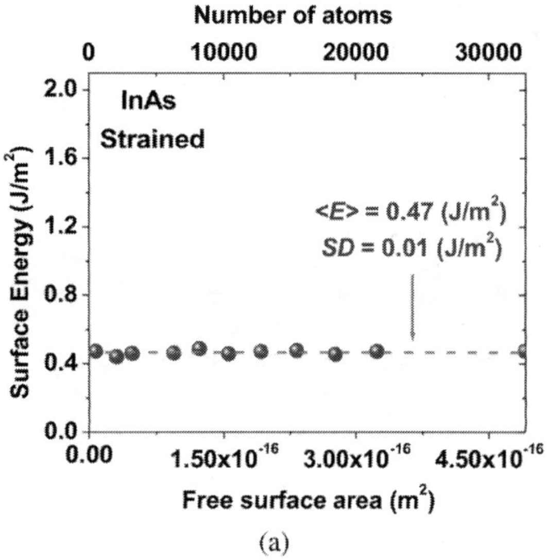

(a)

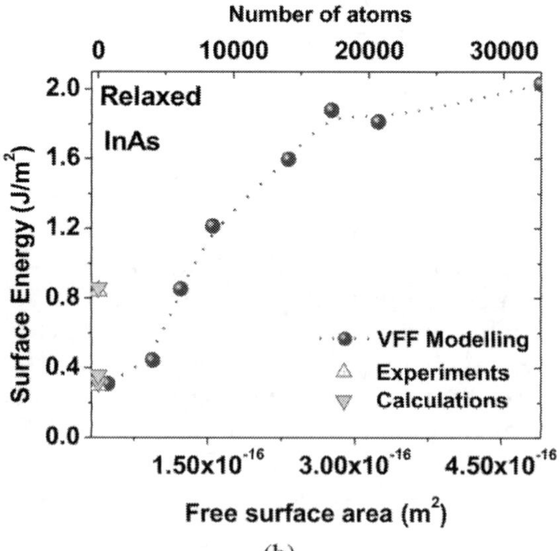

(b)

Figure 3. Surface energy as functions of free surface area and number of atoms in modelling crystals (blue circles) as well as average surface energy $<E>$ with standard deviation SD (red dashed line). Green downward triangles (∇) are *ab initio* simulations and theoretical calculations and yellow upward triangles (Δ) are experimental results [41-43].

In Figure 3 and Figure 4 it can be seen a good agreement simulations with experiments and other calculations [41-43] for small number of atoms. However, apparently small number of atoms of modelling crystals is not enough to estimate surface energy of real nanosystems, especially of systems like nanoneedls, quantum dots etc (in this case we can not apply periodic boundary conditions). So, firstly, necessary to consider information about the free surface area and geometry of the system, for which the data obtained experimentally and/or calculated; secondly, the need to develop new methods for calculating (theoretical and numerical) of the surface energy of different materials and systems.

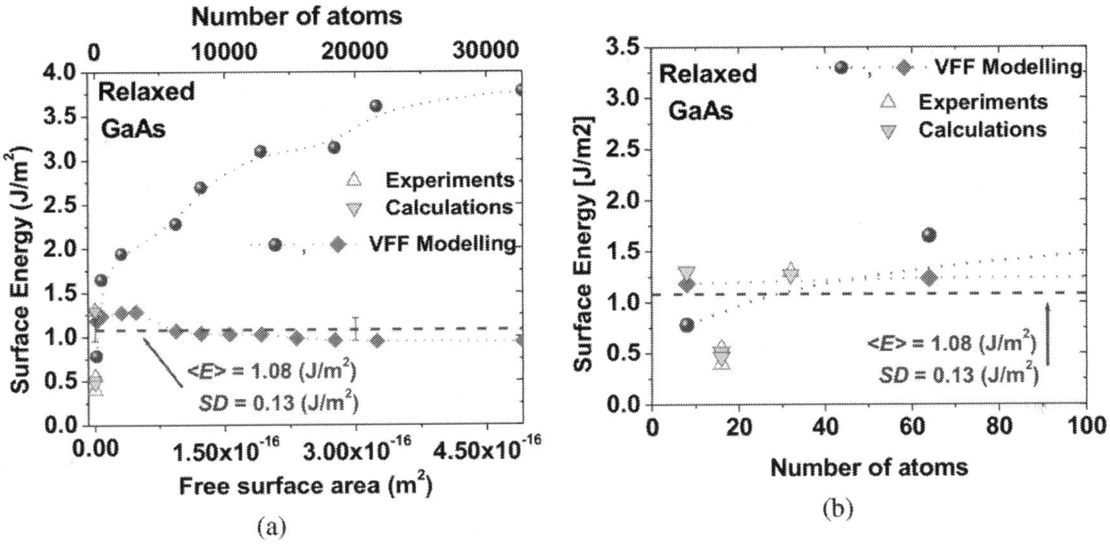

Figure 4. Surface energy as functions of free surface area and number of atoms in modelling crystals: blue circles (o) in the case of "relaxed" GaAs modelling crystals and red diamonds (◇) in case of the "strained" modelling crystals. Average surface energy *<E>* (dark red dashed line) with standard deviation SD. Green downward triangles (∇) are *ab initio* simulations and theoretical calculations and yellow upward triangles (Δ) are experimental results [41-43].

5. Conclusions

Accordingly, now we are developing a computer program for simulation of structural (distribution of atoms and impurities, elastic constants, strain distribution, etc.) and thermodynamic (surface energy, elastic energy, mixing energy, mixing enthalpy, mixing entropy, etc.) properties of new and practically important nanomaterials - low-dimensional nanostructures (e.g. nanowires, nanoneedles etc.) based on III-V, II-VI and IV semiconductors.

Numerical (operational) parameters of simulations and software packages can significantly affect simulation results (quality of energy relaxation). Small number of atoms of modelling crystals is not enough to estimate surface energy of real nanosystems, especially of systems like nanoneedls, quantum dots etc (in this case we can not apply periodic boundary conditions). To compare simulations, calculations and experimental results correctly one should know additional information (free surface area, geometry, simulation parameters etc.)

6. Acknowledgements

This work was partially supported by different grants of the Russian Academy of Sciences, the Russian Foundation for Basic Research and SPb Science Center.

References

[1] Bryllert T, Wernersson L-E, Lowgren T and Samuelson L 2006 *Nanotechnology* **17** S227

[2] Gradecak S, Qian F, Li Y, Park H-G and Liebera C M 2005 *Appl. Phys. Lett.* **87** 173111

[3] Hayden O, Zheng G, Agarwal P and Lieber C M 2007 *Small* **3** 2048

[4] Liang G, Xiang J, Kharche N, Klimeck G, Lieber C and Lundstrom M 2007 *Nano Lett.* 7 642

[5] Pauzauskie P J and Yang P 2006 *Mater. Today* **9** 36

[6] Yang P D, Yan R X, Fardy M 2010 *Nano Lett.* **10** 1529

[7] L W and C M Lieber 2007 *Nat. Mater.* **6** 841

[8] Wagner R S and Ellis W C 1964 *Appl. Phys. Lett.* **4** 89

[9] Givargizov E I 1987 *Highly anisotropic crystals* (Berlin and Heidelberg: Springer)

[10] Ye Y, Dai L, Gan L, Meng H, Dai Y, Guo X and Qin G 2012 *Nanoscale Res. Lett.* **7** 218

[11] Dubrovskii V G, G E Cirlin, I P Soshnikov, A A Tonkikh, N V Sibirev, Yu B Samsonenko and V M Ustinov 2005 *Phys. Rev. B* **71** 205325

[12] Schubert L, Werner P, Zakharov N D, Gerth G, Kolb F M, Long L and Gosele U 2004 *Appl. Phys. Lett.* **84** 4968

[13] Dubrovskii V G, Cirlin G E, Sibirev N V, Jabeen F, Harmand J C and Werner P 2011 *Nano Lett.* **11** 1247

[14] Dubrovskii V G and N V Sibirev 2008 *Phys. Rev. B* **77** 035414

[15] Dubrovskii V G, Sibirev N V, Cirlin G E, Harmand J C and Ustinov V M 2006 *Phys. Rev. E* **73** 021603

[16] Algra R E, Verheijen M A, Feiner L-F, Immink G G W, Vlieg E and Bakkers E P A M 2011 *Nano Lett.* **11** 1259

[17] Ren X, Huang H, Dubrovskii V G, Sibirev N V, Nazarenko M V, Bolshakov A D, Ye X, Wang Q, Huang Y, Zhang X, Guoland J and Liu X 2011 *Semicond. Sci. Technol.* **26** 014034

[18] Zhmakin A I and Podolska N I 2010 Investigation of surface properties effects on liquid behaviour by molecular dynamics *Problems of Mathematical Physics and Applied Mathematics: Proc. of 4th Int. Symp. (Ioffe Institute, Saint-Petersburg, Russia, 29 September 2010)* ed E A Tropp and E V Galaktionov pp. 201-206

[19] Saito T and Arakawa Y 1999 *Phys. Rev. B* **60** 1701

[20] Takayama T, Yuri M, Itoh K, Baba T and Harris J S, Jr 2000 *J. Appl. Phys.* **88** 1104

[21] Takayama T, Yuri M, Itoh K, Baba T and Harris J S, Jr 2001 *J. Cryst. Growth* **222** 29

[22] Saito T and Arakawa Y 2002 *Journal of Crystal Growth* **1172** 237–239

[23] Yang X-F, Chen X-S, Lu W and Fu Y 2008 *Nanoscale Res Lett* **3** 534

[24] Camacho Mojica D C 2010 Modeling of the structural, electronic and optical properties of GaN/AlN nitride nanowires *(PhD theses, Grenoble University, France)*

[25] Fukui T 1984 *Jpn. J. Appl. Phys.* **Part 2 23** L208

[26] Tsao J Y 1993 *Materials Fundamentals of Molecular Beam Epitaxy* (Academic, Boston)

[27] Pryor C, Kim J, Wang L W, Williamson A J and Zunger A 1998 *J. Appl. Phys.* **83** 2548

[28] Bastard G 1988 *Wave Mechanics Applied to Semiconductor Heterostructures* (Les Editions de Physique, Les Ulis)

[29] Chuang S L and Chang C S 1996 *Phys. Rev. B* **54** 4

[30] Slater J C and Koster G F 1954 *Phys. Rev. B* **94**

[31] Delerue C and Lannoo M 2004 *Nanostructures: Theory and Modelling* (Springer, New York)

[32] Chelikowsky J R and Cohen M L 1976 *Phys. Rev. B* **14** 556; 1984 *Phys. Rev. B* **30** 4828(E)

[33] Wang L W and Zunger A 1995 *Phys. Rev. B* **51** 17398

[34] Kleinman L 1962 *Phys. Rev.* **128** 2614

[35] Keating P N 1966 *Phys. Rev.* **145** 637

[36] Tersoff J 1988 *Phys. Rev.* B **37** 6991

[37] Stillinger F and Weber T A 1985 *Phys. Rev.* B **31** 5262

[38] Karpov S Yu, Podolska N I, Zhmakin I A and Zhmakin A I 2004 *Phys. Rev. B* **70** 235203

[39] Podolska N I, Karpov S Yu and Zhmakin A I, 2008 *Technical Physics Letters* **34** 370–372

[40] Podolska N I 2011 Investigation of thermodynamic proprieties of multicomponent $A^{III}B^{V}$ alloys by Valence Force Field *(PhD Thesis, Ioffe Institute, Saint-Petersburg, Russia, 15 December 2011)*

[41] Panse C, Kriegner D, Bechstedt F, 2011 *Phys. Rev. B* **84** 075217

[42] Cahn J.W., Hanneman R.E., 1964 *Surface Science* 1(4) 387

[43] Sibirev N V, Timofeeva M.A, Bol'shakov A D, Nazarenko M V and Dubrovskii V G 2010 *Physics of the Solid State* **52** 1531

15th Russian Youth Conference on Physics and Astronomy (PhysicA.SPb) IOP Publishing
Journal of Physics: Conference Series **461** (2013) 012005 doi:10.1088/1742-6596/461/1/012005

Phase diagram calculation of binary solutions of the peritectic type in the generalized lattice model

D O Barablin and M A Zakharov

Department of Solid State Physics and Microelectronics, Novgorod State University, 173003, Novgorod the Great, Russia

E-mail: ion.exe@bk.ru

Abstract. Thermodynamics of the binary solutions of the peritectic type is studied within the framework of the generalized lattice model accounting for the specific volumes of components. The corresponding equations describing two- and three-phase chemical equilibria are derived on the base of proposed approach. Several of phase diagrams of the peritectic type are calculated. The theoretically calculated phase diagrams are compared with the corresponding experimental diagrams.

1. Introduction

One of the most important problems of thermodynamics of multicomponent systems is description of two-, three- and multi-phase equilibria in solutions and the construction of the corresponding phase diagrams [1, 2]. A lot of theoretical and experimental works are devoted to these problems (see, for example, [3–5]). Among the various types of phase diagrams of binary solutions, the description of peritectic equilibrium causes significant difficulties. At present there is no *ab initio* models describing the nature of peritectic solutions. As a consequence, the number of phase diagrams of the peritectic type is relatively small. In this study, we developed a thermodynamical approach for the calculation of phase diagrams of peritectic type that is based on the generalized lattice model (see, for example, [6, 7]).

2. Model

According to the basic principles of the generalized lattice model, the chemical potentials of a homogeneous binary solution (per mole of a material) can be represented in the form [8, 9]

$$\begin{aligned} \mu_1 &= \mu_{10} + RT\ln x + W\lambda\left(\frac{1-x}{x+\lambda(1-x)}\right)^2, \\ \mu_2 &= \mu_{20} + RT\ln(1-x) + W\left(\frac{x}{x+\lambda(1-x)}\right)^2, \end{aligned} \qquad (1)$$

where μ_{i0} is the standard chemical potential of the ith component, R is the universal gas constant, T is the temperature of the system, x is the mole fraction of the first component, W is an analog of the energy of mixing in the generalized lattice model, $\lambda = \omega_2/\omega_1$, and ω_i is the characteristic atomic volume of the ith component.

Further let us consider the phase diagram of an A–B binary solution of peritectic type. The corresponding phase diagram is schematically presented in Figure 1.

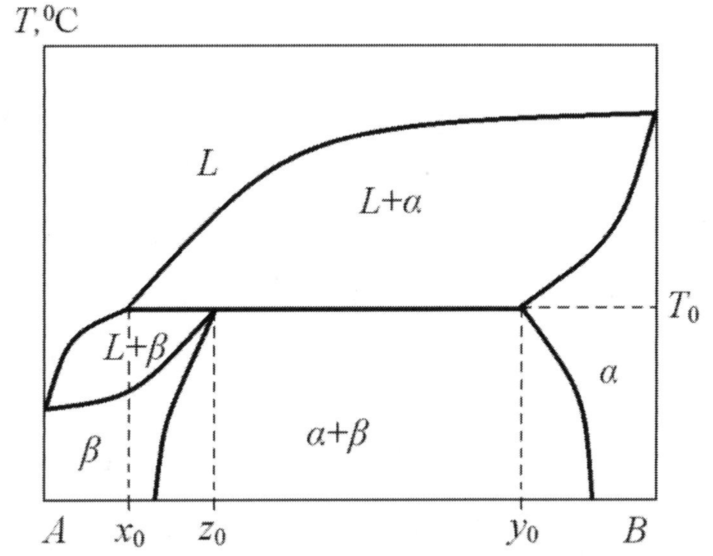

Figure 1. Phase diagram of the peritectic type.

Let x, y and z be the mole fraction of the first component in luiqid phase, in α solid solution and in β solid solution, respectively. We introduce three new parameters W, U and G, where W is an analog of the energy of mixing in the luiqid phase, U is an analog of the energy of mixing in α solid solution and G is an analog of the energy of mixing β solid solution. Then relationships (1) determine the chemical potentials of components in luiqid phase. The chemical potentials of components in solid solutions are considered in a similar way. In these cases for α solid solution we have

$$\begin{aligned} \mu_1^\alpha &= \mu_{10}^S + RT\ln y + U\lambda\left(\frac{1-y}{y+\lambda(1-y)}\right)^2, \\ \mu_2^\alpha &= \mu_{20}^S + RT\ln(1-y) + U\left(\frac{y}{y+\lambda(1-y)}\right)^2, \end{aligned} \tag{2}$$

and for β solid solution we obtain

$$\begin{aligned} \mu_1^\beta &= \mu_{10}^S + RT\ln z + G\lambda\left(\frac{1-z}{z+\lambda(1-z)}\right)^2, \\ \mu_2^\beta &= \mu_{20}^S + RT\ln(1-z) + G\left(\frac{z}{z+\lambda(1-z)}\right)^2. \end{aligned} \tag{3}$$

We can obtain all parameters of generalized lattice model for binary solutions of the peritectic type taking into account the chemical equilibrium in the three-phase system, i.e., equating the chemical potentials of the components in the solid and liquid phases

$$\begin{aligned} \mu_1^L(x_0,T_0) = \mu_1^\alpha(y_0,T_0) = \mu_1^\beta(z_0,T_0), \\ \mu_2^L(x_0,T_0) = \mu_2^\alpha(y_0,T_0) = \mu_2^\beta(z_0,T_0). \end{aligned} \tag{4}$$

Taking into account the chemical equilibrium in the two-phase systems gives

i) Liquid—α solid solution ($x > x_0, y > y_0, T > T_0$):

$$T(x,y) = \frac{q_1T_1 + \widetilde{W}\lambda\left(\dfrac{1-x}{x+\lambda(1-x)}\right)^2 - \widetilde{U}\lambda\left(\dfrac{1-y}{y+\lambda(1-y)}\right)^2}{q_1 - \ln\left(\dfrac{x}{y}\right)} = \frac{q_2T_2 + \widetilde{W}\left(\dfrac{x}{x+\lambda(1-x)}\right)^2 - \widetilde{U}\left(\dfrac{y}{y+\lambda(1-y)}\right)^2}{q_2 - \left(\dfrac{1-x}{1-y}\right)}. \quad (5)$$

ii) Liquid—β solid solution ($x < x_0, z < z_0, T < T_0$):

$$T(x,z) = \frac{q_1T_1 + \widetilde{W}\lambda\left(\dfrac{1-x}{x+\lambda(1-x)}\right)^2 - \widetilde{G}\lambda\left(\dfrac{1-z}{z+\lambda(1-z)}\right)^2}{q_1 - \ln\left(\dfrac{x}{z}\right)} = \frac{q_2T_2 + \widetilde{W}\left(\dfrac{x}{x+\lambda(1-x)}\right)^2 - \widetilde{G}\left(\dfrac{z}{z+\lambda(1-z)}\right)^2}{q_2 - \left(\dfrac{1-x}{1-z}\right)}. \quad (6)$$

iii) α solid solution—β solid solution ($y > y_0, z > z_0, T < T_0$):

$$T(y,z) = \frac{\widetilde{G}\lambda\left(\dfrac{1-z}{z+\lambda(1-z)}\right)^2 - \widetilde{U}\lambda\left(\dfrac{1-y}{y+\lambda(1-y)}\right)^2}{\ln\left(\dfrac{y}{z}\right)} = \frac{\widetilde{G}\left(\dfrac{z}{z+\lambda(1-z)}\right)^2 - \widetilde{U}\left(\dfrac{y}{y+\lambda(1-y)}\right)^2}{\ln\left(\dfrac{1-y}{1-z}\right)}. \quad (7)$$

Here $\widetilde{W} = W/R$, $\widetilde{G} = G/R$, $\widetilde{U} = U/R$, and $q_i = \Delta H_i/RT_i$, where ΔH_i are the latent heats of the liquid–solid transition for the pure components, T_i stands for the melting temperatures of the pure components.

Equations (4)–(7) form a closed system, whose solution allows us to find the concentration dependences of the liquidus branches, solidus branches and solvus branches and, hence, to construct the phase diagrams of peritectic binary systems in the framework of the generalized lattice model.

3. Phase diagram calculation

The table presents the calculated parameters of the generalized lattice model for two of binary systems with the phase diagrams corresponding to the type considered above. The parameters q_i were determined with the use of the reference data on the latent heats of the liquid–solid transition.

The corresponding phase diagrams calculated with these parameters are depicted in Figure 2. It should be noted that the results obtained from the theoretical calculations are in agreement with the experimental data available in the literature [10, 11].

Table 1. Parameters of the generalized lattice model of binary systems.

A–B	q_1	q_2	λ	$\widetilde{W}$, K	$\widetilde{U}$, K	$\widetilde{G}$, K
Ag–Pt	1.282	1.164	2.858	3411	4556	4721
Au–Cr	1.182	1.140	1.705	2831	5091	4109

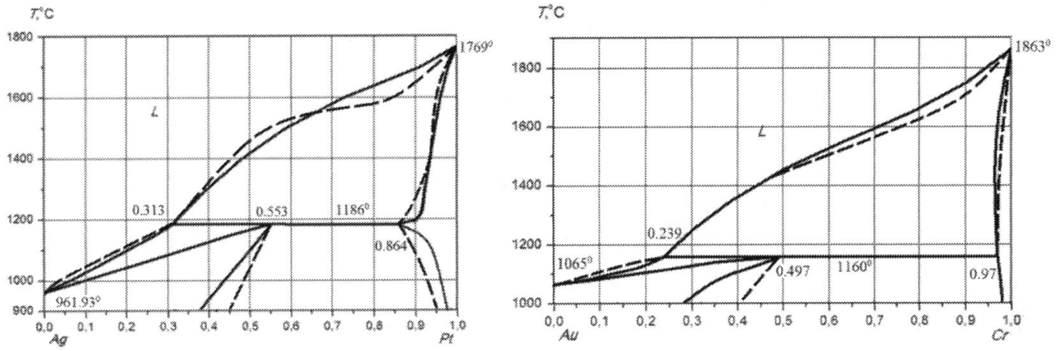

Figure 2. Phase diagrams of the Ag–Pt and Au–Cr systems. Dashed lines are the experimental data taken from [10, 11]. Solid lines indicate the results obtained from theoretical calculations.

References

[1] Gorban A N, Kaganovich B M, Filippov S P, Keiko A V, Shamansky V A and Shirkalin I A 2006 *Thermodynamic Equilibria and Extrema: Analysis of Attainability Regions and Partial Equilibria* (Berlin-Heidelberg-New York: Springer)
[2] Anderson G M 2005 *Thermodynamics of Natural Systems* (Cambridge: Cambridge University Press)
[3] Honig J M 2007 *Thermodynamics. Principles Characterizing Physical and Chemical Processes* (London: Academic Press)
[4] Saunders N and Miodownik A P 2005 *CALPHAD. Calculation of Phase Diagrams* (Oxford: Pergamon)
[5] Haines P J 1995 *Thermal methods of analysis: principles, applications and problems* (London: Blackie Academic & Professional)
[6] Zakharov A Yu and Terekhov S V 1985 *Fiz. Met. Metalloved.* **95** 261
[7] Zakharov A Yu, Zakharov M A and Lebedev V 2005 *Int. J. Quant. Chem.* **104** 133
[8] Korzun E L and Terekhov S V 1987 *Zh. Fiz. Khim.* **61** 1186
[9] Zakharov M A 2007 *Phys. Solid State* **49** 2312
[10] Karakaya L and Thompson W T 1987 *Bull. Alloy Phase Diagrams* **8** 334
[11] Okamoto H and Massalski T B 1985 *Bull. Alloy Phase Diagrams* **6** 223

Influence of radiation on Voltage Terminating Structure of silicon relativistic particle detectors

N Fadeeva[1,2], V Eremin[1], E Verbitskaya[1] and E Terukov[1]

[1]Ioffe Physical-Technical Institute of the Russian Academy of Sciences, Polytekhnicheskaya str. 26, 194021 St. Petersburg, Russia

[2]Saint Petersburg Electrotechnical University, Prof. Popov str. 5, 197376 St. Petersburg, Russia

E-mail: fadeevanadezda@gmail.com

Abstract. Semiconductor radiation detectors are widely applied in high energy physics experiments. The largest particle colliders use silicon detectors with the total area of hundreds square meters that enhances the requirement to their stable long-term operation. This is provided by incorporating the floating p^+ rings surrounding the sensitive area of the p^+-n-n^+ detectors (VTS – Voltage Termination Structure), that prevents an electric breakdown. The physical model of VTS operation in high-resistivity p^+-n-n^+ silicon radiation detectors developed in the study uses an approach of the current injection through the ring spacings of VTS that may occur under a certain electric field distribution. This leads to a strict stabilization of the ring potentials. Investigation of the potential distribution in VTS of silicon detectors irradiated with 1 MeV neutrons up to a fluence of 5×10^{15} n_{eq}/cm^2 was carried out. It was shown that the change of the electric field profile in the detector bulk with increasing radiation fluence is a key factor for the potential distribution. At fluences less than 5×10^{14} n_{eq}/cm^2 the potential distribution in VTS is governed by the punchthrough mechanism, whereas at higher fluences it is controlled by the bulk generation current and interaction with radiation induced deep levels.

1. Introduction

Silicon detectors which are used in high luminosity hadron colliders (such as the Large Hadron Collider (LHC) at CERN (Switzerland), and TEVATRON at the Fermi laboratory (USA)) can require high operational voltages (in excess of 500 V) to maintain full depletion and effective collection of nonequilibrium charge carriers even at irradiation fluences more than 10^{14} n_{eq}/cm^2.

Radiation detectors of low-energy photons and short-range particles are based on the structures with shallow junctions (less than 1 μm) processed by ion implantation. Therefore, detector operation at high reverse bias results in significant electric fields at the p-n junction periphery that is accompanied by abrupt increase in the reverse current and subsequent breakdown.

To prevent low voltage breakdown at the junction periphery, i.e. stabilize the detector I-V characteristics, the junction active region is surrounded by potential dividers with isolated ring p-n junctions (VTS – voltage terminating structures) that gradually reduce the potential from the central active region towards the periphery [1]. Operation of VTS was studied earlier for the diodes processed on silicon with a resistivity ρ of tens and hundreds $\Omega\times$cm and p-n junction depths of tens micrometers [2-7]. Existing models for the VTS calculation are based on the electrostatic approach to the problem in which it is postulated that the electric field is transferred from the inner structural element to an adjacent one when the space charge regions (SCRs) of both elements touch each other.

A principal disadvantage of the electrostatic model becomes clear when it is applied to radiation detectors on high resistivity silicon. For example, at the typical distances of 10-30 μm between the structural elements and n-type silicon resistivity in the range of 3-20 kΩ×cm, SCRs of the individual rings touch each other even at a zero bias due to a built-in potential. According to the electrostatic model, all rings in such structures should have close potentials, i.e. the VTS appears to be inconsistent. This contradicts with the practice of VTS operation in radiation detectors.

It is necessary to maintain stable operation in detectors irradiated to the fluences exceeding 10^{14} n_{eq}/cm^2. One of the aspects of solving this problem is related with the influence of the high fluences of relativistic particles on VTS. Irradiation of silicon with high energy particles leads to the formation of the radiation defects, including those that have energy levels in the band gap, and so are electrically active. In turn, this results in the trapping of nonequilibrium charge carriers during their drift to the collection electrodes, which modifies the distribution of the electric field in the sensitive area of the detector and gives rise to a decrease in the amplitude of the detector signal. Generation of the equilibrium charge carriers via midgap energy levels of radiation induced defects brings to an increase in the leakage current of the detector and a decrease in signal-to-noise ratio.

In this paper we consider a physical model of VTS operation of nonirradiated and neutron irradiated detectors which is based on the injection mechanism of current flow through the ring spacings. The measurements were performed for the samples manufactured within the frames of a project for the development of proton detectors for the TOTEM experiment at CERN. The studies are focused on the development of radiation hard detectors for the LHC upgrade and are also important for the development of silicon detectors for the future experiments within the FAIR program at Research Center GSI (Germany).

2. Experimental

A photograph of p^+-n junction detector (top view) and its fragment with VTS are shown in figure 1. Detectors were processed on n-type FZ Si wafers manufactured by TOPSIL, Denmark. Si resistivity was 7 kΩ×cm. All structures had the same topology with the central p^+ area (pad) 5 × 5 mm^2. The VTS consisted of four closed floating p^+ rings separated by ring spacings. The surface of the ring spacings was passivated with a SiO_2 layer. The ohmic side was an n^+ sheet implant. The thickness of the structures was 300 μm. Detectors were irradiated with 1 MeV neutrons up to a fluence of 5×10^{15} n_{eq}/cm^2, which, in fact, covers the expected range of radiation fluences for Si tracking detectors in the principal experiments in the LHC after a planned increase in the beam intensity.

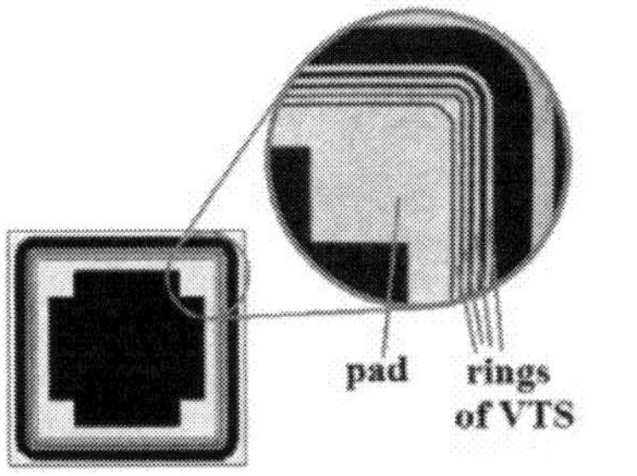

Figure 1. A photograph of the detector.

Measurements of the ring potentials and I-V characteristics of ring spacings of the detectors were performed using a setup with microprobes and a Keithley 487 Picoammeter/Voltage Source. The probe was sensitive to the potential and had the resistance 10^{12} Ω, which ensured the conditions of floating potential for a ring; the central p^+-n junction was reverse biased with the voltage V_{bias} at a grounded n^+ contact. I–V characteristics of ring spacings were measured at bias voltage V_{bias} applied to the pad and the voltage ΔV applied to the ring spacing. All measurements were performed at a temperature 20°C.

3. Experimental results

Figure 2 shows the dependences of the potentials φ in VTS on the reverse bias V_{bias} for nonirradiated detectors. The magnitudes of negative voltages and potentials are given in the figures and text. The solid line shows the potential of the pad, the numbers of curves correspond to the numbers of the rings.

It can be seen that for a nonirradiated detector the potentials of all rings monotonically increase with the voltage V_{bias} applied to the pad. In this case, the potential of an individual ring begins to increase at a certain critical voltage V_{cr} which significantly exceeds the built-in potential at the p^+-n junction (~ 0.6 V), which is sufficient for joining the SCRs of the adjacent p^+-n junctions in the electrostatic model.

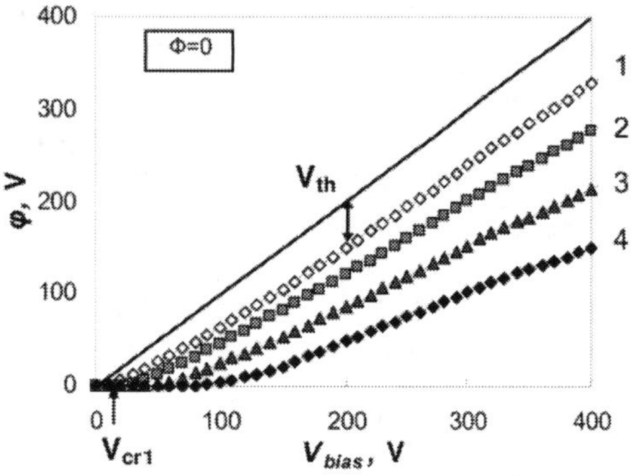

Figure 2. Ring potentials of nonirradiated detector. The solid line shows the potential of the pad, the numbers of curves correspond to the numbers of the rings.

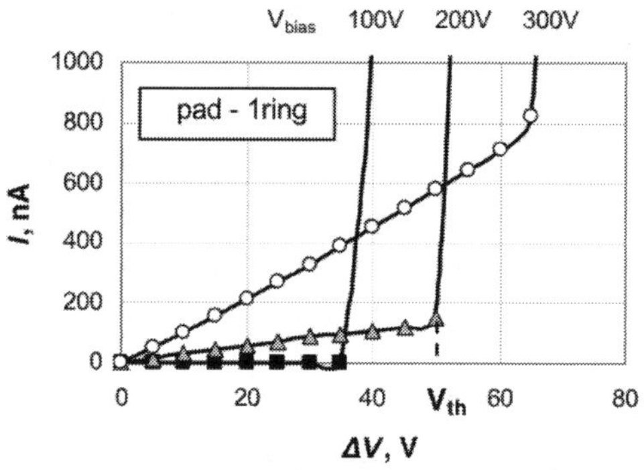

Figure 3. I-V characteristic of the pad-1 ring spacing at bias V_{bias} = 100, 200 and 300 V.

Figure 3 shows the dependences of the pad – 1st ring spacing current I for nonirradiated detector on the voltage ΔV applied to the spacing. Measurements were performed at three various bias V_{bias}, 100, 200 and 300 V, respectively. There are two regions in the characteristics: a region with a weakly increasing low current and a region of a sharp current increase at the threshold voltage V_{th}. The former one is associated with leakages of ring spacing insulation due to the ohmic surface conductivity and generation-recombination processes at the Si/SiO_2 interface [8]. The sharp current increase at the V_{th} is not a breakdown due to the fact it can be reproduced many times, so it is a beginning of the injection current. Indeed, the structure with ring spacings is symmetric and the current increase corresponds to forward biasing of one of the p^+-n junctions (punch-through).

It should be noted that the value of the threshold voltage V_{th} on I-V characteristic of the pad – 1st ring spacing (figure 3) equals to the potential difference between the pad and the 1st ring at a certain bias V_{bias} (figure 2). Therefore, potential distribution in VTS is controlled by the injection current in a ring spacing.

Figure 4 shows the dependences of the potentials φ at VTS rings on the reverse bias V_{bias} for detectors irradiated by neutrons in the range of fluences from 5×10^{13} to 5×10^{15} n_{eq}/cm^2. As the fluence increases the value of the critical voltage V_{cr} for all rings decreases and becomes equal to zero at a dose of 5×10^{13} n_{eq}/cm^2 (figure 4 (a)). For detectors irradiated to a fluence 5×10^{14} n_{eq}/cm^2 (figure 4 (b)) the potentials at all rings increase in the entire range of bias V_{bias} in comparison with ones for detectors irradiated to lower fluences. At low bias V_{bias} (0 – 200 V) the ring potentials are close to each other and effective dividing of the potentials is lacking. As the fluence further increases (figures 4 (c), 4 (d)), the potentials of all rings become the same in the entire range of V_{bias}, while the voltage drop between the pad and the 1st ring considerably increases.

4. The model of the potential distribution in VTS of nonirradiated detectors

The potential distribution in VTS of nonirradiated detectors is described in terms of an injection model of current flows through the ring spacings. At a zero bias SCR corresponding to the built-in potential of the pad junction and the 1st ring are joined and a potential minimum in the central of spacing is observed. As the bias increases, the SCR of the pad junction extends and the potential minimum shifts to the ring, the potential of which remains zero. The existing potential barrier for the holes prevents their transport between the p^+ regions and the ring remains insulated from the pad, that corresponds to the region $\Delta V < V_{cr}$. This state will retain until the potential barrier will decrease to a value allowing hole injection from the ring to the spacing. The ring spacing resistance will sharply drop. The punch-through of the ring spacing occurs and the ring potential becomes different from zero. A further increase in V_{bias} brings about an increase in the ring potential. The conditions required for a punch-through of the ring spacing depend on the electric field distribution in the ring spacing and are determined by the competition between the normal and tangential (parallel to the surface of the ring spacing) vectors of the electric field. Injection is stimulated by the tangential component of the electric field, whereas the normal component inhibits injection by moving the charge carriers to the detector bulk. As V_{bias} increases, the voltage drop between the rings also increases as a result of the normal component increase.

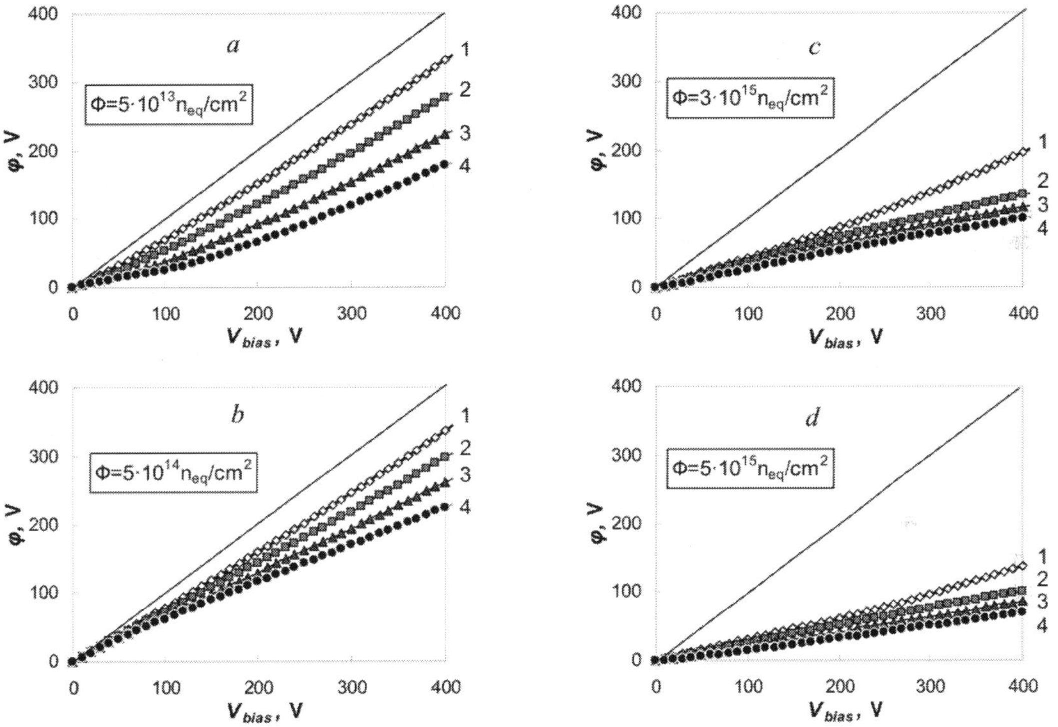

Figure 4. Ring potentials of neutron irradiated detector. Fluence was (a) 5×10^{13}; (b) 5×10^{14}; (c) 3×10^{15}; (d) 5×10^{15} n_{eq}/cm^2. The solid line shows the potential of the pad, the numbers of curves correspond to the numbers of the rings.

Indeed, figure 3 shows that V_{cr} increases with V_{bias}. In the limit, at sufficiently high V_{bias}, the path of hole transport between rings can be broken and injection will be terminated. Injection can be restored at larger V_{cr}. Therefore, an increase in V_{cr} with the bias voltage can be considered as a result of hole runaway from the injection path, which is equivalent to an increase in the potential barrier height.

According to the model, the criterion of the stable potential distribution in VTS is the injection current flow in VTS. In this case, the injection conditions will be individual for each ring spacing. The current flowing in certain ring spacing, being a bulk generation current will also be individual and will increase when approaching the pad due to summing of the currents of outer rings.

5. Influence of neutron irradiations on potential distributions in VTS

As was shown in the study of the mechanisms of the potential distribution in VTS before irradiation, the value of the electric field in the ring spacing is the most important factor, which determines the

Journal of Physics: Conference Series **461** (2013) 012006 doi:10.1088/1742-6596/461/1/012006

potential distributions. Irradiation of silicon detectors brings about the generation of radiation defects with energy levels that affect the effective space charge density N_{eff} in the SCR and the corresponding profile of the electric field. In the case of irradiation of n-Si, compensation with acceptor levels mainly occurs; therefore, at low fluences Φ, the N_{eff} in the SCR decreases. At higher Φ, the sign of the space charge is changed from positive to negative, i.e., space charge sign inversion takes place. The dependence of the effective concentration N_{eff} on the neutron fluence Φ can be calculated using the following parametric equation [9]:

$$N_{eff}(\Phi) = N_{d0} \cdot \exp(-\gamma \cdot \Phi) - \beta \cdot \Phi , \qquad (1)$$

where N_{d0} is the effective space charge density before irradiation with $N_{d0} = N_{eff}(\Phi = 0)$; $\gamma = 8.8\times10^{-14}$ cm^2 is the donor removal rate; and $\beta = 0.022$ cm^{-1} is the acceptor introduction rate. According to equation (1), inversion occurs at $\Phi = 1\times10^{13}$ n_{eq}/cm^2 for the silicon with the resistivity 7 kΩ×cm ($N_{d0} = 6\times10^{11}$ cm^{-3}) used in the study. As the radiation fluence is further increased, the magnitude of N_{eff} increases.

The operation of the VTS radically changes at a fluence of 5×10^{13} n_{eq}/cm^2 (figure 4 (a)). V_{cr} becomes equal to zero for all rings, while the dependences of the ring potentials on V_{bias} are significantly linearized. The space charge is negative and the electric field maximum is located near the n^+ contact; at a voltage lower than the full depletion voltage for the structure, the rings on the p^+ side are found to be in the zero electric field. Indeed, in a partially depleted detector, the conducting base near the p^+ contact connects all the rings to each other; the leakage current of the n^+-n junction flows over the edge of the detector and then through the VTS, thus increasing the potential of the outermost ring and then of all rings. Therefore, in the case of full depletion of the detector (more than 75 V), a drop in the potential arises between the rings according to the standard punch-through mechanism, however, now at a level of nonzero potential of the outermost ring.

At a fluence of 5×10^{14} n_{eq}/cm^2, an increase in the potentials of all rings is observed; as a result, the drop in the voltage at the periphery of the detector reaches a value of more than 50 % of the voltage V_{bias}. It is noteworthy that, in this case, the potentials of the rings increase virtually linearly in the entire range of voltages; an appreciable difference between the potentials of neighboring rings arises at $V_{bias} > 200$ V (figure 4 (b)). This value is found to be several times smaller than the full depletion voltage, which, according to [10], is equal to about 750 V. This fact is a direct evidence of the existence of two maxima for the electric field in heavily irradiated p^+-n-n^+ detectors; these maxima are located near the opposite contacts [11]. The main depleted region, on which a larger portion of the applied voltage drops, is situated near the n^+ contact. An additional region of the electric field at the p^+ contact is formed due to the capture of holes thermally generated in the bulk of the structure and is the cause of the appearance of a potential difference between the rings. It is evident that a difference in the ring potentials is insignificant since the potential drop at SCR of the p^+ contact is several times smaller than that at SCR of the n^+-contact.

At higher fluences, $3\times10^{15} - 5\times10^{15}$ n_{eq}/cm^2, the trend in the potentials variation becomes reverse (figures 4 (c), 4 (d)). As a fluence increases, the potentials of the rings decrease and an effective division of the potentials is lacking. For example, at a fluence of 5×10^{15} n_{eq}/cm^2, 80 % of the potential drops between the pad and the 1st ring. The reason for such a large difference between the potential of the pad and the 1st ring may be the dominance of the current related mechanism of the electric field formation in the bulk of the detector. The leakage current in the structure increases in proportion to the radiation fluence and, at $\Phi \sim 10^{15}$ n_{eq}/cm^2, reaches tens of $\mu A/cm^2/300$ μm. At a high generation rate in the bulk and a significant concentration of radiation defects, which are trapping centers for free charge carriers, the electric field profile is formed by the trapped charge of electrons and holes, i.e., this field mainly exists in the regions where the current flows. On the p^+ side, these include the region adjacent to the sensitive p^+-n junction where the main current of the detector is collected. Thus, the VTS rings surrounding this region and isolating it from the pad do not collect current and are situated away from the path of holes drifting toward the p^+ contact. VTS is situated in the region with a low electric field and the potential is close to that of the ohmic n^+ contact. This is completely consistent with the conditions of the performed experiment since, at the highest fluence (5×10^{15} n_{eq}/cm^2), the thickness of the SCR on the n^+ side does not exceed 100 μm at V_{bias} = 400 V. Therefore, the electric field distribution is determined by the path of the current flow in the base; this current forms the space charge by the mechanism of charge carrier trapping. Qualitatively, the relation between the electric field distribution and the flowing current is given by the following expression:

$$divE \sim qN_{tr} \sim f(\tau)(j_p - j_n), \quad (2)$$

where q is the elementary charge; N_{tr} is the difference between the concentrations of captured electrons and holes; $f(\tau)$ is a function describing the probability of the capture of electrons and holes; and j_h and j_n are the electron and hole current densities, respectively. Since the hole current is dominant at the p^+ contact, the value of $divE$ is always positive; consequently, the maximum electric field is bound to be located at the surface of the p^+ contact, which collects holes. The electric field strength near the isolated rings is bound to be at a minimum and, as experiments show, their potential is close to that of the n^+ contact (figure 4 (d)).

6. Conclusions

The developed physical model of the potential distribution in VTS considers both the current flow in the SCR and electrostatic aspects. The model can be applied to the development of floating ring structures on high purity materials.

The radiation induced degradation of a detector bulk led to a disruption in the operation of the VTS. The dominant mechanism in the case of the fluence increase is the variation in the electric field distribution in the detector bulk and the replacement of the electrostatic mechanism for field formation in the SCR by a mixed mechanism with a dominant influence of the density of the electron-hole generation current, which flows in the bulk. In this situation, the potential that dropped in the VTS is reduced and becomes substantially lower than the voltage applied to the structure, i.e., the VTS loses the properties of a system that optimizes the potential distribution.

In silicon radiation detectors, which operate at high radiation fluences and high voltages, the observed stability of their characteristics (i.e. without breakdown) is governed by the space charge limited current, i.e., by suppression of the electric field in regions where the subthreshold high electric field gives rise to the injection of charge carriers from the contact to the detector bulk. This conclusion confirms this mechanism, which was suggested previously by V. Eremin, et al. [12].

Acknowledgments

This study was supported by the grants of the Scientific School of the President of the Russian Federation (grants no. NSh_3306.2010.2, 3308.2012.2), by the program "Experimental and Theoretical Studies of Basic Interactions Related to Experiments at the Accelerator Complex CERN", and is in the frames of the scientific program of CERN-RD50 collaboration.

References

[1] Koa Y C and Wolley E D 1967 Proc. IEEE 55 1409-14
[2] Adler M S, Temple V A K,.Ferro A P and Rustay R C 1997 IEEE Trans. Electron Devices ED-24 107-13
[3] Baliga B J 1990 Sol. St. Electron. 33 485-8
[4] Suh K D, Hong S W, Lee K and Kim C 1990 Sol. St. Electron. 33 1125-9
[5] Brieger K P, Gerlach W and Pelka J 1983 Sol. St. Electron. 26 739-45
[6] Bae D G and Chung S K 2000 Sol. St. Electron. 44 2109-16
[7] Boisson V, Helley M Le and Chante J P 1985 IEEE Trans. Electron Devices ED-32 838-40
[8] Eremin V, Verbitskaya E, Ilyashenko I, Eremin I, Safonova N, Tuboltsev Yu, Egorov N, Golubkov S and Konkov K 2009 Semiconductors 43 796-800
[9] Li Z, Dezillie B, Chen W and Bruzzi M 2002 Nucl. Instrum. Meth. A 476 628-38
[10] Sze S M 1969 Physics of Semiconductor Devices (New York: Wiley-Interscience)
[11] Eremin V, Verbitskaya E and Li Z 2002 Nucl. Instrum. Meth. A 476 537-49
[12] Eremin V, Verbitskaya E, Zabrodskii A, Li Z and Härkönen J 2011 Nucl. Instrum. Meth. A, 658 145-51

Journal of Physics: Conference Series **461** (2013) 012007 doi:10.1088/1742-6596/461/1/012007

Time-dependent model of particle acceleration in the vicinity of approaching magnetohydrodynamic flows

Gladilin P.E.[1,2], Bykov A.M.[1,2] and Osipov S.M.[1,2]

[1]Ioffe Physical-Technical Institute of the Russian Academy of Sciences, Saint-Petersburg, Russia,

[2]Saint-Petersburg State Polytechnical University, Saint-Petersburg, Russia

E-mail: peter.gladilin@gmail.com

Abstract. Supernova explosions and powerful stellar winds with strong shocks can convert a sizeable part of the kinetic energy release into energetic particles. The starforming regions are argued as a favorable site of energetic particle acceleration and could be efficient sources of nonthermal emission. We present the analytic solution for the time-dependent diffusion-convection equation for the case of two approaching shocks. We also present here time-dependent model of particle acceleration in the vicinity of two closely approaching fast magnetohydrodynamic (MHD) shocks. Such MHD flows are expected to occur in rich young stellar cluster where a supernova is exploding in the vicinity of a strong stellar wind of a nearby massive star. We find that the spectrum of the high energy particles accelerated at the stage of two closely approaching shocks can be harder than that formed at a forward shock of an isolated supernova remnant. The presented method can be applied to model particle acceleration in a variety of systems with colliding MHD flows.

1. Introduction

Diffusive shock acceleration (DSA) mechanism thought to be operating at the fast shocks of young supernova remnants (SNRs) is likely responsible for the galactic cosmic ray (CR) acceleration up to the knee-region energies (see e.g. [1, 2]) and may even exceed 10^{17} eV as it was advocated by [3] for the case of nuclei accelerated in an isolated young Type IIb SNRs. The basic features of the DSA process have been revealed in the pioneering papers of [4], [5], [6] and [7] where the high efficiency of the acceleration mechanism was shown. Effects of nonlinear backreaction of the accelerated particles on the structure of the supersonic shock flow were discussed later by [8], [9], [10], [11], [12], [13] and [14]. It has been found that the pressure of accelerated particles can modify the bulk plasma flow in the shock upstream and that may result in a substantial increase of the flow compression and flattening in the particle spectra at the maximal energies.

The maximum energy of accelerated particles strongly differs for different types of supernovae (see e.g.[3]), it depends on the circumstellar medium around the supernova progenitor star. Moreover, core-collapsed supernovae produced by massive stars often occur in OB-star associations where the intense radiation of hot massive stars, powerful stellar winds and supernova shocks strongly modify the interstellar environment, producing large hot cavities of a few tens of parsec size, called superbubbles.

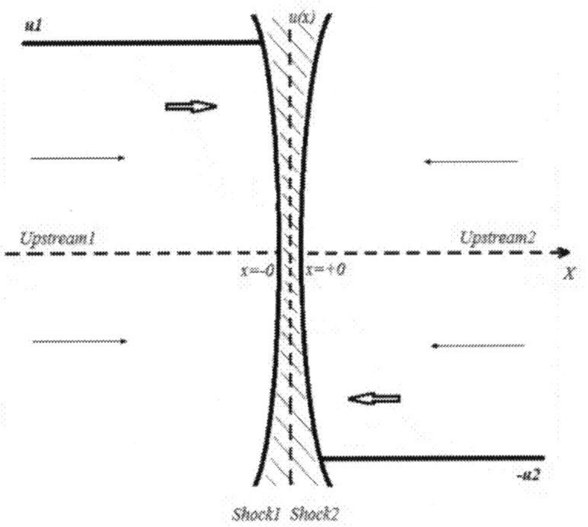

Figure 1. Simplified scheme of velocity profiles for the approaching shocks system.

In the present paper we model a class of a few parsec size particle accelerators associated with collision of a young supernova shock with a fast stellar wind of a massive star. Some basic features of such systems were noticed in the previous paper [15]. The modeled stage starts a few hundred years before the supernova shock collides with the wind termination shock. At this stage the maximal energy particles accelerated via DSA at the SNR shock reach the fast wind termination shock and are scattered back by magnetic fluctuations carried by the fast stellar wind. Therefore, the high energy particles that have mean free path $\Lambda(p)$ larger than the distance between the two shocks L_{12} start to be accelerated by the converging fast flows. This is the most favorable circumstance for the efficient Fermi acceleration. While the structure of the MHD flow in the vicinity of a supernova shell colliding with the stellar wind termination shock is rather complex, it is possible to consider a simplified model. When the magnetic field fluctuations are strong enough to provide the so-called Bohm diffusion regime with $\Lambda(p) = \xi r_g(p)$, where p is the particle momentum, $r_g(p)$ is the particle gyroradius in the mean magnetic field and $\xi \gtrsim 1$, the high energy particles bouncing between the converging flows are likely to have a spectrum harder than that produced at an individual shock and may contain a sizeable fraction of the total kinetic energy of the converging MHD flows.

2. Analytic time-dependent solution

Consider a model describing a population of high energy CR particles with $\Lambda(p) > L_{12}$ in a vicinity of two approaching shocks with $R_{\rm sh} = R_{\rm sw} \gg L_{12}$, so planar geometry can be applied. In Figure 1 we illustrate a simplified scheme of the local flow with two approaching shocks. At $x < 0$ there is the upstream region of the shock 1 and at $x > 0$ there is the upstream region of the shock 2.

To derive the distribution function at the shock in the case of the two colliding shock fronts we employ a time-dependent diffusion-convection equation with the CR particle injection rate:

$$\frac{\partial f_i(t,x,p)}{\partial t} = \frac{\partial}{\partial x} D_i(x,p) \frac{\partial f_i(t,x,p)}{\partial x} - u_i \frac{\partial f_i(t,x,p)}{\partial x} + \frac{p}{3}\frac{\partial u_i}{\partial x}\frac{\partial f_i(t,x,p)}{\partial p} + Q_0 \delta(x)\delta(p-p_0)H(t-t_0) \qquad (1)$$

Journal of Physics: Conference Series **461** (2013) 012007 doi:10.1088/1742-6596/461/1/012007

where index $i = 1$ corresponds to the $x < 0$ region and $i = 2$ for $x > 0$ (see Fig.1), Q_0 - is the injection rate of the particle, $\delta(x)$ - is Dirac's delta-function, $H(\tau)$ - is the Heaviside step function, $D_i(x,p)$ - is the diffusion coefficient in the upstream area 1 and 2 (for simplicity we will assume D_i to be independent of x).

Velocity profile has been chosen as

$$u_i = \begin{cases} u_1, & \text{if } x < 0; \\ -u_2, & \text{if } x > 0; \end{cases}$$

The distribution function of the particles $f(t,x,p) = 0$ at $t = 0$. Boundary conditions at the point $x = 0$ can be obtained by integration of the Eq.1 by x:

$$f_1 = f_2, \tag{2}$$

$$D_1\frac{\partial f_1}{\partial x} + \frac{u_1}{3}p\frac{\partial f_1}{\partial p} = D_2\frac{\partial f_2}{\partial x} - \frac{u_2}{3}p\frac{\partial f_2}{\partial p} + Q_0\delta(p-p_0)H(t-t_0). \tag{3}$$

On taking the Laplace transform with respect to time:

$$g_f(s,x,p) = \int_0^\infty exp(-st)f(t,x,p)dt, \tag{4}$$

equation (1) and boundary conditions (2),(3) become:

$$sg_{fi} + u_i\frac{\partial g_{fi}}{\partial x} = D_i\frac{\partial^2 g_{fi}(s,x,p)}{\partial^2 x}, \tag{5}$$

$$D_1\frac{\partial g_{f1}}{\partial x} + \frac{u_1}{3}p\frac{\partial g_{f1}}{\partial p} = D_2\frac{\partial g_{f2}}{\partial x} - \frac{u_2}{3}p\frac{\partial g_{f2}}{\partial p} + \frac{Q_0}{s}\delta(p-p_0). \tag{6}$$

In order to satisfy boundary conditions $g_{fi}(s,x,p) \to 0$ as $x \to \pm\infty$ we look for the solution $g_{fi} \propto exp(\beta_i x)$, where

$$\beta_i = \frac{u_i}{2D_i}\left(1 + \sqrt{1 + \frac{4D_i\,s}{u_i^2}}\right). \tag{7}$$

Thus we have:

$$D_1\beta_1 g_{f0} - D_1\beta_2 g_{f0} + \frac{1}{3}p\frac{\partial g_{f0}}{\partial p} = \frac{1}{s}Q_0\delta(p-p_0). \tag{8}$$

where $g_{f0} = g_f(s,0,p)$ - Laplace transform at $x = 0$.

In general case inverse transform of the equation (8) cannot be evaluated. But for physically important values $t \gg D_i/u_i^2$ one can easily perform inverse Laplace transform by taking into account that for this case $\sqrt{1 + \frac{4D_i\,s}{u_i^2}} \approx (1 + \frac{2D_i\,s}{u_i^2})$.

Therefore, solution of the equation (1) reads:

$$f_i(x,p,t) = \frac{3Q_0}{u_1+u_2}\left(\frac{p}{p_0}\right)^{-3} H(p-p_0)H(t-\tau_a)exp\left(-\frac{u_i|x|}{D_i(p)}\right) \propto \frac{1}{p^3}, \tag{9}$$

and the particle spectrum:

$$\frac{\partial N}{\partial p} \propto p^2 f \propto \frac{1}{p}, \tag{10}$$

where

$$\tau_a(p^*) = \int_{p_0}^{p^*} \frac{3}{(u_1+u_2)}\left(\frac{D_1}{u_1} + \frac{D_2}{u_2}\right)\frac{dp}{p} \tag{11}$$

is the CR acceleration time up to the momentum p^*.

Presented acceleration mechanism can provide efficient creation of a nonthermal particle population with a very hard energy spectrum, containing a substantial part of the kinetic energy released by the supernova. The high energy nonthermal emission of the sources is characterized by a very hard spectral energy distribution peaked at the maximal photon energies and have the apparent properties of the dark accelerators.

3. Time-dependent model simulations

Via time-dependent DSA simulations we calculated the energy spectra of cosmic ray protons and electrons at a SNR shock approaching a strong wind of a nearby early type star. For relativistic electrons/positrons we account for the energy losses due to synchrotron and inverse Compton (IC) radiation. We solve one-dimensional diffusion-advection equations (see [16, 17, 18]) for the pitch-angle-averaged phase space distribution function of protons, $f_p(x,p,t)$, and electrons, $f_e(x,p,t)$, given by

$$\frac{\partial g_p}{\partial t} + u_i \frac{\partial g_p}{\partial x} = \frac{1}{3}\frac{\partial u_i}{\partial x}\left(\frac{\partial g_p}{\partial y} - 4g_p\right) + \frac{\partial}{\partial x}\left(D(p)\frac{\partial g_p}{\partial x}\right) \tag{12}$$

$$\frac{\partial g_e}{\partial t} + u_i \frac{\partial g_e}{\partial x} = \frac{1}{3}\frac{\partial u_i}{\partial x}\left(\frac{\partial g_e}{\partial y} - 4g_e\right) + \frac{\partial}{\partial x}\left(D(p)\frac{\partial g_e}{\partial x}\right) + \exp(y)\frac{\partial}{\partial y}\left[b\exp(-2y)g_e\right] \tag{13}$$

where $g_p = p^4 f_p$, $g_e = p^4 f_e$, $y = ln(p)$. The dimensionless particle momentum is expressed in the units of $m_p c$. The cooling term $b(p) = -dp/dt$ describes the electron synchrotron and IC losses. These equations were solved with integration-interpolation algorithm and the standard Crank-Nicolson scheme. The model allows to calculate $f_{e,p}(x,p,t)$ at any position between the shocks and in the post-shock flows.

In Fig. 2 we present the results of calculations of the proton distribution function $f_p(x,p,t)$ that were obtained by time-dependent colliding shocks flow model and proton spectrum $dN_p(p)/dp$ at the moving left shock for the corresponding times comparing to the spectrum predicted by analytic test-particle solution (Eq. 9) .

The left shock is propagating in the positive direction of the x- axis and the right shock is moving in the opposite direction. In Fig. 2 the CR proton distribution function $f(x,p,t)$ in the phase space (p,x) is shown at different moments that correspond to the distances between the shocks of 0.6 pc, 0.5 pc, 0,3 pc, 0.1 pc. Spectral evolution of the proton distribution function between the colliding shocks is clearly seen. The CRs are initially concentrated around the shocks, and proton spectrum is close to the spectrum of the identical SNR shock $dN_p(p)/dp \propto 1/p^2$. But as the inter-shock distance reduces the proton spectrum gets harder and CRs concentrate in between the shocks. It can be seen from the Fig. 2 that when the distance between the shocks becomes 0.1 pc proton spectrum almost coincides with the analytic solution (Eq. 9).

Assuming that the amplified magnetic field near the shock is $\sim 100\ \mu G$ as it was inferred from young SNR observations recently reviewed by [19], the distance between the shock fronts $L_{12} \sim 1$ pc, we used $D_{is} = 100 \times D_{sn}$ and $D_{sw} \simeq D_{sn}$ for the Bohm-type diffusion, where $D_{is}\, D_{sn}\, D_{sw}$ - are interstellar, SNR-upstream and SW-upstream diffusion coefficients respectively. These parameters are convenient for the efficient particle acceleration in that system (Bykov A., Gladilin P., Osipov S., 2012, to be published).

4. Conclusion

The model describing the time-dependent particle acceleration in the vicinity of two colliding shocks is presented. This model predicts an unusual hard spectrum of CR protons and electrons confined in the flow and allows to accelerate particles up to the 10^{14} eV. We concentrated here on

Journal of Physics: Conference Series **461** (2013) 012007 doi:10.1088/1742-6596/461/1/012007

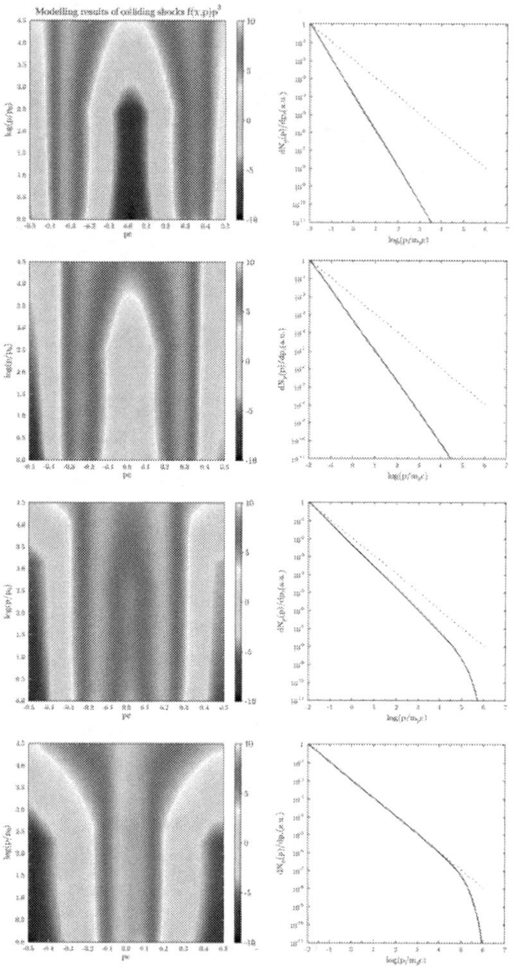

Figure 2. On the left: proton distribution function $f(x, p)\, p^3$ as a function of the CR momentum and the observer position x, presented for four different distances between the shocks - from top to bottom: 0.6 pc, 0.5 pc, 0,3 pc, 0.1 pc. On the right: proton spectrum dN_p/dp at the moving left shock for a corresponding times; dashed line - spectrum predicted by the test-particle solution $dN_p/dp \propto 1/p$ (eq. 9,10), solid line - simulated spectrum.

the test-particle time-dependent solutions without taking into account CR pressure modification of the shocks fronts. The non-linear effects of the described system will be discussed in the following paper (Bykov A., Gladilin P., Osipov S., 2012, to be published).

The stage of the closely approaching SNR and stellar wind flows that is favorable for particle acceleration typically lasts for 200-1,000 years depending on the stellar wind velocity and the termination shock radius. Therefore, such sources can likely contribute to the galactic CRs population.

Acknowledgments
The work was supported in part by the Russian government grant 11.G34.31.0001 to the Saint-Petersburg State Polytechnical University, and also by the RAS Programs (P21 and OFN 16), and by the RFBR grant 11-02-12082-ofi-m-2011. The numerical simulations were performed at SC at Ioffe Institute.

References

[1] Hillas A M 2005 *Journal of Physics G Nuclear Physics* **31** 95–131
[2] Aharonian F, Bykov A, Parizot E, Ptuskin V and Watson A 2012 *Space Sci. Rev.* **166** 97–132 (*Preprint* 1105.0131)
[3] Ptuskin V, Zirakashvili V and Seo E S 2010 *ApJ* **718** 31–36 (*Preprint* 1006.0034)
[4] Axford W I, Leer E and Skadron G 1977 *International Cosmic Ray Conference* vol 11 pp 132–137
[5] Krymskii G F 1977 *Akademiia Nauk SSSR Doklady* **234** 1306–1308
[6] Bell A R 1978 *MNRAS* **182** 147–156
[7] Blandford R D and Ostriker J P 1978 *ApJ* **221** 29–32
[8] Blandford R and Eichler D 1987 *Phys. Reports* **154** 1–75
[9] Berezhko E G and Krymskij G F 1988 *Uspekhi Fizicheskikh Nauk* **154** 49–91
[10] Bell A R 1987 *MNRAS* **225** 615–626
[11] Malkov M A and O'C Drury L 2001 *Reports on Progress in Physics* **64** 429–481
[12] Blasi P 2004 *Astroparticle Physics* **21** 45–57 (*Preprint* arXiv:astro-ph/0310507)
[13] Amato E and Blasi P 2005 *MNRAS* **364** 76–80 (*Preprint* arXiv:astro-ph/0509673)
[14] Vladimirov A E, Bykov A M and Ellison D C 2008 *ApJ* **688** 1084–1101 (*Preprint* 0807.1321)
[15] Bykov A M, Gladilin P E and Osipov S M 2011 *MemSAIt* **82** 800 (*Preprint* 1111.2587)
[16] Skilling J 1975 *MNRAS* **172** 557–566
[17] Toptygin I N 1985 *Cosmic rays in interplanetary magnetic fields (Moscow: Izdatel'stvo Nauka)*
[18] Kang H 2011 *Journal of Korean Astronomical Society* **44** 49–58 (*Preprint* 1102.3109)
[19] Vink J 2012 *Astron. Astroph. Reviews* **20** 49 (*Preprint* 1112.0576)

Journal of Physics: Conference Series **461** (2013) 012008 doi:10.1088/1742-6596/461/1/012008

Relation Between Initial Cosmological Conditions and the Properties of Dark Matter Haloes

Vadim Semenov[1,2]
[1]Astro Space center of Lebedev Physical Institute of Russian Academy of Sciences, Profsojuznaja st. 84/32, 117997 Moscow, Russia
[2]Moscow Institute of Physics and Technology, Institutskij per. 9, 141700 Dolgoprudny, Russia

E-mail: vsemenov@asc.rssi.ru

Abstract. The core-cusp problem is one of the essential issues in modern cosmology. The Entropy Theory of haloes Evolution recently suggested by Lukash, Doroshkevich and Mikheeva is one of the possible solutions to this problem. This work compares some results of numerical simulation of Large-Scale Structure with the conclusions of the Entropy Theory in order to verify this theory. The numerical simulation was performed in a volume 100 Mpc/h in a side using $\sim$ 17 million particles. Dark matter particles, which then form virialized haloes, were found in the initial perturbation field. This work investigates the distribution of these dark matter particles and measures the velocity dispersion profiles. It also traces evolution of haloes entropy profiles. On the whole, simulation results correspond to Entropy Theory of haloes evolution.

1. Introduction

In the standard (ΛCDM) cosmological model the energy density of dark matter is 5 times higher than that of baryon matter. Therefore, it is the dark matter that governs the processes of clustering and formation of the Large-Scale Structure of the Universe. Numerical simulation of the dark matter distribution is a powerful method to investigate the properties of the Universe structure on the range of scales and verify or exclude existing physical models. Even though nowadays results of numerical simulations are mainly in a good agreement with observations [1], some drawbacks exist. One of them is a divergence of dark matter haloes' density behaviour near the center of halo in observations and numerical simulations, the so-called core-cusp problem. Observations of the velocity curves of galaxies with low surface brightness (LSB) show that there is constant density in dark matter haloes, whereas haloes in numerical simulations have singular density in the center ($\rho(r) \propto r^{-1}$).

There are several plausible solutions to this problem. First of all, cusps might be produced by inadequacy of the simulation technique itself, where one deals with the truncated power spectrum of initial perturbation due to limited size of cells and the box. It worth noting that cosmological numerical simulation is an intensively developing brunch of science. It has been being refined for several last decades. The computation power of modern supercomputers is becoming higher and higher. Nevertheless, the density profiles of dark matter haloes still have singular cusps. The other solution to the problem is to modify the theory. For instance, the core-cusp problem might be solved by including the exotic dark matter in the theory. However, authors of [2] believe that there is no reason in such significant modifications and the core-cusp problem can be solved within the ΛCDM model. They suggest the Entropy Theory of dark

Journal of Physics: Conference Series **461** (2013) 012008 doi:10.1088/1742-6596/461/1/012008

matter halo formation. According to this theory cusps in numerical simulations are produced by the underestimating of initial small-scale entropy of dark matter particles.

In order to verify some conclusions of the Entropy Theory a numerical simulation of dark matter distribution was performed. This report provides the first results of this simulation.

2. The numerical simulation of the Large-Scale Structure of the Universe

To investigate properties of the inner structure of dark matter haloes the numerical simulation was run in a cubical volume 100 Mpc/h in a side. 253^3 probe dark matter particles were used and corresponding mass resolution was $4.96 \cdot 10^9$ $M_\odot$. The simulation was performed with the aid of GADGET-2 code [3, 4]. The used values of the main parameters are specified in the table 1.

Table 1. Parameters of the model.

Variable	Value	Variable	Value
Box Size	100 Mpc/h	Dark energy density (Ω_Λ)	0.7
Number of particles	256^3	Matter density (Ω_0)	0.3
Mass of particles	$4.96 \cdot 10^9$ $M_\odot$	Hubble constant (H_0)	70 Mpc/(km/s)
Force resolution	7 kpc/h	Spectrum index (n)	1
Initial redshift (z_{init})	79		

To find haloes of dark matter widely applied spherical overdensity criterium [5] was used. The approach starts with finding of the densest clusters of particles via Minimal Spanning Tree Method. This method is very similar to "Friend-of-Friend" Method [6] and provides ability to find all clusters which are denser than given threshold density (specified by threshold length l_{th}) and threshold mass (i.e. the number of particles within a cluster M). The threshold length defines the boundary density of the cluster, where separation of particles is about l_{th}, whereas within the cluster separation is lower than l_{th}. In this work l_{th} was selected to be about the length of force resolution. Apart from finding clusters, this choice of l_{th} showed that the length of force resolution was reasonable. Found clusters were associated with the inner parts of dark matter haloes. It may be suggested that the center of the halo, that corresponds to some cluster, lays within this cluster, because found clusters are the densest parts of haloes.

On the first step the center of each halo can be roughly associated with the center of mass of the corresponding dark matter particles cluster. In order to define the size of the particular halo, one can place a probe sphere and choose the size of this sphere so that the average density inside the sphere would exceed the critical density by a factor of $\Delta_c \approx 200$. For virialized systems in standard cosmological model dependence of Δ_c on z is given in [7].

In order to obtain appropriate profiles of density and entropy, one should accurately define the center of the halo. While there are several definitions of the halo's center [8], in this paper the density maximum is used as one. All the haloes were further fitted with universal Navarro-Frenk-White (NFW) profile [9] with dependence of concentration on a halo's mass proved to be very similar to the one obtained in [10].

3. Evolution of the individual halo

3.1. Initial clouds of particles

With data on the position and velocity of each particle in different moments of time, one can study in details processes in the evolution of the individual halo. This data also provides means

Journal of Physics: Conference Series **461** (2013) 012008 doi:10.1088/1742-6596/461/1/012008

for investigating the distribution of particles on the early stages of evolution and even in the initial conditions. It is interesting to study initial properties of particles that form halo. These particles form extended "clouds" in the initial distribution of dark matter particles. However, the investigating of such clouds is complex due to entanglement of particles, which will form a halo, with particles, which will be thrown out during halo formation. This work shows that profile of a halo's particle fraction is almost the same for all clouds and in the center of the cloud there are almost all particles will get into the halo. Furthermore, distribution of these particles from the central part of the cloud in the halo depends on process of the halo evolution. In case of monolithic collapse these particles are concentrated in the center of the halo. In more general case of hierarchical clustering these particles are spread over the halo.

3.2. Evolution of central density

It is the central part of the halo where one has surplus of particles in comparison to the observation data. Therefore, it is interesting to study the evolution of the central density of the halo. Typical behaviour of mass within the sphere of physical radius equals to $r_s/2 \approx 100$ kpc/h in the center of the halo is shown on figure 1.

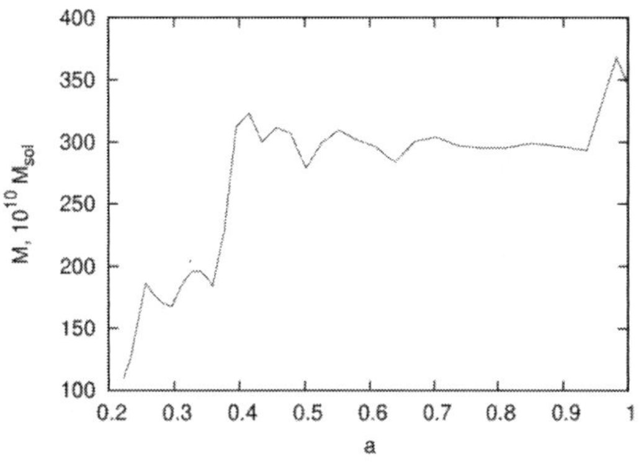

Figure 1. Solid line represents evolution of mass within the central region of size $r_s/2 \approx 100$ kpc/h with scale factor a.

Mass changes significantly during the first collapse and then remains almost constant. Actually, after formation of the central part smaller sub-haloes continue falling onto the main halo. However, this process cannot change central mass significantly due to high velocities of dark matter particles and sub-haloes. Falling particles come through the central part and it results in slight oscillations of the mass (figure 1, $a > 0.4$).

3.3. Spatial asymmetry of haloes

In analytical models of the spherical cloud collapse one can obtain density profile with quite a steep slope in the center: $\rho(r) \propto r^{-\alpha}$, where $\alpha \approx 1.6 - 2.0$ [11, 12]. However in numerical simulations haloes density profiles can be approximated with NFW profile with slope index $\alpha = 1$. The obvious difference between spherical initial conditions and real conditions in simulations of the Large-Scale Structure is the spatial asymmetry of real distribution of dark matter particles. In this research the criterion of asymmetry was chosen as the ratio of the largest eigenvalue of the inertia tensor to the lowest one. During the evolution this ratio increases because sub-haloes collapse and then decreases with falling of sub-haloes on the center. Therefore, maximum of this ratio for the individual halo can be associated with the moment of formation of sub-haloes.

As for initial and final values of asymmetry, the ratio of eigenvalues changes significantly. For the main part of initial clouds of particles this ratio covers the range from 1.2 to 1.6, while for the major part of final haloes this ratio lies between 1.3 and 3.0. It is worth noting that the minimal values of the ratio for initial and final states are 1.12 and 1.25, respectively. It gives an indication of maximal symmetry in the simulation.

4. The Entropy Theory of haloes formation

The Entropy Theory suggested in [2] provides a possible solution to the core-cusp problem. It claims that cusps in numerical simulations appear due to underestimation of low-scale initial perturbation. To make the analysis clearer authors suggest to convert density profiles $\rho(r)$ into entropy profiles $E(M)$. Here M is the mass within sphere of radius r and E is an entropy function, defined as $E = \sigma^2 \cdot \rho^{-2/3}$, where σ is the dispersion of particles' velocities. The entropy function behaviour close to the center can be approximated as $E \propto M^\beta$. Critical value $\alpha = 1$ corresponds to $\beta = 5/6$. In this research inner slopes of all haloes proved to be $\beta > 5/6$, with averaged inner slope being about $\beta = 1$.

4.1. Initial entropy profiles

In the initial state distribution of dark matter particles is almost uniform, thus entropy function depends only on velocities dispersion: $E \propto \sigma^2$. To obtain analytical shape of initial clouds entropy profiles one can consider the perturbation of the cloud as a set of plane waves falling into the center of the cloud. In this case velocities dispersion is given by:

$$\sigma^2(r) = \frac{1.7aH_0^2\Omega_m}{4\pi^2} \int\limits_0^\infty P(k) \left[1 - \frac{\sin(2kr)}{2kr}\right] dk,$$

where $P(k)$ is a power spectrum of perturbation.

Article [2] provides solutions obtained with other methods. These solutions, the solution given above and numerical results are given in figure 2. The article also predicts a large scatter of σ, thus the results are in a good agreement.

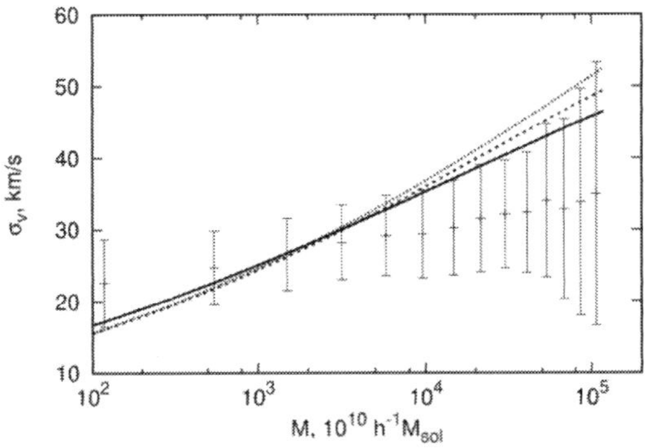

Figure 2. Dispersion profiles in the initial conditions. Solid line represents solution obtained in this research, dashed lines represent solutions from [2]. The averaged entropy profile of particles clouds in the initial conditions ($z = 79$) is given by error bars.

4.2. Evolution of entropy profiles

Figure 2 shows that entropy profiles are initially plane, whereas in the haloes they have steep slope $\beta \approx 5/6$. Tracing of entropy profiles evolution shows the existence of two different stages. Initially, on the linear stage of evolution, entropy profiles remain plain and then radically change their shapes during the hierarchical collapse. It is the non-linear stage of evolution when entropy profiles attain the slope $\beta > 5/6$. For that reason density profiles were converted into entropy ones. It makes analysis clearer.

Unfortunately, the efforts to find direct dependence of slopes on initial entropy magnitudes were not successful (fig. 3). The case needs more profound and refine analysis. For instant, the next step might be distinguishing between different times of collapse.

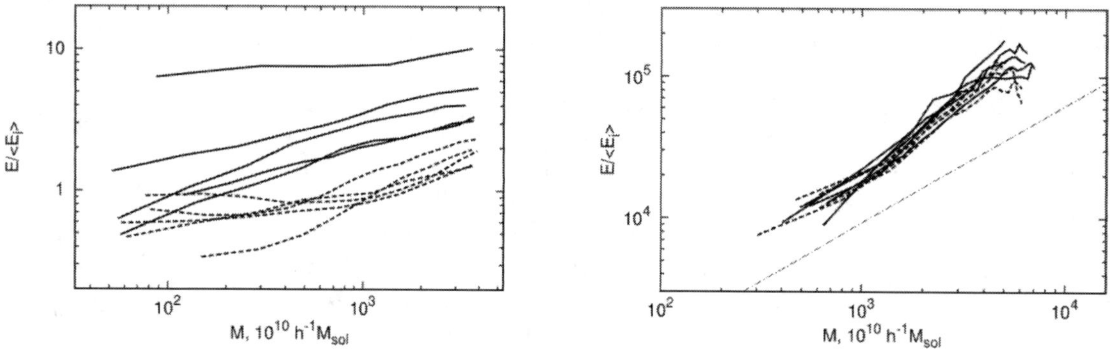

Figure 3. Entropy profiles in the initial conditions (left panel, $z = 79$) and in the results of simulation (right panel, $z = 0$). Solid lines represent haloes with high initial entropy, while dashed ones represent haloes with low initial entropy. The dash-dotted line on the right panel shows the slope equals to 5/6.

5. Following steps

Finding actual processes that govern formation of haloes inner structure is a complicated problem. Now we have variety of means to study evolution of haloes in details. There are still plenty of factors which need more profound analysis: minor/major merging, velocity anisotropy, angular momentum, etc. Furthermore, investigation of modified spectra with enhanced or suppressed perturbations of various ranges is very instructive, since it enables to examine how perturbations of different scales affect inner structure of dark matter haloes. Thus, this report contains merely intermediate results of extensive research aimed to shed light on the core-cusp problem and working upon the problem is going on.

Acknowledgments

I express my deepest gratitude to Dr. A.G. Doroshkevich for his invaluable guidance and to S.V. Pilipenko for discussions and constant support. I am also very grateful to Dr. V.N. Lukash for general formulation of the problem and discussions. This work has been supported by the Ministry of Education and Science of the Russian Federation (contract 8422), the "Scientific Personnel" Federal Targeted Program (contract 16.740.11.0460) and the "Dynasty" Foundation.

References

[1] Springel V, Frenk C S and White S D M 2006 *Nature* **440** 1137
[2] Doroshkevich A G, Lukash V N and Mikheeva E V 2012 *Phys. Usp.* **55** 3
[3] Springel V 2005 *Mon. Not. R. Astron. Soc.* **364** 1105
[4] Springel V, Yoshida N and White S D M 2001 *New Astron.* **6** 51
[5] White M 2001 *Astron. and Astrophys.* **367** 27
[6] Einasto J, Klypin A A, Saar E and Shandarin S F 1984 *Mon. Not. R. Astron. Soc.* **206** 529
[7] Brean G L and Norman M L 1998 *Astrophys. J.* **495** 80
[8] Kravtsov A V and Borgani S *Annual Rev. of Astron. and Astrophys.* **50** 353
[9] Navarro J F, Frenk C S and White S D M 1995 *Mon. Not. R. Astron. Soc.* **275** 720
[10] Klypin A, Trujillo-Gomez S and Primack J 2011 *Astrophys. J.* **740** 102
[11] Fillmore J A and Goldreich P 1984 *Astrophys. J.* **281** 1
[12] Gurevich A V and Zybin K P *Sov. Phys. JETP* **67** 1

Simulation of pumping mechanism of H_2O-masers in circumstellar envelopes of late-type stars

A V Nesterenok
Ioffe Physical-Technical Institute, 26 Polytechnicheskaya St., 194021, Saint Petersburg, Russia
St. Petersburg State Polytechnic University, 29 Polytechnicheskaya St., 195251, Saint Petersburg, Russia

E-mail: alex-n10@yandex.ru

Abstract. The radiative transfer problem is considered in dense gas-dust clouds. Physical parameters are adopted in simulations corresponding to gas-dust clouds in the circumstellar envelopes of late-type stars. An one-dimensional plane-parallel slab geometry of the cloud is considered. The radiative transfer problem is solved using the accelerated lambda iteration technique. The rotational levels of the five lowest vibrational levels of H_2O-molecule are considered in the simulations. The level populations of H_2O-molecule are computed as a function of distance in the cloud. The conditions of the level population inversion occurrence are investigated. In particular, the factors controlling the population inversion in the 22.2 GHz maser line are considered. A radiative excitation of the molecules by external radiation field of a star is considered together with the collisional pumping of the maser.

1. Introduction

Water maser emission has been detected in many astrophysical objects – active galactic nuclei, star formation regions, circumstellar envelopes of late-type stars. In this paper we will focus on the water masers in the envelopes of the late-type stars.

Stars with masses up to a few solar masses evolve onto the Asymptotic Giant Branch after the hydrogen fuel has been exhausted in its interior. The stars become very large (> 1 AU) and cool (≈ 2500 K). Stars with higher mass become Red Supergiants and have ten times the size of AGB stars. The evolution of late-type stars is dominated by the mass loss. The photosphere and expanding circumstellar envelope are effective producers of a variety of molecular species. Oxygen-rich stars produce SiO, H_2O and OH masers with spatially very compact, spectrally narrow lines. The high intensity and the narrowness of spectral features of a maser allow one to obtain detailed maps of the distribution of maser sources. The H_2O masers around late-type stars are found at the distances of about 5-50 stellar radius from the central star [1]. They provide a very high resolution probe of the structure, kinematics and conditions in the circumstellar envelope.

In this study a simulation is done of the pumping process of water masers in a circumstellar envelope of late-type star. Two mechanisms of molecule excitation are considered – collisional excitation of the molecules and excitation by the radiation of the central star.

2. The method

2.1. Radiative transfer and population balance

Each energy level i of the molecule is characterized by its statistical weight g_i, energy ε_i and the population n_i. The energy relations $\varepsilon_i > \varepsilon_j$ and $\varepsilon_i < \varepsilon_j$ are abbreviated by $i > j$ and $i < j$, respectively. The equations of statistical equilibrium for the level populations may be written

$$n_i \sum_{j \neq i}^{M} \left(R_{ij} + C_{ij}\right) - \sum_{j \neq i}^{M} n_j \left(R_{ji} + C_{ji}\right) = 0 \tag{1}$$

where C_{ij} are the collisional rate coefficients and R_{ij} represent radiative rate coefficients, M is the total number of energy levels involved in the model. The values of R_{ij} are given by $R_{ij} = A_{ij} + B_{ij}\bar{J}_{ij}$ for $i > j$ and $R_{ij} = B_{ij}\bar{J}_{ij}$ for $i < j$, A_{ij} and B_{ij} are the Einstein coefficients, $\bar{J}_{ij}$ is the line-weighted mean intensity, defined by

$$\bar{J}_{ij} = \frac{1}{4\pi} \int_0^{4\pi} d\Omega \int_0^{\infty} d\nu\, \varphi(\Omega,\nu) I(\Omega,\nu)$$

where $I(\Omega,\nu)$ is the intensity at frequency ν and angle Ω, $\varphi(\Omega,\nu)$ is the normalized line profile function. In the plane parallel geometry the transfer equation for the intensity is

$$\mu \frac{\partial I(\mu,\nu)}{\partial z} = -\chi(\mu,\nu) I(\mu,\nu) + \eta(\mu,\nu) \tag{2}$$

where z is the coordinate along the normal to the cloud plane, $\mu = l/cos\,\theta$, θ is the angle between the normal to the cloud plane and a given direction, $\chi(\mu,\nu)$ and $\eta(\mu,\nu)$ are the total opacity and emissivity at frequency ν and direction μ. The emissivity and opacity in the neighbourhood of a given line $i \to j$ depend on the optical properties of the dust and level populations n_i and n_j [2]. The problem of calculating the populations of energy levels consists in solving the set of rate equations (1) coupled to the equations of radiative transfer (2).

Various techniques have been developed to yield a self-consistent set of level populations and radiation fields [3]. We have adopted the accelerated lambda iteration (ALI) method for molecular level calculation. The detailed description of the method was given by [2]. In brief outline, the equations are solved iteratively starting with any initial level populations. The mean intensities computed with the aid of equation (2) are substituted to the statistical equilibrium equations (1) and new level populations are calculated. The procedure is repeated until the relative change in the level populations or mean intensities between two successive steps falls below a given convergence criterion. The concept of the lambda operator is introduced. The lambda operator represents the entire procedure of computing the line-weighted mean intensities $\bar{J}_{ij}$ from the source function

$$\bar{J}_{ij} = \Lambda_{ij}\left[S^{\dagger}(\Omega,\nu)\right]$$

where $S^{\dagger}(\Omega,\nu)$ is the source function from the previous iteration step. The acceleration of iteration scheme is achieved by splitting the lambda operator onto the local contribution to the intensity and the remainder [2]. The calculation of the line-weighted mean intensity at each iteration step is modified

$$\bar{J}_{ij} = \left(\Lambda_{ij} - \Lambda^{*}_{ij}\right)\left[S^{\dagger}(\Omega,\nu)\right] + \Lambda^{*}_{ij}\left[S(\Omega,\nu)\right] \tag{3}$$

This is inserted into the equations of statistical equilibrium (1) and the equations are solved for the new level populations. Note, that the equation (3) contains the new populations in the last term which have to be found in the end of iteration step. In addition to this operator splitting technique, one can apply iteration improvement schemes such as the Ng acceleration [2].

2.2. Collisional rate coefficients and spectroscopic data

We calculate the populations for the 350 rotational levels of the five lowest vibrational levels of H_2O molecule. The energy of the uppermost level is about 4800 cm^{-1}, or 6900 K in units of temperature. The spectroscopic data for the H_2O molecule is taken from the HITRAN 2008 database [4]. The rate coefficients for H_2O collisional transitions in inelastic collisions of H_2O with H_2 were taken from [5]. The rate coefficients for H_2O collisional transitions in collisions of H_2O with He were taken from [6]. The data [6] are restricted to 45 lower H_2O levels. For higher H_2O levels collisional rate coefficients are taken to be scaled rate coefficients H_2O-H_2 with the scaling factor being 0.74. The factor accounts for the different reduced mass when He is the collisional partner. The helium-to-hydrogen ratio was assumed to be He/H = 0.1. The rate constants of inelastic collisions involving H_2O molecule and atomic hydrogen H are assumed to be a factor 1.15 larger than He-H_2O rate coefficients. This value takes both the smaller mass and the smaller cross section of H atom into account.

2.3. Physical model

The physical parameters adopted in the calculations are representative for the gas clumps in the wind of AGB stars [1]. The distance between the masing cloud and the star is taken to be $D = 10R_S$ where R_S = 4×10^{13} cm – the star radius. An one-dimensional plane-parallel slab geometry of the cloud is considered. The cloud thickness H is chosen to be 5×10^{13} cm. The cloud is divided into N = 100 layers, using a distance scale defined by

$$z_{k+1} - z_k = a(z_k - z_{k-1}),\ z_{k+1} - z_k = z_{N-k} - z_{N-k-1},\ k = 0,\dots,N/2-1,$$

where k is the depth point number, a is the coefficient. For the boundary layers the size is chosen such that the maximum optical depth in any line is low, $z_1 - z_0 = z_N - z_{N-1} = H/10^6$.

The mass loss rate of the star is adopted to be $\wp$ = 3×10^{-6} of solar masses per year. One may estimate the gas density of the outflow based on the mass conservation equation

$$\rho = \wp\left(4\pi r^2 \upsilon\right)^{-1}$$

where ρ is the gas density, r is the distance from the star, υ is the outflow velocity. Taking the gas outflow velocity of υ = 10 km/s, for the hydrogen atom concentration (both unbound and bound in molecular hydrogen) we have $n_H = 4\times10^7$ cm^{-3} at the distance in question. It's suggested that the densities of H_2O maser clouds are higher than the average wind densities [1]. We have used three values of the total hydrogen atom concentration in the maser cloud – 4×10^7 cm^{-3}, 1.2×10^8 cm^{-3} and 4×10^8 cm^{-3}. The ratio of concentrations of molecular and atom hydrogen is adopted to be 1. The H_2O abundance is taken to be equal to the average solar system abundance of oxygen – 5.4×10^{-4} [7]. The velocity of turbulent motions of the gas is adopted to be 1 km/s, velocity gradient in the cloud – 10^{-9} s^{-1} [1].

We consider the fixed size of the dust grains 0.1 μm. We use the complex dielectric function for the dust material presented in [8]. The absorption efficiency is computed from Mie theory for spherical grains (program BHMIE from [9]). The dust to gas mass ratio is taken to be 0.01.

The gas and dust temperatures in the cloud depend on cooling and heating processes such as gas-dust collisions, radiative cooling and heating by starlight. We use the calculation results from [10,11]. The dust and gas temperatures are taken to be equal 400 K and 600 K, respectively.

Journal of Physics: Conference Series **461** (2013) 012009 doi:10.1088/1742-6596/461/1/012009

The stellar temperature is taken to be 2500 K. The radiation field intensity of the star is assumed to have a blackbody spectrum. The stellar radiation is taken into account in the boundary conditions for the radiation transfer differential equation (2).

Using the ALI method with a local operator and Ng acceleration, the system is iterated towards a solution with a convergence criterion of 10^{-4} in relative change in level populations for successive iterations.

3. Calculation results

The H_2O maser line at 22.2 GHz corresponds to the transition between rotational levels of the ground vibrational state of ortho-H_2O molecule $6_{16} - 5_{23}$. The Figure 1 presents the calculated gain at the line profile center for the maser line 22.2 GHz. The spectral gain profile broadening due to the hyperfine splitting is taken into account [12]. The average values of the gain are 1.1×10^{-13}, 0.9×10^{-13} and 0.8×10^{-13} cm^{-1} for the values of total hydrogen atom concentration 4×10^7, 1.2×10^8 and 4×10^8 cm^{-3}, respectively. The average value of the gain decreases with increasing gas density.

The effect of the stellar radiation field on the maser pumping process is studied. The figure 2 presents the calculation results for the gain in two cases – with and without taking into account the stellar radiation in the radiative transfer. The total hydrogen atom concentration is taken to be 1.2×10^8 cm^{-3}. The average values of the gain are 0.9×10^{-13} and 1.3×10^{-13} cm^{-1}, with the gain being lower when the stellar radiation is taken into account. The result contradicts to the generally accepted conclusion that the stellar radiation has a minimal effect on the pumping process of the 22 GHz maser [13].

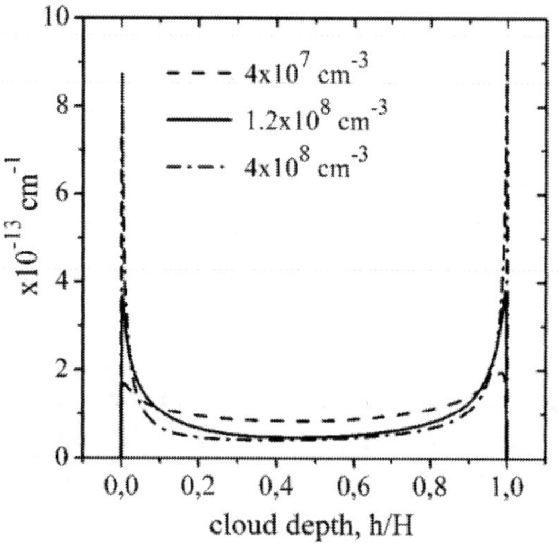

Figure 1. The gain in the 22 GHz maser line as a function of the cloud depth. The results for three values of total hydrogen atom concentration are presented.

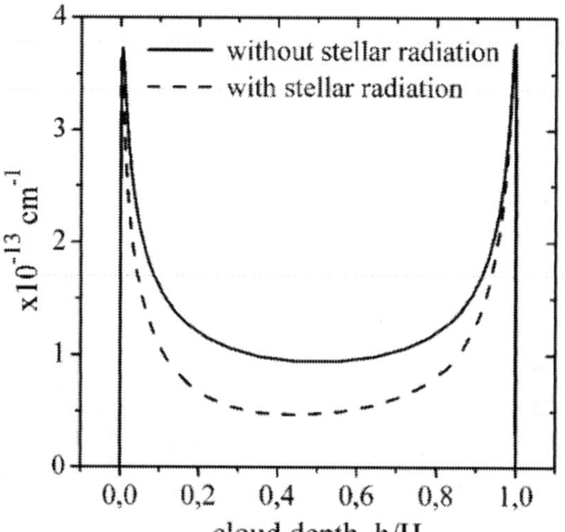

Figure 2. The gain in the 22 GHz maser line as a function of the cloud depth. The results are given for the calculations with and without taking into account stellar radiation in the radiative transfer.

4. Conclusions

In the present work, accelerated lambda iteration method has been used to calculate the level populations of H_2O molecule in the dense gas-dust clouds. The physical conditions encompassing in the inner part of circumstellar envelopes are adopted in the simulations. A radiative excitation of the molecules by radiation field of a star is considered in the model. The gain in the maser line of ortho-H_2O molecule 22.2 GHz is calculated. It is found that the effect of the stellar radiation on the gain in the maser line is substantial. The radiation field of the star has to be taken into account in the simulations of the maser pumping process.

Acknowledgments
This work was supported by the Russian Foundation for Basic Research (project no. 11-02-01018a), the Program of the President of Russia for Support of Leading Scientific Schools (project no. NSh-4035.2012.2), by Ministry of Education and Science of Russian Federation - Agreement No.8409, 2012 and contract # 11.G34.31.0001 with SPbSPU and leading scientist G.G. Pavlov.

5. References

[1] Richards A M S, Etoka S, Gray M D, Lekht E E, Mendoza-Torres J E, Murakawa K, Rudnitskij G and Yates J A 2012 *Astron. Astrophys.* **546** A16
[2] Rybicki G B and Hummer D G 1991 *Astron. Astrophys.* **245** 171–81
[3] van Zadelhoff G-J, Dullemond C P, van der Tak F F S, Yates J A, Doty S D, Ossenkopf V, Hogerheijde M R, Juvela M, Wiesemeyer H and Schoier F L 2002 *Astron. Astrophys.* **395** 373–84
[4] Rothman L S et al. 2009 *J. Quant. Spectr. Radiat. Transfer* **110** 533–72
[5] Faure A and Josselin E 2008 1991 *Astron. Astrophys.* **492** 257–64
[6] Green S, Maluendes S and McLean A D 1993 *Astrophys. J. Suppl. Ser.* **85** 181–5
[7] Lodders K, Palme H and Gail H-P 2009 in *Landolt-Börnstein: Group VI Astronomy and Astrophysics Numerical Data and Functional Relationships in Science and Technology, Solar System* ed Trümper J E (Berlin: Springer) **4B** (*Preprint* arXiv:0901.1149)
[8] David P and Pegourie B 1995 *Astron. Astrophys.* **293** 833–9
[9] Bohren C F and Huffman D 1983 Absorption and scattering of light by small particles (New York: John Wiley)
[10] Babkovskaya N and Poutanen J 2006 *Astron. Astrophys.* **447** 949–62
[11] Decin L, De Beck E, Brunken S, Muller H S P, Menten K M, Kim H, Willacy K, de Koter A and Wyrowski F 2010 *Astron. Astrophys.* **516** A69
[12] Nesterenok A V and Varshalovich D A 2011 Astronomy Letters **37** 456–67
[13] Cooke B and Elitzur M 1985 *Astrophys. J.* **295** 175-82

Magnetic field generation in galactic molecular clouds

A Kiselev

Moscow Institute of Physics and Technology, Institutskii per., 9, Dolgoprudny, Moscow Region, Russia

E-mail: kiselevalexs@mail.ru

Abstract. We discuss magnetic field generation in weakly ionized turbulent gas. We consider the case of nonzero mean magnetic field. We assume that fluctuations are isotropic, and derive evolution equation for pair correlation function of magnetic fluctuations. Stationary solution of this equation is presented.

1. Introduction

A few years ago Pamela satellite has observed an excess of positrons in cosmic rays at 10-100 GeV energy range. Several theoretical explanations of this effect were presented, among them generation of positrons in pulsars or in dark matter annigalation process. We propose another source of positrons: cosmic rays acceleration in galactic giant molecular clouds and secondary particle production there. This mechanism has been studied earlier in [1],[3]. In these papers an exess of positrons in cosmic rays has been predicted. Appearance of appropriate observations requires a detailed investigation of this mechanism, with taking into account modern data.

Molecular clouds are clusters of molecular hydrogen with a complex inhomogeneous structure. According to observations, gas in them is strongly turbulent. The turbulence has a power-law Kolmogorov-like spectrum. In addition, gas is partially ionized, $N_i/N_n = 10^{-8} - 10^{-5}$. In such a system stochastic magnetic field arises, which can accelerate charged particles. Polarization observations [2], which were carried out for dozens molecular clouds, have shown that magnetic field directions in distant points of cloud may be similar. Probably, mean homogeneous magnetic field exists in clouds along with stochastic field.

In this study we investigate magnetic field generation in weakly ionized turbulent gas with nonzero mean magnetic field.

2. Magnetohydrodinamics eguations for weakly-ionized gas

In our problem for description of gas motion one can use double-fluid hydrodynamic equations. We denote $\mathbf{v}$ and $\mathbf{u}$ velocity of neutral and ionized gas components, ρ_i - dencsity of ionized component and μ_{in} - ion-neutral collision rate. We consider inertial range of scales $L_\nu < L < L_0$, where L_0 is determined by the size of the system, and L_ν corresponds to the viscous scale. Typical values for molecular clouds - $L_0 = 10^{19}$ cm, $Re = 10^8$, hence for the Kolmogorov turbulence $L_\nu = 10^{13}$ cm. In this range of scales viscosity can be neglected. We will also assume gas to be incompressible.

Since the concentration of ions is very small, neutral gas does not feel them. Ions motion, by contrast, is completely determined by the motion of the neutral gas. Thus we can treat the motion of the neutral component known - it coincides with the ordinary hydrodynamic turbulence. For ionized components only two forces are essential - the force of friction on the neutral gas and the Lorentz force. Therefore the equation of motion of the ionized component becomes

$$\frac{(\nabla \times \mathbf{B}) \times \mathbf{B}}{4\pi\rho_i} - \mu_{in}(\mathbf{u} - \mathbf{v}) = \mathbf{0} \tag{1}$$

Well-known induction equation for magnetic field is

$$\frac{\partial \mathbf{B}}{\partial t} = \nabla \times (\mathbf{u} \times \mathbf{B}) \tag{2}$$

Let us denote

$$a = \frac{1}{4\pi\rho_i\mu_{in}} \tag{3}$$

As far as we assume gas to be incompressible, a = const. Expressing velocity of ionized component and substituting it in the induction equation (2), we obtain

$$\frac{\partial \mathbf{B}}{\partial t} = \nabla \times (\mathbf{v} \times \mathbf{B}) - a\nabla \times (\mathbf{B} \times (\nabla \times \mathbf{B}) \times \mathbf{B}) \tag{4}$$

This equation gives the dependence of the magnetic field on the velocity $\mathbf{v}$ of neutral gas. See [1] for details of derivation of (4).

3. Correlator's evolution equation

We will use a model, introduced by Kazantsev [4]. To describe the turbulent motion of the neutral gas, we assume its velocity $\mathbf{v}(t)$ to be a Gaussian stochastic process with zero mean $< \mathbf{v} >= 0$. All information about it is contained in the pair correlation function $< v_i(\rho, t)v_j(\rho', t') >$. The angle brackets here and below denote averaging over an ensemble of realizations. We will consider the simplest case of a delta-correlated in time stochastic process

$$< v_i(\rho, t)v_j(\rho', t') >= v_{ij}(\rho - \rho')\tau_c\delta(t - t') \tag{5}$$

Neutral gas in our problem is a homogeneous isotropic medium. We assume it to be incompressible, hence $\operatorname{div} \mathbf{v} = \mathbf{0}$, consequently $\partial_i v_{ij} = \partial_j v_{ij} = 0$, which leads us to the general form of the correlation tensor

$$v_{ij}(r) = 2V(r)\delta_{ij} + rV'(r)(\delta_{ij} - \frac{r_ir_j}{r^2}). \tag{6}$$

The magnetic field we also consider to be a Gaussian stochastic process, but with nonzero mean. Denote its mean and fluctuating components by $\mathbf{H}$ and $\mathbf{b}$ respectively:

$$\mathbf{B} = \mathbf{H} + \mathbf{b}, \qquad < \mathbf{B} >= \mathbf{H} \tag{7}$$

We believe mean component to be constant in space and time $\mathbf{H} = \text{const}(\mathbf{r}, t)$. In the presence of the mean magnetic field appears a preferred space direction, so the correlator of the magnetic field can be anisotropic. But for simplicity we will assume it to be isotropic. This can be done because, as we will see below, in the case of small mean field H the amplitude of the fluctuations $|b| \simeq H^{1/2}$ is greater than H.

Maxwell equation $\operatorname{div} \mathbf{b} = 0$ is similar to the incompressibility equation, so tensor structure of the magnetic field correlator is

$$< b_i(\mathbf{x}, t)b_j(\mathbf{x} + \mathbf{r}, t) >= 2Q(r)\delta_{ij} + rQ'(r)(\delta_{ij} - \frac{r_ir_j}{r^2}) \tag{8}$$

Journal of Physics: Conference Series **461** (2013) 012010 doi:10.1088/1742-6596/461/1/012010

Using equation (4), one can derive an evolution equation for the magnetic field correlator, that is to compute $\partial Q/\partial t$. To do this, we have to split the arising correlators like $< vb^2 >$ and $< b^4 >$. Since all our random processes are Gaussian, to split the first correlator we use Furutsu-Novikov formula, and to split the second we present it as a product of pair correlators. To obtain an isotropic tensor, we have to replace

$$h_i h_j \Longrightarrow \frac{1}{3}\delta_{ij}$$

where $h_i = H_i/H$ - unit vector in the direction of the mean field. After some tensor algebra we get an equation

$$\frac{1}{2\tau_c}\frac{\partial Q(r)}{\partial t} = \left(V(0) - V(r) + \lambda + \frac{2a}{\tau_c}\frac{H^2}{3}\right)(Q'' + \frac{4Q'}{r}) - V'Q' - \frac{1}{r}(4V' + rV'')(Q + \frac{1}{6}H^2) \quad (9)$$

where we denote an unknown parameter

$$\lambda = \frac{4a}{\tau_c}Q(0) \quad (10)$$

4. Stationary correlation function

Let us reduce our equation to dimensionless one. For distance unit we take the size of the molecular cloud L_0, for unit of magnetic field - the value $\sqrt{\frac{\tau_c}{2a}V(0)}$, a for unit of time the value

$$\frac{L_0^2}{2\tau_c V(0)} = \frac{\tau_{max}^2}{\tau_c}$$

where τ_{max} - eddy turnover time at scale L_0.

Thus, we reduce equation (9) to the form

$$\frac{\partial Q(r)}{\partial t} = \left(V(0) - V(r) + \lambda + \frac{1}{3}H^2\right)(Q'' + \frac{4Q'}{r}) - V'Q' - \frac{1}{r}(4V' + rV'')(Q + \frac{1}{6}H^2) \quad (11)$$

where $\lambda = 2Q(0)$, and $V(0) = 1$.

For typical parameters of molecular clouds: concentration of neutrals $N_n = 10^3\,\mathrm{cm}^{-3}$, and ions $N_i = 10^{-2}\,\mathrm{cm}^{-3}$, $L_0 = 10^{19}$ cm, $L_\nu = 10^{13}$ cm, the unit of measurement of the magnetic field is equal to 5μG.

We choose a function $V(r)$, which describes the motion of the neutral gas, in the form

$$V(r) = \begin{cases} 1 - r^\alpha & \text{when } r < 1, \\ 0 & \text{when } r > 1 \end{cases} \quad (12)$$

Below we consider only the case of $\alpha = 2/3$, which corresponds to Kolmogorov turbulence.

Let us find stationary solutions of (11). We have linear second-order equation for the function $Q(r)$. It has two independent solutions. When $r \to 0$ ($r_\nu \ll r \ll \lambda^{1/\alpha}$) one can find the asymptotic behavior of solutions, and when $r > 1$ one can obtain an explicit solution

$$r \to 0 \;:\quad Q^{(1)} \sim 1; \qquad Q^{(2)} \sim r^{-3} \quad (13)$$

$$r > 1 \;:\quad Q^{(1)} = 1; \qquad Q^{(2)} = r^{-3} \quad (14)$$

We are interested in a solution, bounded at zero and decreasing at infinity. One can show that such a solution does exist, and it fix the value of the parameter λ (for each value of the

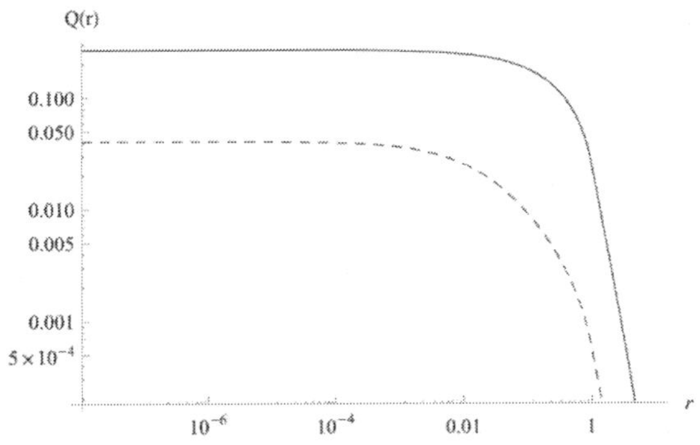

Figure 1. Correlation function $Q(r)$ for $H = 1$ - solid line, and for $H = 0.1$ - dotted line

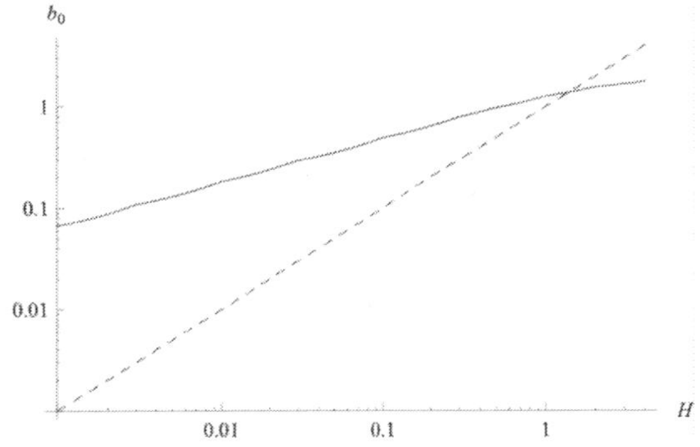

Figure 2. Fluctuations amplitude b_0 against the value of mean field H. For comparison, line $b_0 = H$ is shown dashed

average magnetic field H). Graph of this solution for the values of $H = 1$, $H = 0.1$ is shown in Figure 1. Since

$$< b_i(\mathbf{r}) b_i(\mathbf{r}) >= b_0^2 = 3\lambda \tag{15}$$

where b_0 - the amplitude of the fluctuations of the magnetic field, we obtain the dependence of b_0 on H, shown in Figure 2. In the case $H < 1$ one can see that $b_0 \sim \sqrt{H}$. Therefore, when $H \ll 1$, we obtain $b_0 \gg H$ - random field exceeds the mean field.

It should be noted that when $H \to 0$ the value of the random field b_0 decreases to zero, and in the limit $H = 0$ our solution becomes identically zero. The fact is that if $H = 0$, there are no (non-zero) solutions, bounded at $r \to 0$ and decreasing at infinity.

5. Discussion

In the papers [5], [6] similar problem of a turbulent dynamo in a conducting fluid is studied for the case $\nu \ll \eta$, where ν, η - kinematic and magnetic viscosity, respectively. The role of our unknown parameter λ plays η, which is considered to be known.

In [5] authors try to take into account the presence of the mean magnetic field, but they derive evolution equation for correlators only for zero mean field. Therefore, in paper mentioned, our general equation (11) was solved only in special case $H = 0$. In this case, for the Kolmogorov turbulence $V(r) = 1 - r^\alpha$ with $\alpha = 2/3$ there are no solutions, growing with time, see [5] (and there are no limited solutions of the stationary equation). To avoid this problem, Rogachevskii et al and Malyshkin et al use rather artificial method - they suppose the correlation time τ_c to be a non-constant function of scale. In this approach, Kolmogorov turbulence corresponds to $\alpha = 4/3$ and growing solutions appear (and for stationary equation - limited solution appears). In this work we have demonstrated that when mean feild is taken into account, limited solution can exist even at $\alpha = 2/3$.

References

[1] Dogiel V A, Gurevich A V, Istomin Ya N and Zybin K P 1987 *MNRAS* **228** 843
[2] Tassis K, Dowell C *et al* 2009 *MNRAS* **399** 1681
[3] Dogiel V A, Gurevich A V, Istomin Ya N and Zybin K P 2004 *Astrophysics & Space Science* **297** 201
[4] Kazantsev A P (1967) Enhancement of a magnetic field by a conducting fluid, *JETP*, **26** 1031
[5] Rogachevskii I and Kleeorin N 1997 *Phys. Rev.* E **56** 417
[6] Malyshkin L M and Boldyrev S 2010 *Phys. Rev. Lett.* **105** 215002

Journal of Physics: Conference Series **461** (2013) 012011 doi:10.1088/1742-6596/461/1/012011

The Peculiarities Of The Initial Phase Of The 24-th Solar Activity Cycle

P A Otkidychev, N N Skorbezh
GAS GAO RAN, Kislovodsk, Russia
E-mail: otkid@list.ru

Abstract. A set of peculiarities of the 24-th solar activity cycle is considered on the base of the data of Kislovodsk Mountain Astronomical Station (www.solarstation.ru) as well as NASA/Marshall data. The peculiarities of the 24-th cycle are: the asymmetry between sunspot areas of northern and southern solar hemispheres; the difference between sunspot latitudes in northern and southern solar hemispheres; low activity of the first years of the cycle; large relative amount of single sunspots.

1. Introduction

The Sun observations in white light at the Kislovodsk Mountain Astronomical Station (GAS GAO RAN) carry out from the October 1954 [1]. Thus, the set of GAS GAO RAN observations includes the majority of the 19-th cycle, complete 20–23 cycles and the beginning of the 24-th cycle. At the present time observations carry out on CCD by the instrumentality of the MEADE telescope ($D = 15$ cm, $f = 120$ cm).

2. The asymmetry between northern and southern solar hemispheres

There is considerable difference between monthly average areas of sunspot groups in northern and southern solar hemispheres. Adopting the beginning of the 24 cycle as January 2009, we obtain the next mean value of monthly average sunspot areas by Kislovodsk data: 334.0 mhm in the north and 185.7 mhm in the south (here and farther the data are given for the state of August 2012); thus the difference is 1.9 times (although in July 2012, on the contrary, sunspot area in the south was 8.5 times greater than in the north). The correlation coefficient between the north and the south is only 0.29. Sunspot areas shown in the figure 1: the total area (thick line), area of northern (thin line) and southern (dotted line) solar hemisphere.

Wolf numbers in the north also overcome ones in the south. Monthly average meanings of Wolf numbers by Kislovodsk data are: 30.0 in the north and 16.3 at the south. The difference between them is 1.8 times. Wolf numbers shown in the figure 2 similarly to sunspot areas.

The asymmetry exists also in latitude distributions of sunspots. Mean yearly latitudes of sunspot groups by NASA/Marshall data for each year from 2009 are given in the table 1 (calculated as the sum of products of latitudes by sunspot areas, divided by the overall sum of sunspots areas for each year and for each hemisphere). Mean latitudes of sunspot groups in the south are appreciably more than in the north. Besides, mean latitudes at the south in 2012 almost haven't changed in comparison with 2011. This fact is in disagreement with Sporer's low. Then, the rate of the change of average sunspot latitude must be about 1.6 degree per year [2].

Journal of Physics: Conference Series **461** (2013) 012011 doi:10.1088/1742-6596/461/1/012011

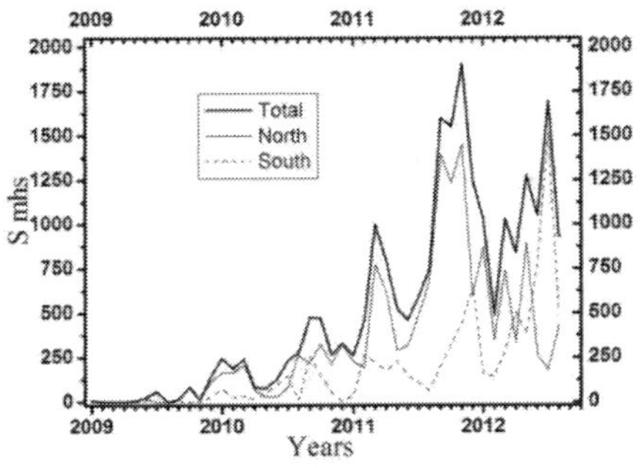

Figure 1. Mean monthly sunspot areas.

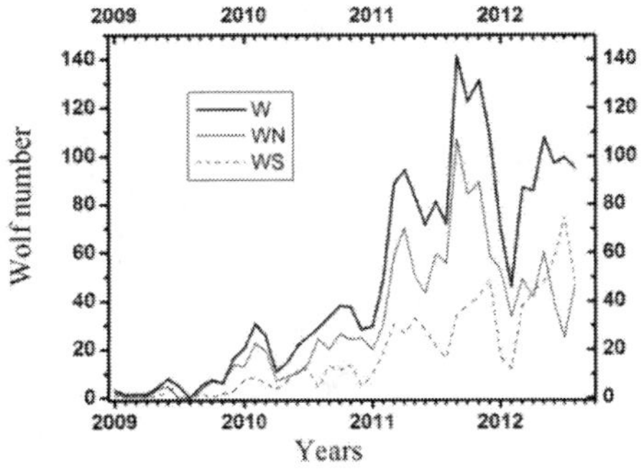

Figure 2. Mean monthly Wolf numbers.

This condition is approximately true only for 2009-10 years. The data of table 1 represented in the figure 3 (thick line for the North and thin line for the South).

Table 1. Mean yearly sunspots latitude of northern and southern solar hemispheres

Year	North	South
2009	22.2	25.8
2010	20.4	24.5
2011	16.7	18.8
2012	15.6	18.6

Mean monthly latitudes (calculated similarly to mean yearly latitudes) represented in the figure 4 (thick line for the North and thin line for the South too). It is worth paying attention to the fact, that sunspots in southern hemisphere "have switched over" to new cycle only in May 2012.

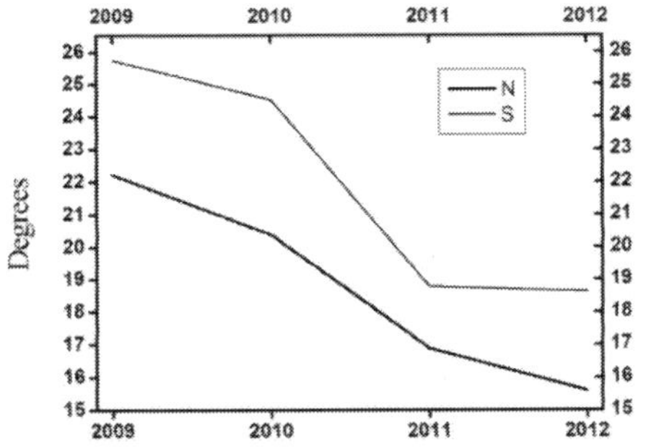

Figure 3. Mean yearly sunspot latitudes.

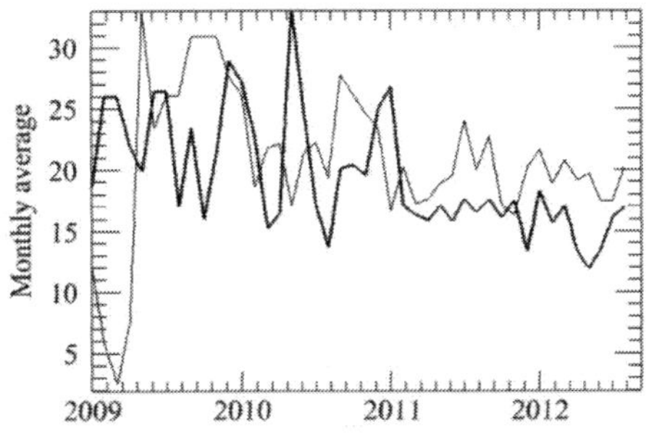

Figure 4. Mean monthly sunspot latitudes.

As for the extremal meaning of latitudes, the least latitude was registered in 24.12.2011 at the group number 11383: 2.3° North, at that one of the sunspots had 1.9° North. Also it should to note group number 11560: for the period 02-06.09.2012 it was crossing the equator, having mean latitude about 3° North. Such low latitudes are typical of the end of the cycle.

3. The initial phase activity of the 24 cycle

The initial phase of the 24 cycle has low activity in comparison with previous cycles. The cycle activity from 20 to 24 cycle is submitted in the table 3: there are mean sunspot area S_{mean} and Wolf number W_{mean} for the first year of each cycle and for the first three years (by GAS data).

Table 2. The cycle activity from 20 to 24 cycle by GAS GAO RAN data

Cycle number & beginning	S_{mean} for the 1-st year	S_{mean} for the 1-st 3 years	W_{mean} for the 1-st year	W_{mean} for the 1-st 3 years
20 (1964/10)	83.2	564.2	17.3	61.6
21 (1976/06)	158.6	926.5	18.5	83.4
22 (1986/09)	325.2	1279.7	30.7	91.1
23 (1996/05)	113.5	567.9	12.9	51.9
24 (2009/01)	30.6	403.0	5.5	40.3

NASA/Marshall data permit to carry out the comparison of the first years of the 24-th cycle with the first years of the weak cycles in the past. From the cycles, which represented in the NASA/Marshall database, 12-th and 14-th are the weakest. The cycle activity of 12-th and 14-th cycles is submitted in the table 3: there are mean sunspot area S_{mean} (by NASA/Marshall data) and Wolf number W_{mean} (by Solar Influences Data Analysis Center, Royal Observatory of Belgium) for the first year of each cycle and for the first three years.

Table 3. 12-th and 14-th cycles by NASA/Marshall & SIDC data

Cycle number & beginning	S_{mean} for the 1-st year	S_{mean} for the 1-st 3 years	W_{mean} for the 1-st year	W_{mean} for the 1-st 3 years
12 (1878/12)	33.7	369.8	5.4	29.6
14 (1901/09)	32.2	237.8	2.8	19.1

The comparison shows that 24 cycle is similar by activity with two weakest cycles from the beginning of the RGO observations, namely with 12 and 14 ones.

The data of tables 3 and 3 has combined in the figures 5, 6 (areas) and 7, 8 (Wolf numbers).

4. The relative quantity of single sunspots

The data of Kislovodsk Station enable to calculate the total and the relative amount of single sunspots (i.e. groups, composed of only one sunspot). The relative quantity of single sunspots (calculated for each year as number of groups, composed of only one sunspot, divided by number of all groups for this year) from 1955 to 2012 represented in the figure 9.

5. Conclusions

As follows from the figure 9, the minima of the relative quantity of single sunspots fall on the cycle minima, with the exception of the minimum between 19-th and 20-th cycles. 24-th cycle demonstrates very high level of the relative amount of single sunspots by comparison with previous years. This fact as well as low cycle activity and spot asymmetry between the north and the south makes this cycle extraordinary. This is the weakest cycle for last hundred years.

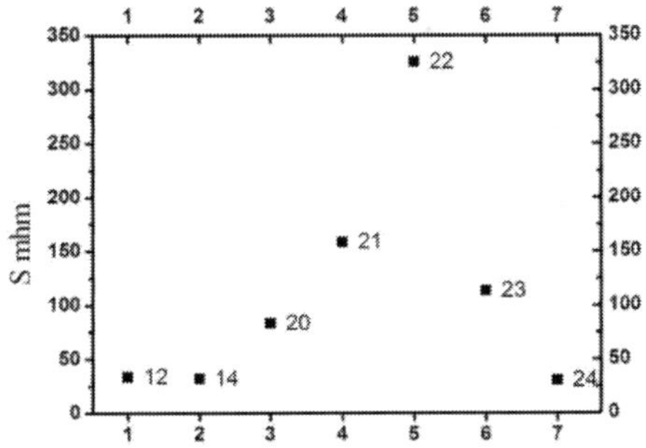

Figure 5. Mean sunspot areas for the first year from the cycle beginning.

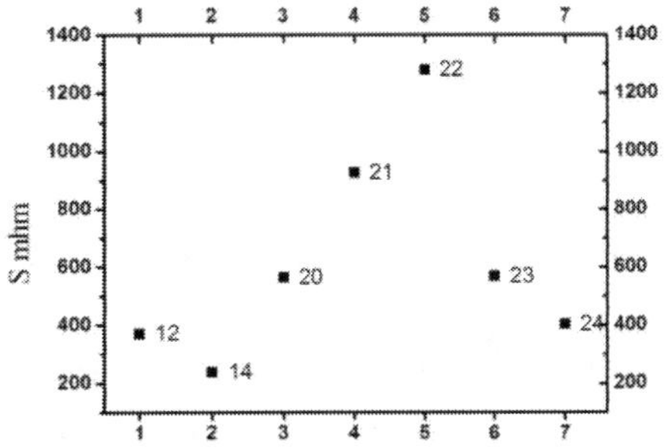

Figure 6. Mean sunspot areas for the first three years from the cycle beginning.

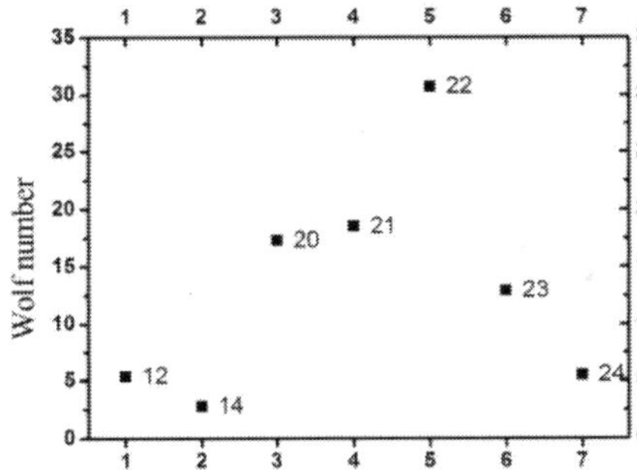

Figure 7. Mean Wolf numbers for the first year from the cycle beginning.

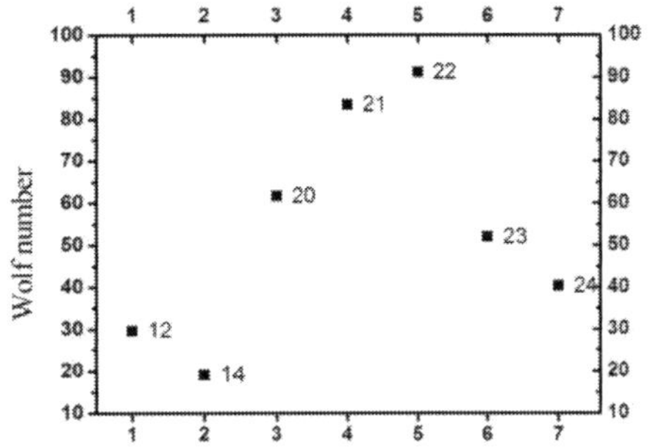

Figure 8. Mean Wolf numbers for the first three years from the cycle beginning.

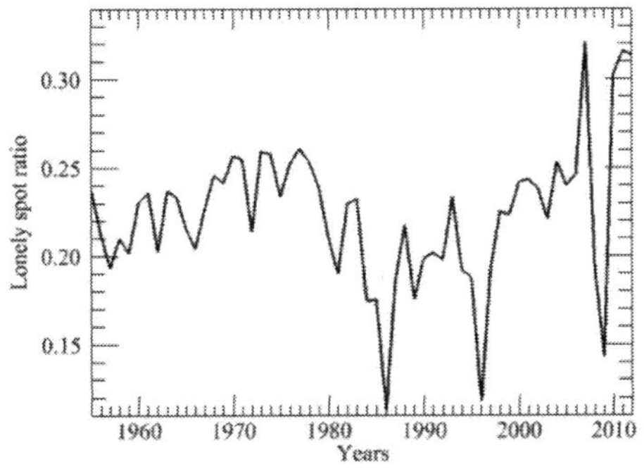

Figure 9. The relative quantity of single sunspots.

Acknowledgments
The work is supported by the grant of Russian Foundation for Basic Research 12-02-31128.

References

[1] Gnevyshev M N and Makarov V I 1989 *Kislovodsk Mountain Astronomical Station GAO AN SSSR* (Leningrad: Nauka) p 46
[2] Li K J, Yun H S and Gu X M 2001 *The Astronomical Journal* **122** 2115-17

Journal of Physics: Conference Series **461** (2013) 012012 doi:10.1088/1742-6596/461/1/012012

The influence of core superfluidity on the neutron stars long-term rotation evolution.

D P Barsukov[1,2], O A Goglichidze[1] and A I Tsygan[1]

[1]Ioffe Physical-Technical Institute of the Russian Academy of Sciences, st. Polytekhnicheskaya, 26, Saint-Petersburg, 194021, Russian Federation.
[2]Saint-Petersburg State Polytechnical University, st. Polytekhnicheskaya, 29, Saint-Petersburg, 195251, Russian Federation.

Abstract. We investigate the evolution of neutron star rotation taking into account the superfluidity of the neutrons in the neutron star core. The neutron star is treated as a two-component system consisting of a charged component (including the crust and the core protons, electrons and normal neutrons) and a core superfluid neutron component. The components are supposed to interact through the mutual friction force. We assume that the charged component rotates rigidly. The neutron superfluid velocity field is calculated directly from linearized hydrodynamical equations. It is shown that the superfluid core accelerates the evolution of inclination angle and makes all pulsars evolve to orthogonal state. But as it is known from observations the rate of the angle evolution is not very high. Therefore, a small size of superfluid cores is more likely. These facts may allow to examine superfluid models.

1. Introduction

Neutron stars cores consist mostly of the neutrons with small fractions of protons, electrons and probably more exotic particles like hyperons. The neutrons in some regions of the star must be in superfluid state at that time as the protons must be in superconducting state. Superfluidity may appear in the rotational dynamics in several ways and several time-scales. Present work is devoted to studying how the superfluidity affects the long-term rotation evolution.

2. Superfluid hydrodynamics

It is well known that for the uncharged superfluids like neutron liquid the velocity field $\vec{v}$ must satisfy the equation $\mathrm{rot}\,\vec{v} = 0$. However, such superfluids can rotate by forming an array of vortices. The velocity field near each vortex has the form:

$$\vec{v} = \frac{\hbar}{m_{cp}\tilde{r}}\vec{e}_\phi, \tag{1}$$

where m_{cp} is the mass of Cooper pair, $\tilde{r}$ is the distance from the vortex core, $\vec{e}_\phi$ is the azimuthal unit vector. Vorticity is located only in the vortex cores, in which superfluidity breaks. However, one can average $\vec{v}$ over the volume greater than the intervortex space. This procedure allows to use the hydrodynamical equations instead vortices dynamics consideration even when the rotating superfluids are studied (see [1] for example).

In a mixture of superfluids so-called entrainment effect takes place (see [2]). For the neutron-proton mixture it leads to the following formulas:

$$\vec{J}_n = (\rho_n - \rho_{nn} - \rho_{np})\vec{v}_{ex} + \rho_{nn}\vec{v}_n + \rho_{np}\vec{v}_p, \tag{2}$$

$$\vec{J}_p = (\rho_p - \rho_{pp} - \rho_{pn})\vec{v}_{ex} + \rho_{pp}\vec{v}_p + \rho_{pn}\vec{v}_n, \tag{3}$$

$$\vec{p}_n = (\rho_n - \rho_{nn} - \rho_{np})\vec{v}_{ex} + (\rho_{nn} + \rho_{np})\vec{v}_n, \tag{4}$$

$$\vec{p}_p = (\rho_p - \rho_{pp} - \rho_{pn})\vec{v}_{ex} + (\rho_{pp} + \rho_{pn})\vec{v}_p, \tag{5}$$

where $\vec{J}_\alpha$ is the mass current of the α constituent ($\alpha = p, n$), $\vec{p}_\alpha$ are the proton and neutron momentum densities, $\vec{v}_{ex}$ is the thermal excitation velocity (it is assumed that the excitation velocity is the same for both fluids), $\rho_{\alpha\beta}$ is the mass density matrix. It can be shown that $\rho_{np} = \rho_{pn}$. The superfluid velocities here are determined just like for the ordinary superfluids:

$$\vec{v}_n = \frac{\hbar}{2m_n}\nabla S_n, \quad \vec{v}_p = \frac{\hbar}{2m_p}\nabla S_p - \frac{e}{m_p c}\vec{A}, \tag{6}$$

where S_α are the phases of the complex order parameters, $\vec{A}$ is the electromagnetic vector potential. So in the superfluid mixtures the mass currents corresponding to each fluid are no longer parallel to they velocities and don't equal to corresponding momentum density. Note, however, that

$$\vec{J}_n + \vec{J}_p = \vec{p}_n + \vec{p}_p. \tag{7}$$

3. The Model

We treat a neutron star core as a two-component system. The first component consists of the superfluid neutrons and moves with smooth averaged velocity $\vec{v}_s$. Protons, electrons, and normal neutrons are coupled to each other on the small time-scales and form the second component which we denote as "charged". All particles forming the charged component are supposed to move with the same velocity $\vec{v}_c$:

$$\vec{v}_e = \vec{v}_{ex} = \vec{J}_p/\rho_p = \vec{v}_c. \tag{8}$$

The star crust rotates with angular velocity $\vec{\Omega}$ which is identified with the observed angular velocity of the pulsar. The external torque $\vec{K}$ acts on the crust. We will consider the torques depending only on $\vec{\Omega}$ and the magnetic field configuration which supposed to be frozen in the crust. In this case, the torque $\vec{K}$ is a very slowly evolving vector in the frame co-rotating with the crust.

Except the entrainment effect, superfluid and charged components interact through the so-called mutual friction, which arises when the charged component particles are scattered by the neutron vortices. The system of hydrodynamical equations describing the mixture of superfluid neutrons, superconducting protons and degenerate electrons taking into account the neutron-proton entrainment, mutual friction and gravitational force was developed by Mendell and Lindblom (see [3] and [4]). We are based on these equations.

Introducing $\vec{u}_\alpha = \vec{v}_\alpha - [\vec{\Omega} \times \vec{r}]$ which is a velocity field of the α constituent measured in the frame co-rotating with the crust, we suppose that

$$\Omega r \gg u_s \gg u_c, \tag{9}$$

so we keep only the linear in u_s terms and neglect all terms containing u_c. The second part of inequality (9) requires a sufficiently effective physical mechanism which damps the differential motions of the charged component and connect this component to the crust. It can be ensure by viscosity or by the magnetic field (the last requires the second type superconductivity for the protons). Terms containing rot $\vec{\lambda}_n$ in Mendell-Linblom equations arise from vortices self-acting. These terms do not participate in transferring the angular momentum from the superfluid component to the charged one, so in the linear approximation they can be ignored. We also neglect the gravitational potential perturbation.

Vector $\vec{\Omega}$ as well as the external torque $\vec{K}$ evolves very slowly in co-rotation frame in compare with the mutual friction time-scales Thus, the long-term evolution corresponds to the quasistationary solution. It means that at the large time-scales

$$\partial_t \vec{u}_s \approx [\vec{\Omega} \times \vec{u}_s] - \left([\vec{\Omega} \times \vec{r}] \cdot \nabla \right) \vec{u}_s, \tag{10}$$

and equations for $\vec{u}_s$ take the form:

$$2\Omega \frac{\varrho^2}{\rho_{pp}\rho_s} \left(\delta [\vec{e}_z \times \vec{u}_s] - \beta \vec{e}_z \times [\vec{e}_z \times \vec{u}_s] \right) + \nabla \tilde{\mu}_1 = -[\dot{\vec{\Omega}} \times \vec{r}] \tag{11}$$

$$\operatorname{div} \left(\frac{\varrho^2}{\rho_{pp}} \vec{u}_s \right) = 0, \tag{12}$$

$$\left[\frac{\varrho^2}{\rho_{pp}} (\vec{u}_s \cdot \vec{r}) \right]_{r=r_{core}} = 0, \tag{13}$$

where $\delta = (1 - \beta')$, β and β' are the mutual friction coefficients, $\varrho^2 = \rho_{nn}\rho_{pp} - \rho_{np}^2$, r_{core} is the radius of sphere in which the superfluidity breaks, $\dot{\vec{\Omega}} = d_t \vec{\Omega}$, $\tilde{\mu}_1 = -\frac{1}{2}[\vec{\Omega} \times \vec{r}]^2 + \tilde{\mu}_n + \Phi_G$, $\tilde{\mu}_n$ is the neutrons chemical potential, Φ_G is the non-perturbed gravitational potential. Note that all mass densities here are not perturbed too.

The solution satisfying equations (11), (12), and (13) has the form

$$\vec{u}_s = \frac{\rho_{pp}\rho_s}{\varrho^2} [\vec{\varpi} \times \vec{r}] - \frac{\rho_{pp}\rho_s}{\varrho^2} \frac{\dot{\Omega}_{||}}{2\Omega} \frac{\delta - \beta\psi}{\delta^2 + \beta^2} \vec{\tilde{r}} + \vec{e}_z \frac{\rho_{pp}}{\varrho^2} \frac{\dot{\Omega}_{||}}{2\Omega} \int\limits_0^z \frac{1}{\tilde{r}} \frac{\partial}{\partial \tilde{r}} \left(\tilde{r}^2 \rho_s \frac{\delta - \beta\psi}{\delta^2 + \beta^2} \right) dz', \tag{14}$$

$$\vec{\varpi} = -\frac{\dot{\vec{\Omega}}_{||}}{2\Omega} \frac{\beta + \delta\psi}{\delta^2 + \beta^2} - \frac{\beta}{\delta^2 + \beta^2} \frac{\dot{\vec{\Omega}}_{\perp}}{\Omega} + \frac{\delta}{\delta^2 + \beta^2} \left[\vec{e}_z \times \frac{\dot{\vec{\Omega}}_{\perp}}{\Omega} \right], \tag{15}$$

$$\tilde{\mu}_1 = -yz\dot{\Omega}_{\perp} - \dot{\Omega}_{||} \int\limits_0^{\tilde{r}} \psi(\tilde{r}')\tilde{r}' d\tilde{r}', \tag{16}$$

$$\psi(\tilde{r}) = \left[\int\limits_0^{z_b} \frac{\rho_s \delta}{\delta^2 + \beta^2} dz' \right] \left[\int\limits_0^{z_b} \frac{\rho_s \beta}{\delta^2 + \beta^2} dz' \right]^{-1}, \quad z_b = \sqrt{r_{core}^2 - \tilde{r}^2}. \tag{17}$$

Here r_{core} is the radius of the superfluid core. The definitions of used coordinates and vectors are given in fig. 1.

In general case, the flow has a complex form. It consists of a differential rotation about the local axis $\vec{\varpi}$ and a poloidal flow. The last, however, is a consequence of the dependence of β and β' on the r and it vanishes if we make β and β' to be constant. Neglecting, in addition, the entrainment effect ($\rho_{pp}\rho_s/\varrho^2 \approx 1$), we obtain a rigidly rotating superfluid.

Protons of course are not required to be superconducting everywhere inside the superfluid core. Entrainment vanishes where the superconductivity breaks (see [5]). Obtained solution, however, remains to be valid there. One just needs to replace $\frac{\rho_{pp}\rho_s}{\varrho^2} \to 1$ and $\frac{\rho_{pp}}{\varrho^2} \to \frac{1}{\rho_s}$. Except this formally replacements the superconductivity breaking apparently leads to decrease by several orders of magnitude of the mutual friction coefficients.

4. The equations of motion

Possessing the expression for the superfluid velocity field, one can calculate the angular momentum transfer rate from superfluid core to the charged component. From the angular momentum conservation law we have

$$\dot{\vec{M}} = \vec{K}, \tag{18}$$

where $\vec{M}$ is the full angular momentum of the star which obviously is a sum of angular momenta of each component, so one can write

$$\vec{M} = \vec{M}_c + \vec{M}_s = \int \vec{r} \times (\vec{p}_c + \vec{p}_s) dV = \int \vec{r} \times \vec{J}_c dV + \int \vec{r} \times \vec{J}_s dV \tag{19}$$

Here we have used (7). By the assumption $\vec{J}_c \approx \rho_c[\vec{\Omega} \times \vec{r}]$ and $\vec{J}_s \approx \rho_s[\vec{\Omega} \times \vec{r}] + \frac{\varrho^2}{\rho_{pp}} \vec{u}_s$. Using (10), (14), (15), we obtain:

$$I_c \dot{\vec{\Omega}} = S_1 I_s \dot{\vec{\Omega}} - S_2 I_s \dot{\vec{\Omega}}_{||} + S_3 I_s [\vec{e}_z \times \dot{\vec{\Omega}}] + \vec{K}, \tag{20}$$

where

$$S_1 = \frac{8\pi}{3I_s} \int \frac{\delta\beta' - \beta^2}{\delta^2 + \beta^2} \rho_s r^4 dr, \; S_2 = \frac{8\pi}{3I_s} \int \frac{\delta}{\delta^2 + \beta^2} \rho_s r^4 dr, \tag{21}$$

$$S_3 = \frac{8\pi}{3I_s} \int \frac{\beta}{\delta^2 + \beta^2} \rho_s r^4 dr, \; I_\alpha = \frac{8\pi}{3} \int_0^R \rho_\alpha r^4 dr. \tag{22}$$

If we introduce basis vectors $\vec{\varepsilon}_x$, $\vec{\varepsilon}_y$, and $\vec{\varepsilon}_z$, anchored in the star crust, where $\vec{\varepsilon}_z = \vec{m}/m$ and vectors $\vec{\varepsilon}_x$ and $\vec{\varepsilon}_y$ are perpendicular to $\vec{\varepsilon}_m$ and to each other, the orientation of $\vec{\Omega}$ can be determined by two angles χ and φ_Ω (see fig. 2). Angle χ is the pulsar inclination angle, and the variation of φ_Ω relates with star precession. Without making any additional restriction external torque $\vec{K}$ can be represented as

$$\vec{K} = K_\Omega \vec{e}_z + K_m \vec{\varepsilon}_z + K_\perp [\vec{e}_z \times \vec{\varepsilon}_z] \tag{23}$$

Equation (20) can be solved for $\dot{\vec{\Omega}}$ and rewritten as three scalar equations:

$$\dot{\Omega} = \frac{K_\Omega + K_m \cos\chi}{I_s + I_c}, \tag{24}$$

$$\dot{\chi} = -\frac{1}{\Omega} \frac{(I_c + S_1 I_s) K_m - S_3 I_s K_\perp}{(I_c + S_1 I_s)^2 + S_3^2 I_s^2} \sin\chi, \tag{25}$$

$$\dot{\varphi}_\Omega = -\frac{1}{\Omega} \frac{(I_c + S_1 I_s) K_\perp + S_3 I_s K_m}{(I_c + S_1 I_s)^2 + S_3^2 I_s^2}. \tag{26}$$

Up to now the developed formalism does not require any specification of the profiles of the mutual friction coefficients and the mass densities as well as K_Ω, K_m and $K_\perp$ can be arbitrary small in magnitude functions. Next we apply the particular models (simplistic at some points) in order to demonstrate that the choice of the model may significantly affect the rate of the rotation evolution.

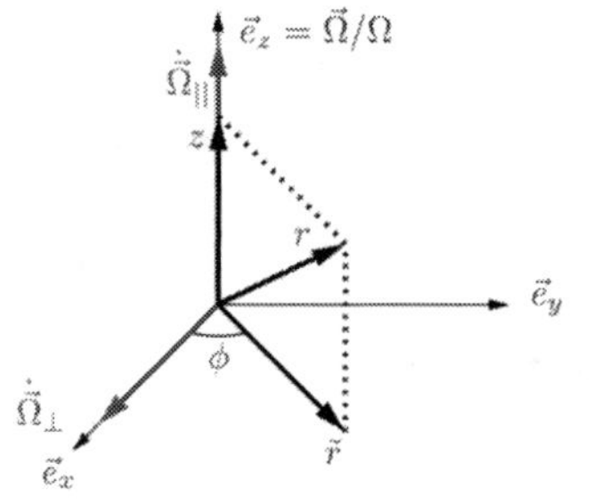

Figure 1. Coordinates and vectors used in section 3.

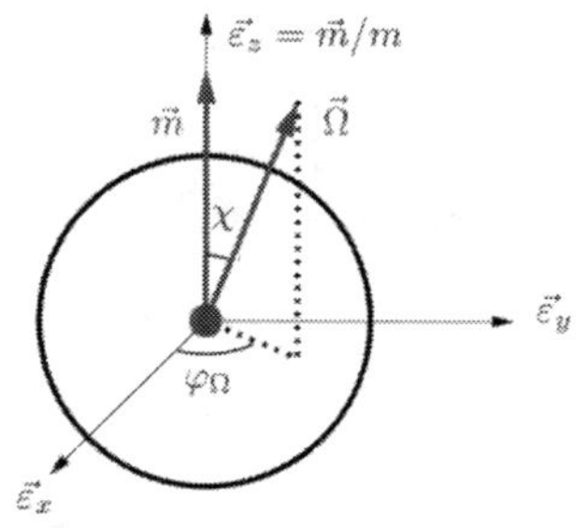

Figure 2. Coordinates and vectors used in section 4.

5. Inclination angle evolution

We suppose that the most effective mutual friction mechanism is based on the electrons scattering on the vortices magnetic field proposed by Alpar and Sauls (see [6]).

$$\beta = \frac{\sigma}{\sigma^2+1}, \quad \beta' = \frac{\sigma^2}{\sigma^2+1}, \quad \sigma = 1.3\times 10^{-2}\frac{x}{1-x}(x\rho_{14})^{1/6}\left(\frac{m_p}{m_p^*}\right)^{1/2}\left(\frac{\rho_{np}}{\rho_{pp}}\right)^2, \tag{27}$$

where $x = \rho_c/\rho$, $\rho_{14} = \rho/10^{-14}$ g cm^{-3}, m_p^* is the effective proton mass.

Superfluid mass density matrix $\rho_{\alpha\beta}$ with taking into account temperature effects has been calculated by Gusakov and Haensel (see [5]). Densities $\rho_{\alpha\beta}$ depend on the parameters $\tau_\alpha = T/T_{c\alpha}$, where T is the core temperature, $T_{c\alpha}$ are the temperatures of the protons and neutrons phase transition temperature. We model $T_{c\alpha}$ profiles by the parabolas (see fig 3):

$$\log_{10}\left(\frac{T_{c\alpha}}{T_{c\alpha}^{max}}\right) = -4\log_{10}^2\left(\frac{\rho}{\rho_\alpha^{max}}\right). \tag{28}$$

It is not quite realistic of course but the variation of T_{cn}^{max} or T_{cp}^{max} allows us to demonstrate the sensitivity of the results to the choice of critical temperature profiles. We determine r_{core} as a radius of the sphere on which $T = T_{cn}$ and suppose that $T = 5\times 10^7$K.

As for the torque $\vec{K}$, we use the one proposed by Barsukov, Polyakova and Tsygan (see [7] for the detail) for which $K_\Omega = -K_0$, $K_m = K_0(1-\alpha(\chi,\varphi_\Omega))\cos\chi$, $K_\perp = K_0 R\cos\chi$, $K_0 = \frac{2\Omega^3 m^2}{3c^3}$, $R = \frac{9c}{10\Omega r_{ns}}$, $\alpha(\chi,\varphi_\Omega)$ is the function ~ 1 which describes the structure of the small-scale magnetic field in vicinities of neutron star magnetic poles.

One can average the equations over the precession period and divide (25) over (26), taking into account that $P = 2\pi/\Omega$ is the pulsar period:

$$\frac{d\chi}{dP} \approx -\frac{1}{P}\frac{I_c(I_c+I_s)}{I_c^2+S_3^2 I_s^2}\frac{\sin\chi\cos\chi}{\sin^2\chi+\bar{\alpha}(\chi)\cos^2\chi}\left[1-\bar{\alpha}(\chi)-S_3\frac{I_s}{I_c}\frac{9}{20\pi}\left(\frac{c}{r_{ns}}\right)P\right]. \tag{29}$$

The evolution trajectories obtained for different initial χ and P and different T_n^{max} are given in figures 4, 5 and 6.

6. Discussion

The presence of the superfluid core increases the rate of inclination angle evolution and makes all pulsars evolve to the orthogonal state. This facts together with observations may allow to examine superfluid models.

We do not take into account any pinning phenomena which may play significant role in neutron star rotational dynamics. Also we do not take into account the thermal evolution of the neutron star. We will include these factors in our future developments.

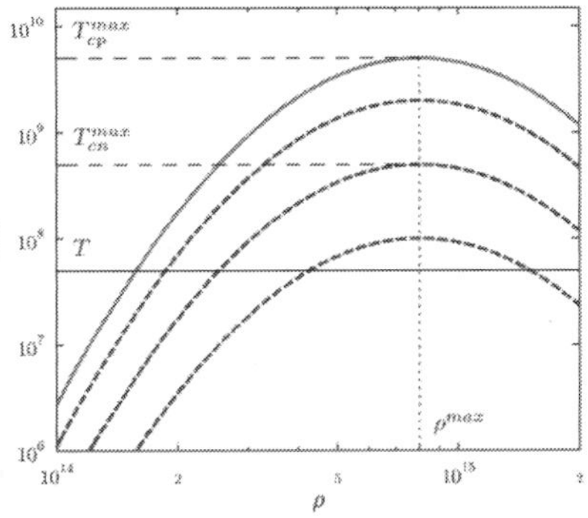

Figure 3. Critical temperature profiles for protons (solid line) and neutrons (dashed lines) used in the calculations. $\rho_n^{max} = \rho_p^{max} = 8 \times 10^{14}\mathrm{g/cm^3}$, $T_{cp}^{max} = 5 \times 10^9$K, $T_{cp}^{max} = 1,\ 5,\ 20 \times 10^8$K.

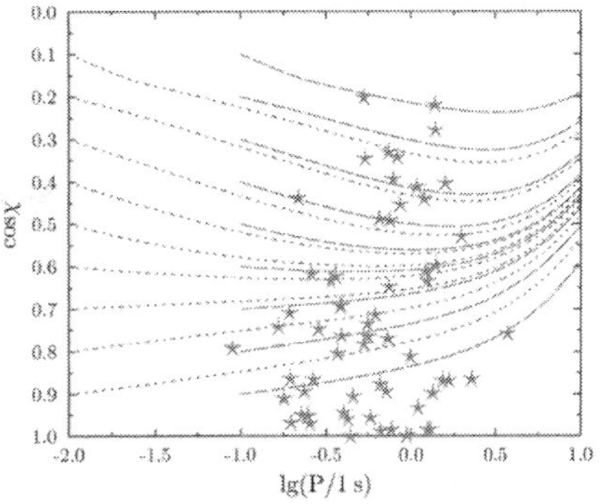

Figure 4. The evolution trajectories obtained for non-dipolarity parameter $\nu = 0.5$ and $T_{cn}^{max} = 1 \times 10^8$K ($I_s/I_c = 0.08$) for pulsars with initial periods equal to 10 msec (dotted lines) and 100 msec (dashed lines). Stars demonstrate the observation data for 62 pulsars from [8].

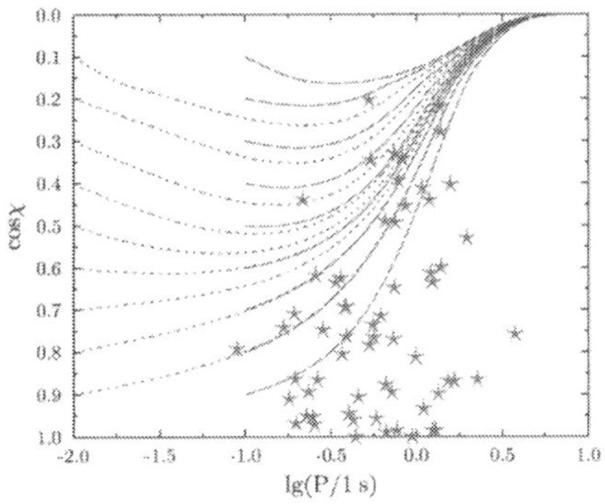

Figure 5. The same as before but $T_n^{max} = 5 \times 10^8$K ($I_s/I_c \approx 0.5$)

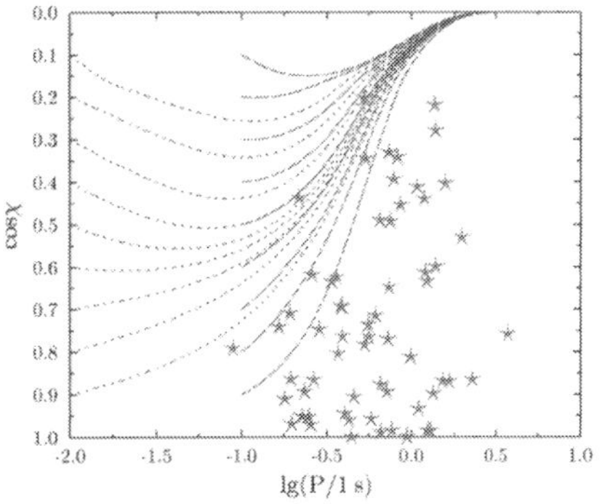

Figure 6. The same as before but $T_n^{max} = 2 \times 10^9$K ($I_s/I_c \approx 1$)

Acknowledgements
The authors are grateful to M E Gusakov and E M Kantor for discussion. This work was supported by the Russian Foundation for the Basic Research (project 13-02-00112), the Program of the State Support for Leading Scientific Schools of the Russian Federation (grant NSh-4035.2012.2) and Ministry of Education and Science of Russian Federation (Agreement No.8409, 2012 and contract No. 11.G34.31.0001).

References
[1] Baym G, Chandler E 1983 *Low Temp. Phys.* **50** 57
[2] Andreev A F, Bashkin E P 1976 *JETP*, **42** 164
[3] Mendell G Lindblom L 1991 *An. of Phys.*, **205** 110
[4] Mendell G 1991 *Astrophys. J.* **380** 515
[5] Gusakov E M, Haensel P *Nucl. Phys. A* **761** 333
[6] Alpar M A, Langer S A, Sauls J A 1984 *Astrophys. J.* **282** 533
[7] Barsukov D P, Polyakova P I, Thygan A I 2009, *Astron. Rep.* **53** 1146
[8] Malov I F 2004 *Radio Pulsars* (Moskow: Nauka) [in Russian]

Journal of Physics: Conference Series **461** (2013) 012013 doi:10.1088/1742-6596/461/1/012013

Spectral index map of the Crab Nebula in the optical range

N I Dolindo[1] and Yu A Shibanov [1,2]

[1] St. Petersburg State Polytechnical University, Polytechnicheskaya, 29, St. Petersburg, 195251, Russia

[2] Ioffe Physical Technical Institute, Politekhnicheskaya 26, St. Petersburg, 194021, Russia

E-mail: n.dolindo@gmail.com

Abstract. Information about spatial distribution of the spectral index of the pulsar wind nebulae is important for understanding the structure of the nebulae and physical processes responsible for their formation. We present preliminary results on construction of a detailed index map of the Crab Nebula based on archival optical data obtained with the Hubble Space Telescope. Our map has much better spatial resolution than the previous one obtained by Veron-Cetty and Woltjer in 1993. It allows us to study in great details how the spectral slope of the synchrotron emission of the Crab varies along its torus and jet-like structures.

1. Introduction

Pulsar wind nebulae (PWNe) are the nebulae formed by the interaction between the interstellar medium and relativistic pulsar wind. The Crab nebula (NGC 1952, Taurus A) powered by the pulsar B0531+21 is a classical example of such nebulae. It lies at a distance about 2 kpc from the Earth and is well-observed from radio to gamma-rays. The optical radiation of the Crab is composed of a nonthermal continuum and emission lines from a net of thin filaments covering the nebula. In radio and X-gamma-rays only continuum emission is observed which is described by a power law spectrum with different spectral indices in various spectral domains. The nebula has a spectacular torus-like shape with jets along the symmetry axis of the torus. Information about spatial distribution of the spectral index of the continuum is important for understanding the PWNe structure and physical processes responsible for their formation by comparison of observations with theoretical models. While there are many observations of the Crab, only a few studies of the spectral index variation was performed. The X-ray index map with a $\sim$2.5"–5" shows that the X-ray spectrum is hardest in the central area, close to the pulsar [3],[2]. The spectral index map in the radio with a resolution of $\sim$16" was presented by Bietenholz [4]. It shows a general steepening of the radio spectrum with the distance from the pulsar. In the mid-infrared the index map has a similar spatial resolution of $\sim$1" as in X-rays [5]. While in the optical the available map has significantly coarser spatial resolution of $\sim$10" [1], which does not allow to resolve properly the torus and jets. Here we present a preliminary results on the optical spectral index map constructed based on the archival Hubble Space Telescope (HST) data on Crab Nebula obtained with highest up to data subarcsecond spatial resolution.

2. Observations and data reduction

We used images of the Crab which were obtained with Wide Field and Planetary Camera 2 (WFPC2) on the HST. It images a 150" x 150" "L"-shaped region with a spatial sampling of 0.1" per pixel, and a smaller 34" x 34" square field with 0.046" per pixel[1]. The data using the wide and medium band filters F300W, F547M, and F814W were obtained in 1995 August 14 (Program ID: 6129, PI Jeff Hester) with 3200s, 2000s, and 1400s total exposure times, respectively. These filters do not cover most bright emission lines from the Crab, hence, they are useful for continuum emission analysis. However, some emission lines can enter filter passbands. For construction spectral index map it is necessary to use a single set of observations because of high time variability of the Crab PWN features. Observations in 1995 are one of the first detailed broadband observations of the Crab with the HST. Besides that, we retrieved data for the narrow band filter F673N obtained on 1995 January 02 (Program ID: 5354, PI Kris Davidson) with 2300.0 exposure duration to determine Crab filaments emitting in S II line (λ6731). Also we use Chandra ACIS-S data (Observation ID: 02001, PI Jeff Hester) obtained on 2001 January 30 to compare the optical data with the X-ray image.

Individual optical exposures obtained from the archive has been pipeline reduced, and we only combined them removing cosmic rays. After that we performed astrometric referencing of the resulting images. For data processing we used the IRAF, including STSDAS and DAOPHOT packages.

3. Construction of the spectral index map

To construct the index map we followed the same procedure as has been used by Temim et al. [5] for the mid-infrared. The continuum emission from the Crab Nebula can be described by a power law spectrum of the form $L_\nu \propto \nu^{-\alpha}$, where L_ν is the synchrotron specific luminosity and α is the spectral index. We compute index maps using the equation

$$\alpha = \frac{ln(F_1/F_2)}{ln(\lambda_1/\lambda_2)},$$

where F_1 and F_2 are the surface brightnesses in the images obtained in two different filters, and λ_1, λ_2 are the pivot wavelengths of filters. All data were background subtracted, calibrated to the physical flux units, and derrended using R=3.1 and A_V=1.61 [8]. $A(\lambda)$ was obtained using analytic formula by J.A.Cardelli, G.C.Clayton, and J.S.Mathis [7]. The background was taken from the Crab Eastern Bay region. Arithmetical operations with the images following the above expression were performed using *imcalc* utility from the IRAF/STSDAS.

4. Results and discussion

In Fig. 1 we compare the soft X-ray and optical images of the Crab PWN obtained with Chandra ACIS and with the HST in F547M filter, respectively. As seen, the WFPC2 field of view (FOV) does not allow one to image the whole nebula. However, most important feature of the PWN, including the torus and southern jet structures, eastern bay, pulsar itself, anvil, moving wisps [6], are within the FOV. Some of them are marked in this figure.

At the left panel of Fig. 2 the same FOV is presented but in the F814W filter. The PWN morphology in this filter is identical to that in F547M. At the right panel of Fig. 2 we show the spectral index map created from these images. Bright features with a high α extended from east to west and from south to north at the south edge of the FOV are filaments where the emission in lines is dominated. They are not related to the continuum properties (see below). Excluding these features, generally the Crab spectrum in the PWN center, particularly, within the torus and jet regions, is harder and spectral index grows to the edge of the nebula. This is compatible

[1] *http://documents.stsci.edu/hst/wfpc2/documents/handbooks/cycle17/wfpc2_cover.html*

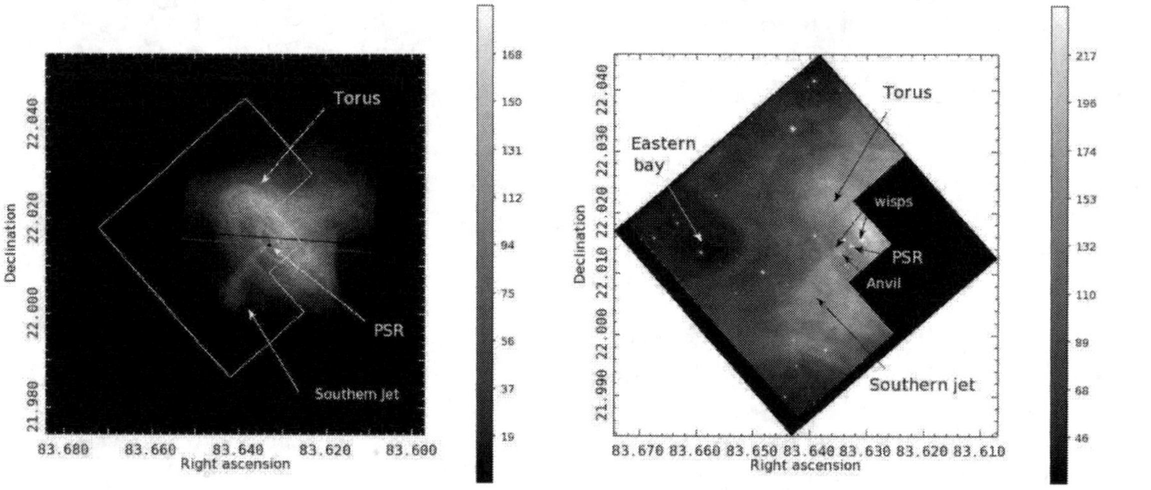

Figure 1. *Left:* The Chandra X-Ray image of the Crab PWN. North is up and east is left. The white region shows the FOV of HST/WFPC2. *Right:* The HST/WFPC2 image of the field in the F547M filter. Main features of the PWN are marked by arrows. Grey bars show the emission intensity scale in counts.

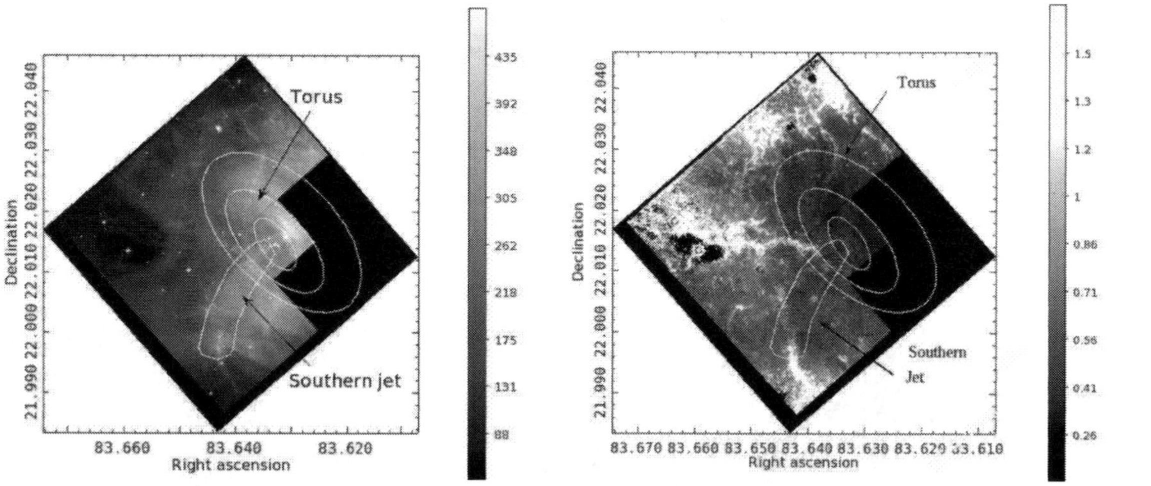

Figure 2. *Left:* The same as at the right part of Fig.1 but in F814W filter. *Right:* Spectral index map constructed using the data of the F547M and F814W filters. The torus, inner ring and jet are indicated for orientation. The grey bar in the *left* panel show the emission intensity scale in counts, while in the *right* panel it shows the scale of the spectral index α .

with the result of Veron-Cetty [1], but on their map the spectrum is systematically softer, i.e., α is in the range of about 0.57–1.0 vs our 0.3–1.0. A possible reason for this discrepancy is that they used a smaller interstellar extinction towards the Crab.

High spatial resolution of the HST allows us to resolve fine structures of the PWN in our index map. In Fig. 3 we show the immediate vicinity of the pulsar. As seen, the most bright extended features in the direct image, such as the anvil, wisp and the nearest part of the torus have much harder spectrum than their surroundings. The wisp and anvil are spatially associated with the internal X-ray PWN ring, which is interpreted as a termination shock of the pulsar wind.

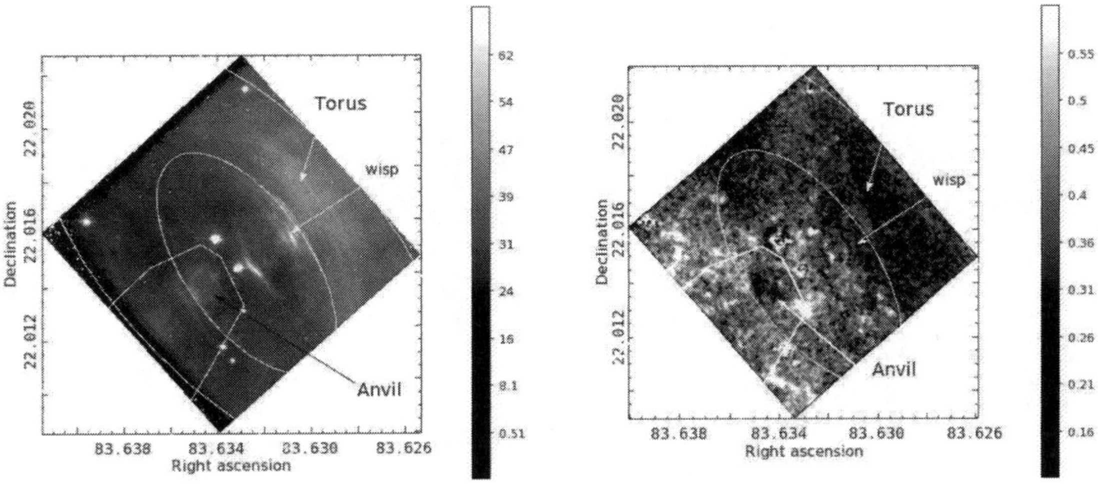

Figure 3. Enlarged image of the central part of the PWN shown in Fig.2. It contains the pulsar, beginning of the jet and the inner ring structures.*Left:* The direct image. *Right:* Spectral index map.

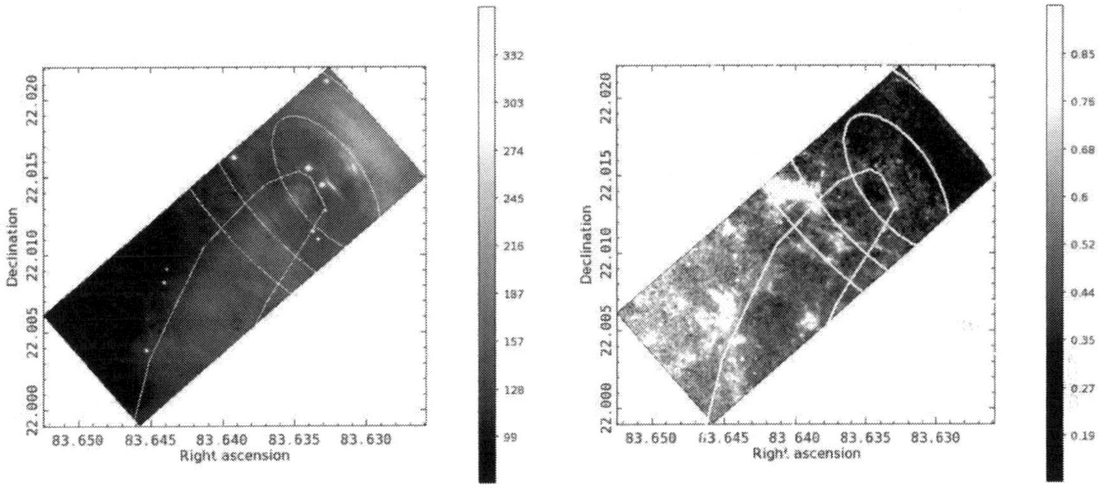

Figure 4. *Left:* HSR/WFPC2 central part of FOV in F547M filter focused on the jet. *Right:* Spectral index map of this part constructed using the data of F547M and F814W filters. Same features as at Fig.2 are marked.

The southern jet region is enlarged in Figs. 4-5. We can see that the spectral index is lower in the jet than in its nearest environments, but the emission of the jet is softer than emission from the torus. The jet itself is nonhomogenious both in the brightness and the index. It might has a spiral structure barely resolved in the X-ray and the direct optical continuum images. Brighter parts of the jet have apparently harder spectra. Comparing the index and F673N image we find that the jet is bounded by a filament which likely spiral with the jet.

We have obtained the index map images using F300W images as well. The structures in that map are almost same as presented here.

To conclude, the described method of obtaining the optical spectral index maps from HST images appears to be quite productive and allows one to study structures of the Crab PWN in great details that has not been possible since the time of publishing of the first index map by

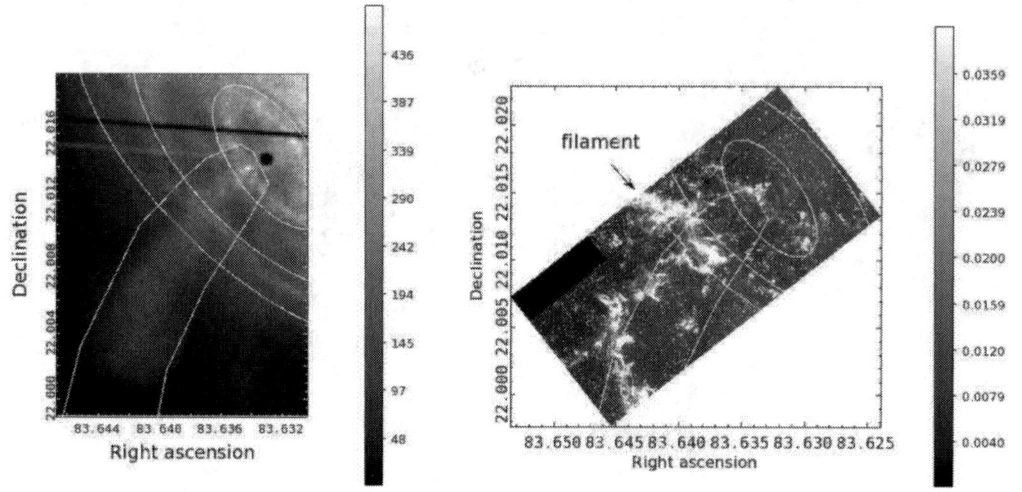

Figure 5. *Left:* Chandra image of the same region as at Fig.4 *Right:* Image in the filter F637N aimed to see Crab filaments emitting in S II line $\lambda 6731$

Veron-Cetty and Woltjer in early 90th. More detail description of our results will be published in a separate paper.

Acknowledgments

We would like to thank E.N.Nekrasova for useful discussions of the method of obtaining the optical spectral index maps, A.A.Danilenko for comments on the text and D.P.Barsukov for technical support.

The work was partially supported by the Russian Foundation for Basic Research (grants 11-02-00253 and 11-02-12082), RF Presidential Program (Grant NSh 4035.2012.2), and the Ministry of Education and Science of the Russian Federation (Contract No. 11.G34.31.0001 and Agreement No.8409, 2012).

This research has made use of the USNOFS Image and Catalogue Archive operated by the United States Naval Observatory, Flagstaff Station (http://www.nofs.navy.mil/data/fchpix/). It is based on observations made with the NASA/ESA Hubble Space Telescope and obtained from the Mikulski Archive for Space Telescopes (MAST) at the Space Telescope Science Institute, which is operated by the Association of Universities for Research in Astronomy, Inc., under NASA contract NAS 5-26555. These observations are associated with programs ID 6129 (PI Jeff Hester) and 5354 (PI Kris Davidson). Data obtained from the Chandra Data Archive and the Chandra Source Catalog was used. IRAF is distributed by the National Optical Astronomy Observatory, which is operated by the Association of Universities for Research in Astronomy (AURA) under cooperative agreement with the National Science Foundation. STSDAS is product of the Space Telescope Science Institute, which is operated by AURA for NASA.

References

[1] Veron-Cetty M P and Woltjer L 1993 *A&A* **270** 370-378
[2] Weisskopf M C et al 2000 *ApJ* **536** L81-L84
[3] Mori K, Burrows D N, Hester J J, Pavlov G G, Shibata S and Tsunemi H 2004 *ApJ* **609** 186-193
[4] Bietenholz M F, Kassim N, Frail D A, Perley R A, Erickson W C and Hajian A R 1997 *ApJ* **490** 291
[5] Temim T, Sonneborn G, Dwek E, Arendt R G, Gehrz R D, Slane P and Roellig T L 2012 *ApJ* **753** 72
[6] Hester J J et al 1995 *ApJ* **448** 240-263
[7] Cardelli J A, Geoffrey C C and Mathis J S 1989 *ApJ* **345** 245-256
[8] Sollerman J 2003 *A&A* **406** 639-644

Journal of Physics: Conference Series **461** (2013) 012014 doi:10.1088/1742-6596/461/1/012014

Roto-translation motion of the stars in close binary systems

A.A. Medvedeva

Sternberg Astronomical Institute, Moscow State University, Universitetsky pr., 13, Moscow 119991, Russia

E-mail: mizar-alcora@rambler.ru

Abstract. This article has to show that the model of p-h which is used to determine the change of the semi major axis of the relative orbit stars is incorrect and leads to large errors in the determination of semi-major axis. The new model, suitable for the elliptical orbits of the stars. To determine relative motion of stars in a close binary system in this paper uses a numerical integration of the equations of motion with the reactive forces, including the rotational component of attraction between the stars and the stream flows into the substance. The calculations of elliptical orbits of close binary stars show that the effect of the reactive force on the evolution of the orbits of stars may be different. The results can be refined by introducing other disturbing factors and making new assumptions based on observations.

1. Introduction

In the mid-20th century the research of the motion of stars in close binary systems was started. This problem is reflected in the works of: A. Kruszewskij [1], D. Hadzhidimetriu [2] S. Piotrowskij [3] etc. That research led to discovery of the conservative mass exchange in a close binary system was obtained a simplified dependence of variation of the semi-major axis a relative stars orbit from the constant rate of increase of the mass of the host star $\dot{M}_2$ in the form

$$\dot{a} = 2a\dot{M}_2 \left(\frac{1}{M_1} - \frac{1}{M_2} \right). \tag{1}$$

This dependence, which is called Paczynskij-Huangs formula, is used in all researches of close binary systems with conservative mass exchange, when $M_1 + M_2 = M = \text{const}$. This formula has been derived using assumptions of the equation of motion of the stars with variable masses permit an integral of the moment of the motion amount.

$$\mathbf{J} = M_1 \mathbf{R}_1 \times \mathbf{V}_1 + M_2 \mathbf{R}_2 \times \mathbf{V}_2 = \mathbf{const}, \tag{2}$$

where $M_i, \mathbf{R}_i, \mathbf{V}_i$, $(i = 1, 2)$ - the mass, radius-vector and the motion speed of the stars.

The assumption of the existence of the integral of the moment of the motion amount is wrong. To let us prove this suggestion please take a look at, based on the most common form of the equations for the Meshcherskijs motion task of two bodies with variable masses:

$$M_1 \frac{d\mathbf{V}_1}{dt} = G \frac{M_1 M_2}{R^3} \mathbf{R} + \mathbf{Q}_1, \qquad M_2 \frac{d\mathbf{V}_2}{dt} = -G \frac{M_1 M_2}{R^3} \mathbf{R} + \mathbf{Q}_2, \tag{3}$$

where $\mathbf{R} = \mathbf{R_2} - \mathbf{R_1}$, $\mathbf{Q}_1$ $\mathbf{Q}_2$ — the reaction forces acting on the star S_1 and S_2.

The existence of the integral (2) in the task of two bodies with variable masses is possible only in the Meshcherskij-Levi-Civitas problem, the equations of the motion are obtained from (3), if the reaction forces are equal:

$$\mathbf{Q}_1 = -\dot{M}_1\mathbf{V}_1, \qquad \mathbf{Q}_2 = -\dot{M}_2\mathbf{V}_2, \tag{4}$$

where $\dot{M}_1 = -\dot{M}_2 = \text{const}$, $\dot{M}_2 > 0$.

However, using the Meshcherskij-Levi-Civitas model to research the motion of the stars in a close binary system with a conservative exchange content is unreasonable because reaction forces (4) in close binary system does not exist.

Nowadays, the department of celestial mechanics SAI MSU (L.Lukyanov, S. Hasanov) developed a model, which determines the motion of close binary systems in the case of the elliptical orbit. In this work, we consider a further adjustment of the trajectories of the stars with the influence of the possible distribution of the flowing stream parameters.

2. Restricted elliptical three-body task

To determine the motion of the particles flowing masses we will use a flat elliptical restricted three-body task, the equations of motion, which are located in a rotating and pulsating Shapners barycentric coordinate system

$$\begin{aligned} \frac{d^2x}{dv^2} - 2\frac{dy}{dv} = \rho\frac{\partial U}{\partial x}, \\ \frac{d^2y}{dv^2} + 2\frac{dx}{dv} = \rho\frac{\partial U}{\partial y}, \end{aligned} \tag{5}$$

where x axis is always directed at the star-accretor S_2, v — true anomaly stars $\rho = 1/(1+e\cos v)$ — dimensionless radius stardom S_2 relatively S_1. U denotes the Jacobis function in restricted elliptic three-body task

$$U = \frac{x^2+y^2}{2} + p^3\left(\frac{1-m}{r_1} + \frac{m}{r_2}\right) - \frac{p^2}{2}(3+m^2-m), \tag{6}$$

r_1, r_2 — the distance from the particle stream flowing respectively to the centres of mass of the first and second stars

$$r_1 = \sqrt{(x+pm)^2+y^2}, \quad r_2 = \sqrt{(x+pm-p)^2+y^2}, \tag{7}$$

where $1-m = M_1/M$ and $m = M_2/M$ — relative masses of the stars, $p = a(1-e^2)$ — focal parameter of the stars orbit, a and e — axis and eccentricity, respectively.

Well assume that the host star has a spherical

$$(x+pm-p)^2 + y^2 = \frac{P^2}{\rho^2}, \tag{8}$$

shape radius of P which in the process of mass transfer changes according to the dependence.

$$P = P_0\sqrt[3]{\frac{m}{m_0}}. \tag{9}$$

Outflow of the matter from the donor star starts when the energy level of the particles is larger than the energy level of the singular point of Euler L_1. Speed of expiration V_0 of the mass particles from the star S_1 is always directed toward the star S_2, forming the axis Ox an angle $\pm\alpha$ and has a value

$$V_0(v) = V_{00}V_1(v)\,\beta, \quad V_1(v) = m\sqrt{\frac{GM}{p}}\sqrt{1+2e\cos v + e^2}, \tag{10}$$

In the initial conditions, the angle α between the jet and the horizontal axis changes within the limits of $-\pi/2 < \alpha < \pi/2$, and the parameter β – between $0.5 < \beta < 2$.

Numerical integration is performed on the interval $v_0 \le v \le v_0 + \tau$, where the value of the true anomaly $v_0 + \tau$ determined being deposited on the surface of the particles of the second star at $x_2 = x(v_0+\tau), y_2 = y(v_0+\tau)$. Without doubts these results can be corrected by introducing other disturbing factors and making new assumptions based on observations.

As the result of numerical integration the components of the speed of expiration $\mathbf{W}_1$ of the star S_1 and of the speed mass flow $\mathbf{W}_2$ of the star S_2 are obtained

$$W_{1x} = V_c,\ W_{1y} = 0,\ W_{2x} = a_x + b_x(\omega+\Omega),\ W_{2y} = a_y + b_y(\omega+\Omega), \tag{11}$$

where

$$a_x = x'(v_0+\tau)\dot{v},\ b_x = y\,(v_0+\tau),\ a_y = y'(v_0+\tau)\dot{v},\ b_y = -x\,(v_0+\tau) - pm + p$$

reaction forces, influencing the stars S_1 and S_2, are considered to be attached to their center of mass and are given by

$$\mathbf{Q}_1 = \dot{m}M\{-V_c, 0\}, \quad \mathbf{Q}_2 = \dot{m}M\{x'(v_0+\tau)\dot{v}, y'(v_0+\tau)\dot{v}\}. \tag{12}$$

A numerical integration is determined as the mass of the steady stream S_3 and coordinates x_3, y_3 its center of mass

$$M_3 = \frac{\tau}{\dot{v}_c}\dot{m}M, \quad x_3 = \frac{1}{\tau}\int\limits_{v_0}^{v_0+\tau} x(v)dv, \quad y_3 = \frac{1}{\tau}\int\limits_{v_0}^{v_0+\tau} y(v)dv, \tag{13}$$

where

$$\dot{v}_c = \frac{1}{2\pi}\int\limits_0^{2\pi}\dot{v}\,dv = \frac{1}{2\pi}\int\limits_0^{2\pi}\sqrt{\frac{GM}{p^3}}\,(1+e\cos v\,)^2\,dv = \sqrt{\frac{GM}{p^3}}\left(1+\frac{e^2}{2}\right)$$

— the mean angular velocity of the orbital motion.

3. Differential equations stars

We explain stars' motion in the inertial coordinate system on a basis of the model which uses the following equations

$$\begin{aligned} M_1\frac{d\mathbf{V}_1}{dt} &= G\frac{M_1\,M_2}{R^3}\mathbf{R} + \mathbf{Q}_1 + G\frac{M_1M_3}{r_{13}^3}\mathbf{r}_{13}, \\ M_2\frac{d\mathbf{V}_2}{dt} &= -G\frac{M_1\,M_2}{R^3}\mathbf{R} + \mathbf{Q}_2 + G\frac{M_2M_3}{r_{23}^3}\mathbf{r}_{23}, \end{aligned} \tag{14}$$

where $r_{13} = \sqrt{(x_3+pm)^2 + y_3^2}$, $r_{23} = \sqrt{(x_3+pm-p)^2 + y_3^2}$ — the distances between the centres of mass of the stars and the jet.

Differential equations in osculating elements for semi-major axis of the relative orbit a, the eccentricity e and the angular rotational speed ω of the star S_2 are presented as

$$\frac{da}{dt} = \frac{2a^2}{\sqrt{GMa(1-e^2)}}[e\sin v \cdot S + (1 + e\cos v)\cdot T],$$
$$\frac{de}{dt} = \sqrt{\frac{a(1-e^2)}{GM}}\left[\sin v \cdot S + \left(\cos v + \frac{\cos v + e}{1 + e\cos v}\right)\cdot T\right], \qquad (15)$$
$$\frac{d\omega}{dt} = \frac{5}{2}\frac{\dot m}{m}\sin\gamma\frac{W_{otn}}{P} = -\frac{5}{2}\frac{\dot m}{P^2 m}\left[A(\omega+\Omega)+B\right]$$

Perturbing acceleration of S and T is becoming known after the integration of the equations (5):

$$S = \frac{\dot m}{m}\left(a_x + b_x(\omega+\Omega) + \frac{mV_c}{1-m}\right) + GM\tau\frac{\dot m}{\dot v_c}\left(\frac{x_3 + pm - p}{r_{23}^3} - \frac{x_3 + pm}{r_{13}^3}\right),$$
$$T = \frac{\dot m}{m}\left[a_y + b_y(\omega+\Omega)\right] + GM\tau\frac{\dot m}{\dot v_c}y_3 r_3, \quad r_3 = \frac{1}{r_{23}^3} - \frac{1}{r_{31}^3} \qquad (16)$$

By choosing the true stars anomaly v and performing averaging using the independent variable, we obtain

$$\frac{d\tilde a}{d\tilde v} = \frac{1}{2\pi}\int_0^{2\pi}\frac{2a^2}{\dot v\sqrt{GMa(1-e^2)}}[e\sin v\cdot S + (1+e\cos v)\cdot T]dv,$$
$$\frac{d\tilde e}{d\tilde v} = \frac{1}{2\pi}\int_0^{2\pi}\frac{1}{\dot v}\sqrt{\frac{a(1-e^2)}{GM}}\left[\sin v\cdot S + \left(\cos v + \frac{\cos v + e}{1+e\cos v}\right)\cdot T\right]dv \qquad (17)$$
$$\frac{d\tilde\omega}{d\tilde v} = \frac{5}{2\dot v}\frac{\dot m}{m}\sin\gamma\frac{W_{otn}}{P} = -\frac{5}{2}\frac{\dot m}{\dot v P^2 m}\left[A(\omega+\Omega)+B\right]$$

if, in addition to the independent variable m we find according to the formula $\frac{d}{d\tilde v} = \frac{\dot m}{\dot v_c}\frac{d}{dm}$, the equations of motion for the average of elements $\tilde a$, $\tilde e$ $\tilde\omega$ with regard to the expression $\dot v$ will take the form

$$\frac{d\tilde a}{dm} = \frac{2+\tilde e^2}{(1-\tilde e^2)^2}\frac{\widetilde W_{2y}}{m} + 2\tilde a^3\sqrt{1-\tilde e^2}\,\tau y_3 r_3 + \frac{\tilde a^2(2+\tilde e^2)}{\sqrt{GM\tilde a(1-\tilde e^2)}}\frac{\tilde b(\tilde\omega+\Omega)}{m},$$
$$\frac{d\tilde e}{dm} = \frac{2+\tilde e^2}{2\tilde a(1-\tilde e^2)}\frac{\tilde e}{1+\sqrt{1-\tilde e^2}}\frac{\widetilde W_{2y}}{m} - \frac{3}{2}\tilde e\,\tilde a^2\sqrt{1-\tilde e^2}\,\tau y_3 r_3 - $$
$$- \frac{3\tilde a\tilde e(2+\tilde e^2)}{4\sqrt{GM\tilde a(1-\tilde e^2)}}\frac{\tilde b(\tilde\omega+\Omega)}{m}, \qquad (18)$$
$$\frac{d\tilde\omega}{dm} = -\frac{5}{4}\frac{2+\tilde e^2}{(1-\tilde e^2)^{3/2}P^2 m}\left[\widetilde W_{2x}\tilde y + \widetilde W_{2y}\tilde b + (\tilde b^2 + \tilde y^2)(\omega+\Omega)\right]$$

where

$$\tilde x = x\,(v_0+\tau), \quad \tilde y = y\,(v_0+\tau), \quad \widetilde W_{2x} = x'(v_0+\tau), \quad \widetilde W_{2y} = y'(v_0+\tau),$$
$$\tilde b = p - pm - \tilde x = \tilde a(1-\tilde e^2)(1-m) - \tilde x.$$

Journal of Physics: Conference Series **461** (2013) 012014 doi:10.1088/1742-6596/461/1/012014

4. Numerical results

Account of the interaction of gravity of stars, star attraction jet and reactive power with the translational and rotational motion

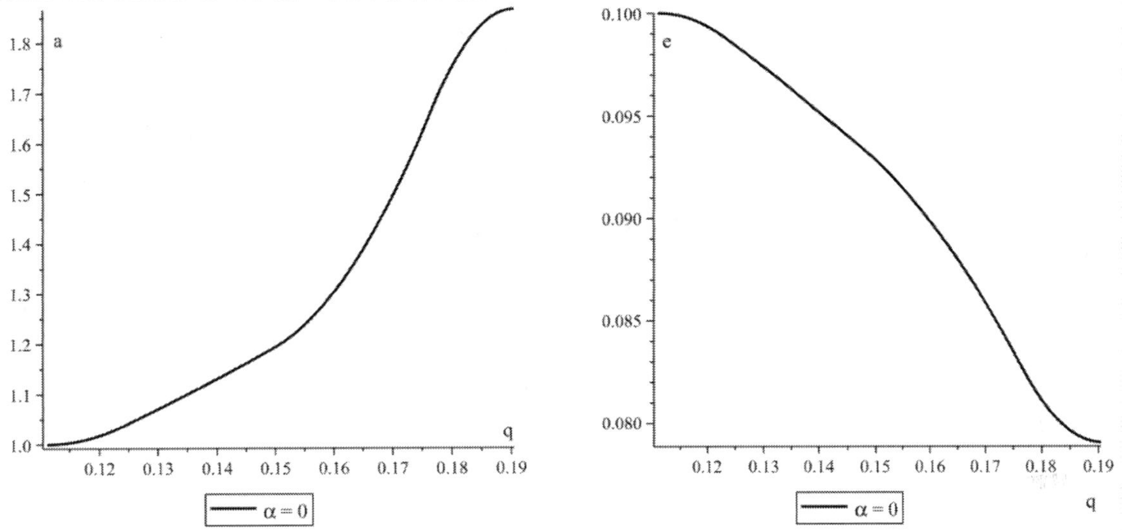

Figure 1. On the left picture is dependence of the semi-major axis of the mass ratio, on the right dependence on the mass ratio of eccentricity

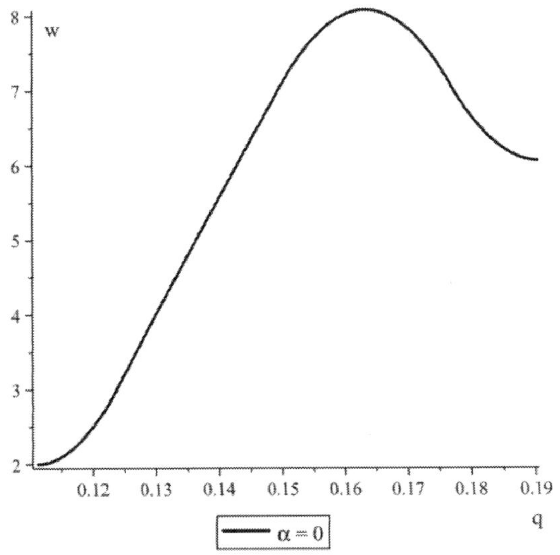

Figure 2. Dependence of the angular velocity of the mass ratio

Here shows the power modules for dependencies.

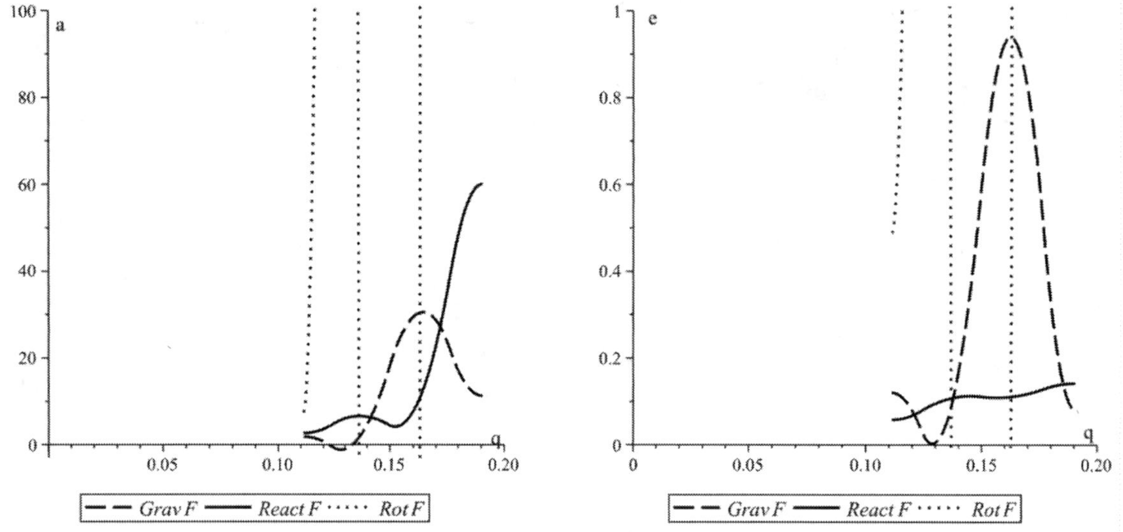

Figure 3. On the left Dependence of the semi-major axis of the mass ratio, on the right dependence on the mass ratio of eccentricity

5. Conclusions

For half a century the incorrect model of Paczynskij-Huang was used to determine the relative orbit of close binary stars. But to this day in the work associated with the close binary systems, the model still has been used. Because we consider this model is wrong, the purpose of this paper was to show fallacy of this model using other methods. To determine the relative motion of stars in a close binary system in this paper we use a numerical integration of the equations of motion with the reactive forces of attraction and stars flowing stream. The calculations of elliptical orbits of close binary stars show that the effect of the reactive force on the evolution of the orbits of stars may be different. There is no doubt that the results can be corrected by introducing other disturbing factors and making new assumptions based on observations.

6. References

[1] Kruszevski A. 1966. *Adv. Astron. and Astrophys.* **4**.P.233.
[2] Hadjidemetriou J. 1969. *Astrophys. and Space Sci.* **3**.P.330.
[3] Piotrovskij S.L. 1967. *Astron. Jorn.* **44**.P.241.

Journal of Physics: Conference Series **461** (2013) 012015 doi:10.1088/1742-6596/461/1/012015

A molecular dynamics and circular dichroism study of a novel synthetic antimicrobial peptide

N P Rodina, A N Yudenko, I N Terterov and I E Eliseev

Laboratory of nanobiotechnologies, St. Petersburg Academic university, Khlopin str., 8/3, St. Petersburg 194021, Russian Federation

E-mail: eliseev@spbau.ru

Abstract. Antimicrobial peptides are a class of small, usually positively charged amphiphilic peptides that are used by the innate immune system to combat bacterial infection in multicellular eukaryotes. Antimicrobial peptides are known for their broad-spectrum antimicrobial activity and thus can be used as a basis for a development of new antibiotics against multidrug-resistant bacteria. The most challengeous task on the way to a therapeutic use of antimicrobial peptides is a rational design of new peptides with enhanced activity and reduced toxicity. Here we report a molecular dynamics and circular dichroism study of a novel synthetic antimicrobial peptide D51. This peptide was earlier designed by Loose et al. using a linguistic model of natural antimicrobial peptides. Molecular dynamics simulation of the peptide folding in explicit solvent shows fast formation of two antiparallel beta strands connected by a beta-turn that is confirmed by circular dichroism measurements. Obtained from simulation amphipatic conformation of the peptide is analysed and possible mechanism of it's interaction with bacterial membranes together with ways to enhance it's antibacterial activity are suggested.

1. Introduction

Antimicrobial peptides (AMPs) are generally defined as short peptides of less than 50 residues with overall positive charge and substantial portion of hydrophobic residues. Natural antimicrobial peptides found in different multicellular eukaryotes and are an important part of the innate immune system that defend host organism against pathogenic microbes[1]. Broad-spectrum activity of antimicrobial peptides against multidrug-resistant bacteria make them potential next-generation antibiotics[2].

Currently more than 2000 AMP sequences can be found in Antimicrobial Peptide Database[3]. Most of the studied antimicrobial peptides with known structure belong to three major structural classes: α-helix, β-sheet or combined α+β-structures. Spatial structure (conformation) of a peptide molecule defines a specific distribution of physico-chemical properties such as an electric charge and hydrophobicity on a molecular surface. Presence of a certain amphipatic conformation is essential for an antimicrobial activity of a peptide. It is widely accepted that antimicrobial peptides target the lipid bilayer of the bacterial cell membrane, and their selectivity is based on higher electrostatic affinity of cationic peptides to negatively charged bacterial membranes than to electrically neutral lipids forming membranes of mammalian cells[4].

Most of the natural antimicrobial peptides are not enough active to be used as therapeutic agents straightforwardly. Thus, new design approaches are required to build synthetic antimicrobial peptides with enhanced activity. Among other modern methods of peptide design there is a method of *single amino acid replacement* in natural antimicrobial peptide sequences, method of *artificial neural network* that had been trained for AMP recognition and ranks a large set of random generated sequences and *linguistic model* where new antimicrobial peptides are generated based on grammatical rules derived from naturally occurring AMPs[5].

Despite the mentioned approaches are rather good for generation of quazirandom peptide sequences that exhibit antimicrobial activity, they do not explicitly take into account the structure of antimicrobial peptides, that is crucial for their activity and can be used in further improvement of their properties. Here we use molecular dynamics simulation in order to complete linguistic model with structural information and give insights into molecular basis of an activity of the designed synthetic peptide D51 (FLFRVASKVFPALIGKFKKK)[5]. Structure obtained by molecular dynamics simulation is used to build hypothesis about peptide interaction with bacterial membranes and propose modifications to enhance it's activity.

2. Materials and methods

2.1. Molecular dynamics simulation

Two MD simulations of the peptide in an explicit solvent were performed using GROMACS 4.5.4 software package[6]. The solvent was modeled using the simple-point charge (SPC) water model. The peptide and peptide-water interactions were described by the GROMOS 43a2 united-atom force field. Simulations were performed using periodic boundary conditions and a time step of 2 fs was used. The temperature was weakly coupled to $T = 300$ K with time constant $\tau_T = 0.5$ ps using the Nose-Hoover thermostat. The pressure was weakly coupled to 1 atmosphere with time constant $\tau_P = 2$ ps using Parrinello-Rahman barostat.

Initial linear structure of the peptide for the first simulation was build with dihedral angles $\phi = \psi = 180°$ for each residue. Final structure of the peptide after the first simulation was used as the starting structure for the second simulation. DSSP[7] algorithm was used for the secondary structure assignment. PLATINUM program[8] was used for the calculation of molecular hydrophobicity potential and PyMol package was used for visualization.

2.2. Peptide preparation

The peptide D51 (FLFRVASKVFPALIGKFKKK-OH, $MW_{theor.} = 2324.96\ Da$) was synthesized using Fmoc (fluorenylmethoxycarbonyl) chemistry with Symphony synthesizer at the Peptide Technologies, Inc. and was supplied as dry powder. The correct mass $MW_{exp.} = 2325.2\ Da$ was observed using Voyager-DE STR MALDI-TOF mass spectrometer (Applied Biosystems) and the purity of the peptide was determined by UPLC using Acquity chromatograph (Waters). For CD measurements the peptide was dissolved in deionized water at the concentration of 0.35 mg/ml.

2.3. Circular dichroism (CD) spectroscopy

CD measurements were performed with the Chirascan CD spectrometer (Applied Photophysics). A 1.0 mm path-length cell was used. CD scans were taken from 190 to 260 nm, with a 1 nm bandwidth and a 1 nm step resolution, three scans were added and averaged for each spectra. In order to estimate secondary structure content averaged CD spectra were analysed using CDNN program[9]. To obtain a melting curve of the peptide structure CD measurements were conducted at different temperatures ranging from 5 °C to 90 °C with a step of 5 °C.

3. Results and discussion

3.1. Peptide folding

Folding dynamics of the peptide were obtained as a result of the molecular dynamics simulation starting from the linear conformation. We investigated a *root mean square deviation* (RMSD) of the C^α atoms that is a measure of distance between different spatial structures. Time evolution of the RMSD of the peptide with it's initial conformation is shown in fig.1. As expected, fast process of structure formation occurred in the first 30 ns of trajectory that is expressed in the large change of RMSD from zero to roughly 1.8 nm. After 30 ns RMSD is nearly constant with fluctuation about 0.25 nm, which is the typical value for the thermal vibrations. A stable structure appears after the 90 ns and remains constant for subsequent calculations.

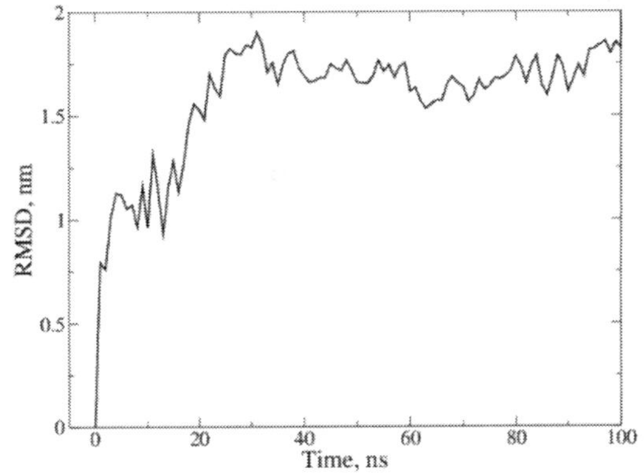

Figure 1. Process of the peptide folding. Root mean square deviation (RMSD) of the peptide C^α atom positions during the first simulation.

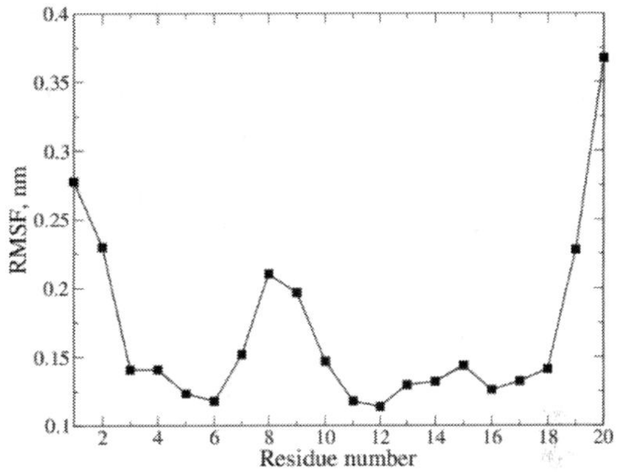

Figure 2. Thermal stability of the peptide structure. Root mean square fluctuation (RMSF) of the peptide C^α atom positions during the second simulation.

The *root mean square fluctuation* (RMSF) of the C^α atoms is a measure of thermal vibrations and rigidity of the peptide structure. RMSF of the D51 calculated from the second simulation is shown on fig.2. Residues 3–6 and 11–18 have the RMSF values below 0.15 nm, so it can be assumed that they belong to the relatively rigid structure motif such as β-sheet. Obviously, the C- and N-terminus of the peptide and also a loop between the two expected β-strands have a large mobility while the atoms in beta-sheets are rather stationary.

3.2. Spatial structure and hydrophobicity

The final peptide conformation after the second simulation was analysed using DSSP program[7]. The results of the analysis of hydrogen-bonding patterns and backbone torsion angles ϕ, ψ and corresponding secondary structure elements are summarized in Table 1.

Table 1. Secondary structure assignment for the final peptide conformation: "E" – residue belongs to β-strand element, "S" – turn, "–" – random coil (unstructured).

Residue number	1	2	3	4	5	6	7	8	9	10	11	12	13	14	15	16	17	18	19	20
Amino acid	F	L	F	R	V	A	S	K	V	F	P	A	L	I	G	K	F	K	K	K
Secondary structure	–	–	–	E	E	E	E	S	S	S	E	E	E	E	S	–	–	–	–	–

It should be mentioned that the secondary structure elements which were found with DSSP algorithm match with rigid regions having lowest RMSF values (fig.2) perfectly. This match is due to hydrogen bonding that occurs between secondary structure elements (β-strands) and reduces their thermal mobility.

Schematic representation of the peptide structure is shown in fig.3(a). Secondary structure elements – antiparallel β-strands are drawn as arrows.

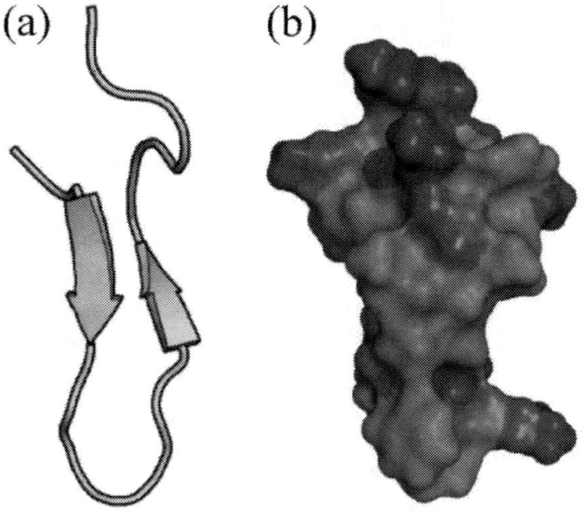

Figure 3. Peptide structure obtained from MD simulations:
(a) – schematic ribbon representation;
(b) – molecular surface representation, red color corresponds to hydrophobic and blue to hydrophilic regions.

Hydrophobic and hydrophilic (amphiphilic) properties are critical for antimicrobial peptides and their ability to bind to bacterial membranes. The surface of the D51 molecule constructed with PLATINUM program[8] using molecular hydrophobicity potential approach is represented in fig.3(b). Blue color corresponds to hydrophilic and red to hydrophobic surface. The most hydrophilic regions correspond to positively charged residues. The highest affinity to the bacterial membrane is usually achieved when spacing between hydrophilic and hydrophobic regions is maximal. In our case the optimal cooperation with the membrane may be interfered by a small hydrophilic area in the main hydrophobic region. This area if formed by polar side chains of S_7 and K_8 amino acids .

3.3. Structure evaluation by CD spectroscopy

In order to evaluate the secondary structure content of the peptide and explore it's thermal stability a number of CD spectra were measured at different temperatures. Six of the obtained curves presented in the fig.4 show gradual variation of spectra shape with increasing temperature. Thermal stability of the peptide was assessed by construction of melting curve using the values of ellipticity at 220 nm for different temperatures. Melting curve of the peptide (fig.5) demonstrates nearly linear decrease of structure or, in other words, unfolding of the peptide with increasing temperature. There is no characteristic temperature or evidence of saturation in the studied range of temperatures. Such a behaviour, meaning very broad range of temperatures where melting occurs, is typical for relatively small biopolymers[10].

The secondary structure content was determined from the CD measurement with the lowest temperature (5 °) and results are listed in Table 2. As can be seen from Table 2 there is a rather good agreement in secondary structure content obtained from molecular dynamics simulation and by CD spectroscopy especially for antiparallel β-sheet. CD measurements also show presence of minor α-helix and parallel β-sheet content that is different to MD results. We assume that this minor components can be neglected because their content is comparable to CD method accuracy.

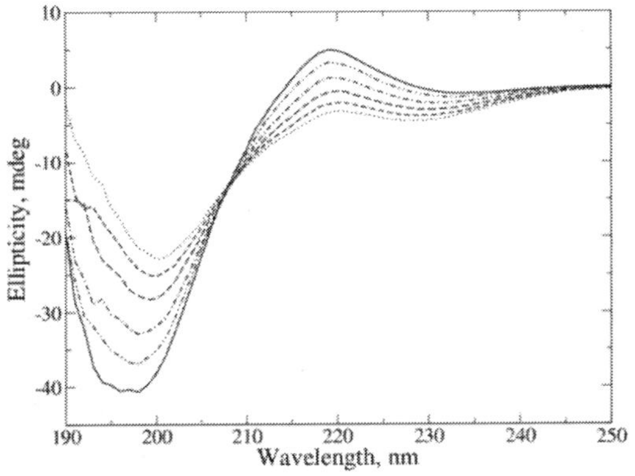

Figure 4. CD spectra at different temperatures: —— 5°C, — · · — 20°C, — · — 35°C, - - - 50°C, - - - - 65°C, · · · · · · 80°C.

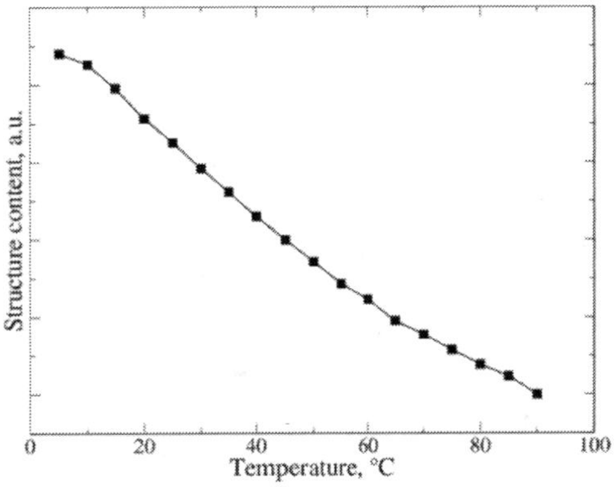

Figure 5. Melting curve of the peptide, measured at 220 nm wavelength.

Table 2. Secondary structure content, %.

Structure element	α-helix	Parallel β-sheet	Antiparallel β-sheet	Turn	Random coil
MD simulation	0	0	40	20	40
CD measurements	11	9	36	15	29

3.4. Hypothesis on the peptide interaction with bacterial membranes

The mechanism of action of cationic AMPs is still not well understood but it is widely accepted that all antimicrobial peptides interact with bacterial membranes[2]. It has been shown that cationic peptides have a higher affinity to LPS in the membrane of the Gram-negative bacteria than native divalent cations such as Mg^{2+} and Ca^{2+}, so peptides displace these cations from the negatively charged LPS leading to a local disturbance in the outer membrane[4]. This facilitates the formation of destabilized areas through which the peptide with it's hydrophobic part integrates into lipid bilayer.

Therefore, for the disruption of bacterial membranes the D51 peptide should first bind to the outer membrane with it hydrophilic areas. Presence of the highly hydrophilic positively charged region (on the top of fig.3(b)) formed by three Lys residues at the C-terminus could account for high electrostatic affinity to LPS. However from the spatial structure it can be seen that there is a charged residue in the hydrophobic area, which can impede further penetration of the peptide into lipid bilayer.

Considering obtained spatial structure and hypothetical mechanism of action of the studied peptide we propose rational modification of it's sequence. One of the most perspective permutation is to move the positively charged K residue to the N-terminus of the peptide: KFLFRVASVFPALIGKFKKK. This permutation leads to further separation of hydrophobic and hydrophilic regions that possibly can improve peptide activity against Gram-negative bacteria.

4. Conclusion

Rational design of new antimicrobial peptides with high antibacterial activity is an important step on the way to the development of new generation antibiotics against multidrug-resistant bacteria. To date, several methods of rational design of AMPs exist, but all of these methods do not take into account spatial structure of peptides which is essential for their action.

In this work we applied methods of molecular dynamics simulation and circular dichroism spectroscopy to give insights into spatial structure organization and possible mechanism of action of previously designed synthetic antimicrobial peptide D51. Structural properties such as secondary structure content and hydrophobicity potential distribution on the molecular surface were determined by molecular dynamics study.

The peptide was chemically synthesized and it's secondary structure was investigated by CD spectrometry, confirming our theoretical results. Considering a possible mechanism of the peptide interaction with bacterial membranes a rational modification of it's structure was proposed.

Acknowledgments

Authors wish to acknowledge technical support by the SPC facility at EMBL Hamburg. Joint Supercomputer Center of the Russian Academy of Sciences (JSCC RAS) is acknowledged for provided computational facilities. This work was financially supported by Presidium of the Russian Academy of Sciences as a part of the "Development of a technology for the design of new antimicrobial peptides with the use of methods of bioinformatics and molecular modeling" project.

References

[1] Wang G (ed) 2010 *Antimicrobial peptides: discovery, design and novel therapeutic strategies* (*Advances in molecular and cellular microbiology* no 18) (Oxfordshire, UK: CABI)
[2] Fjell C D *et al.* 2012 *Nature Reviews* **11** 37–51
[3] Wang G *et al.* 2009 *Nucleic Acids Research* D **37** 933–937
[4] Powers J P S and Hancock R E W 2003 *Peptides* **24** 1681–1691
[5] Loose C *et al.* 2006 *Nature* **443** 867–869
[6] Hess B *et al.* 2008 *J. Chem. Theory Comput.* **4** 435447
[7] Kabsch W and Sander C 1983 *Biopolymers* **22** 2577637
[8] Pyrkov T V *et al.* 2009 *Bioinformatics* **25** 1201–1202
[9] Bohm G *et al.* 1992 *Protein Eng.* **5** 191–195
[10] Finkelstein A V and Ptitsyn O B 2002 *Protein physics: A course of lectures* (Amsterdam: Academic Press)

Propagation of elastic waves in bilayer ferrite-piezoelectric structure

T.A. Galichyan[1, 2] and T.O. Firsova[1]

[1] 173003 Velikiy Novgorod, Russia, Novgorod State University

E-mail: tigrangalichyan@yahoo.com

Abstract. Propagation theory of elastic wave for ferrite-piezoelectric bilayer structures is presented. The expression for the dispersion relation of the longitudinal mode planar oscillations is derived. It is shown that in limited cases this expression come to dispersion relation for ferrite and piezoelectric accordingly.

Keywords: bilayer structure, elastic wave, dispersion relation.

1. Introduction

The mechanism of occurrence of magnetoelectric (ME) effect in ferrite-piezoelectric structure is caused by the elastic interaction of magnetostrictive and piezoelectric subsystems. Alternating magnetic field causes elastic oscillations in the magnetostrictive component which pass into the piezoelectric phase, where an electric field is occurred through polarization. As a consequence of that, the frequency dependence of magneto electric effect in such structures is directly related with the spectrum of elastic waves. The distribution of elastic waves in bilayer medium significantly differs from the distribution in homogeneous medium. Attempts were made earlier [1] to consider ME effect in such structures, but in the same time, a supposition was made, that the amplitude of oscillations was not changing its direction perpendicular to the line of the partition. This suggestion can be used in the description of the effect for rather thin layers, with some accuracy. At more detailed examination this method [1] reduces to the method of effective parameters, which was first used in describing the frequency dependence of ME effect [2,3]. Recently [4], the dispersion relation was obtained for a structure, representing a thin film grown on a semi-infinite substrate. In this paper, the distribution of the elastic waves in bilayer ferrite-piezoelectric structure was investigated given the fact that the amplitude of the wave is changed on the thickness of the sample. The dispersion relation for the elastic waves of acoustic range was obtained. Under the limiting transition it is shown that the dispersion relation, when one of the layers is thinner than the other, passes to a dispersion relation of one, or the other media.

2. Model

As a model, we consider a structure, composed of piezoelectric and magnetostictive phases mechanically interacting on the boundary (figure 1). The alternating magnetic field exits elastic oscillations in the ferrite component transferring through the boundary of the partition to the piezoelectric component, which brings to interconnected oscillations of ferrite and piezoelectric subsystems. The amplitude of the vibrations will be inhomogeneous, perpendicular to the line of the partition, as there is a sharp boundary between the ferrite and piezoelectric layers.

[2] **Corresponding author:** Tigran Galichyan, postgraduate, research fields: condensed matter physics, magnetoelectric effect.

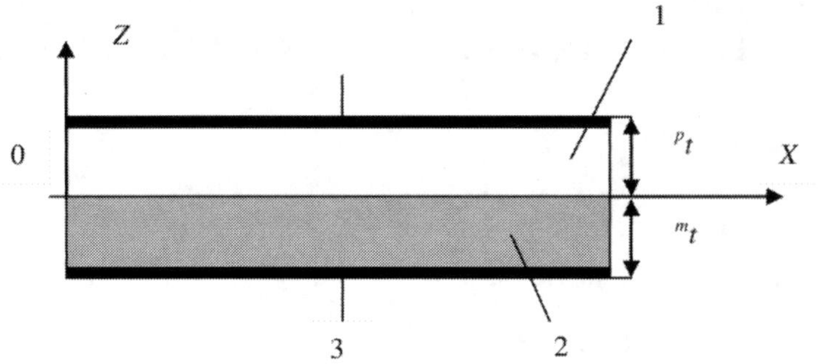

Figure 1. Schematic view of the sample
1- piezoelectric phase with ^{p}t thickness, 2- magnetostictive phase with ^{m}t, 3- Electrodes.

3. Dispersion relation

Taking into account the inhomogeneity along the Z axis, the solution of the equation for the displacement vector of the media will be presented as planar waves, which amplitude is changed on the thickness of the sample which is shown in Eq. (1).

$$^{\alpha}u(x,z)={}^{\alpha}g(z)^{\alpha}A\cos(\omega t-kx)+{}^{\alpha}B\sin(\omega t-kx)\,, \tag{1}$$

where the index α is accordingly m for the ferrite and n for the piezoelectric. $^{\alpha}A$, $^{\alpha}B$ are the constants of integration.

The substitution of this expression into the expression for the motion of the media carries to the equation for function $^{\alpha}g(z)$. Solution of this equation depends on the ration of elastic wave velocities in ferrite and piezoelectric phases. The boundary conditions, namely, the normal components of stress tensor on top and bottom surface are zero, the displacement of ferrite and piezoelectric media on interface are the same, so as the shear stresses. These boundary conditions produce a system of equations, consistency of equations give the dispersion relation in the form shown in Eq. (2).

$$^{m}\chi^{m}Y\mathrm{tg}(^{m}\kappa)={}^{p}\chi^{p}Y\mathrm{th}(^{p}\kappa)\,, \tag{2}$$

where $^{\alpha}\kappa={}^{\alpha}\chi^{\alpha}t$ is a non-dimensional parameter, $^{\alpha}t$ is the thickness of ferrite and piezoelectric phases accordingly $^{m}\chi^{2}=2(1+\nu)\left(k^{2}-\frac{\omega^{2}}{^{m}V_{L}^{2}}\right)$, $^{p}\chi^{2}=-2(1+\nu)\left(k^{2}-\frac{\omega^{2}}{^{p}V_{L}^{2}}\right)$, $\frac{1}{^{m}V_{L}^{2}}=\frac{^{m}\rho}{^{m}Y}$, $\frac{1}{^{p}V_{L}^{2}}=\frac{^{p}\rho}{^{p}Y}$, $^{m}V_{L}$, $^{p}V_{L}$ are the velocities of longitudinal waves, respectively in the ferrite and piezoelectric media, ^{m}Y, ^{p}Y are the Young's modulus of ferrite and piezoelectric phases accordingly.

Equation (2) defines the dependence of angular frequency ω from the wave vector k, in an implicit form, at the distribution of elastic waves in bilayer ferrite-piezoelectric structure.

In limited case, when the thickness of the piezoelectric phase is much less than the thickness of the ferrite phase, we have relation in Eq. (3), which presents the dispersion relation for the ferrite phase.

$$\omega={}^{m}V_{L}k\,. \tag{3}$$

In case, when the thickness of ferrite phase is much less than the piezoelectric phase, we have the Eq. (4), which presents the dispersion relation for piezoelectric phase.

$$\omega={}^{p}V_{L}k \,. \tag{4}$$

For thin ferrite and piezoelectric phases, when ${}^{m}\kappa={}^{m}\chi^{m}t<<1$ and ${}^{p}\kappa={}^{p}\chi^{p}t<<1$, the approximate expression is obtained, for the dispersion relation, which is shown in Eq. (5).

$$^{m}Y({}^{m}\chi)^{2}\,{}^{m}t={}^{p}Y({}^{p}\chi)^{2}\,{}^{p}t \,. \tag{5}$$

Taking into account the expressions for ${}^{m}\chi$, ${}^{p}\chi$ and Eq. (5), after simple transformations, the dispersion relation is obtained in the following form, shown in Eq. (6).

$$\omega=\sqrt{\frac{{}^{m}Y\,{}^{m}t+{}^{p}Y\,{}^{p}t}{{}^{m}\rho\,{}^{m}t+{}^{p}\rho\,{}^{p}t}} \,. \tag{6}$$

Thereby, for small thicknesses of ferrite ${}^{m}t$ and piezoelectric ${}^{p}t$, the linear dependence of angular frequency from the wave vector is retained.

4. Conclusion

The amplitude of the wave is changed on the thickness of the sample, on direction perpendicular to the line of the partition, in the distribution of the elastic waves in bilayer structure. The boundary conditions produce a system of equations which give the dispersion relation for the planar oscillations. Nonlinear relation between the angular frequency ω and the wave vector k is defined from that expression. In limited cases, when one of the layers is thinner than the other, this relation passes to a dispersion relation of one, or the other media, thus the speed of the distribution of elastic waves is less than the speed of the distribution of elastic waves in ferrite, but more than in piezoelectric phase. In limited cases, this dispersion relation passes into dispersion relation for ferrite and piezoelectric accordingly.

References

[1] Bichurin M I, Petrov V M, Averkin S V and Filippov A V 2010 "Electromechanical resonance in magnetoelectric layered structures" *Phys. of the Solid State* **52** 1975-80

[2] Fillipov D A, Bichurin M I, Petrov V M, Laletin V M, Poddubnaya N N and Srinivasan G 2004 "Giant magnetoelectric effect in composite materials in the area electromechanical resonance" *Tech. Phys. Lett.* **30** 15-20

[3] Fillipov D A, Bichurin M I, Petrov V M, Laletin V M and Srinivasan G 2004 "Resonant amplification of the magnetoelectric effect in composite ferrite-piezoelectric materials" *Phys. of the Solid State* **46** 1621-27

[4] Fillipov D A 2012 "Magnetoelectric effect in thin-film magnetostrictive-piezoelectric structures grown on a substrate" *Phys. of the Solid State* **54** 1112-15

Journal of Physics: Conference Series **461** (2013) 012017 doi:10.1088/1742-6596/461/1/012017

Mechanism of charge transport in $Si/Al_2O_3/Al$ structures

T M Borisova [1], R A Castro [1]
[1]Russian State Pedagogical University named after A.I. Hertsen
48, Moyka river Emb., Saint-Petersburg, Russia, 191186
tel: 8(985)266-63-74, e-mail: leonova_tatyana@list.ru

Abstract. The investigation of current – voltage characteristics of structures $Si/Al_2O_3/Al$ on the basis of aluminium oxide layers obtained by a method atomic layer deposition is carried out. It is established, that the possible mechanism of charge transport in structure is the space charge limited currents. The charge carrier concentration (N_t), concentration of traps (n_0) and electron mobility (μ) in Al_2O_3 layer are calculated.

1. Introduction

Aluminium oxide possesses a number of interesting physical and chemical properties - high melting point, heat conductivity and the dielectric parameters, increased radiation stability, high electrical specifications, opacity to alkali ions and stability, wide-gap region, minor external impact resistance, shows no chemical reaction at contacted materials, thus causes its wide application as an active element in modern microtronics and electrooptics. During the recent years films of aluminium oxide have been analyzed as a dielectric layer in condenser structures. Thanks to high dielectric capacitance amorphous films of Al_2O_3 are considered as one of alternatives of replacement of silicon oxide in the process of formation of superfine gate dielectric layers in technology of metal-oxide-semiconductor structure [1].

Nature of films' conductivity depends on film thickness, film coating conditions, electrode system, substratum material, and applied voltage direction. Authors [2] investigated properties of cold cathodes on the basis of nanoflake of Al_2O_3 applied on Indium Tin Oxide. It is revealed that addition of a dielectric layer in thickness up to 11 nm essentially increases emission current. At a field, smaller than $2\cdot10^7$ V/m, prominent mechanisms of emission current formation are trap tunneling and Frenkel-Poole thermoactivation. In the field of $E \leq 6\cdot10^7$ V/m carrier tunneling in vacuo according to Fowler-Nordheim mechanism involving traps is carried out, and at $E \geq 6\cdot10^7$ V/m the emission mechanism changes from resonance to direct Fowler- Nordheim mechanism. Hickmott examined $Al–Al_2O_3–Au$ structures with thickness of a dielectric layer $\approx$ 30 nm [3], [4] and found out that at voltage up to $\approx$ 15 V the Fowler- Nordheim mechanism of tunneling is carried out, and the height of a potential barrier makes up 1.5 eV. At the thickness of the insulation layer of 100 nm and at the voltage of 1 V density of injection current was equal to $8\cdot10^6$ A/cm^2 . Many researchers interpret the conductivity of Al_2O_3 within Frenkel-Poole model [5], [6], [7], [8].

It is known that when using TMA (trimethylaluminum, $Al(CH_3)_3$) and water as reagent in the process of synthesis of amorphous aluminum oxide, crystalline grains of the structure represent modification α-Al_2O_3 (hexagonal, distance between the closest atoms of aluminum and Oxygen is $R_{Al\text{-}O}$ = 0,176$\AA$) [9]. A unit cell of α-Al_2O_3 belongs to a rhombohedral crystallographic system (R3cR) and contains 10 basic elements. Lattice constant: a_1 = 5.160 Å and θ_1=55.286^0. Atoms of oxygen are fourfold coordinated by atoms of aluminum. Atoms of aluminum are coordinated by six atoms of oxygen, and three next atoms at distance of 1.866 $\AA$ and three following the next atoms at distance of 1.983 $\AA$ are available [10], [11]. The valance band of α-Al_2O_3 consists of two subbands divided by an ionic gep, the lower band is formed generally by 2*s* – oxygen states, the top band is created by 2p – oxygen states with a contribution of 3*s*-, 3*p*-states aluminum. Values of forbidden-band width (E_g) is 6.0 – 8.2 eV for α-Al_2O_3 [12].

Excess of aluminum in a crystal latitude of amorphous aluminum oxide that has not been subjected to annealing is caused by existence of its own (structural) defects such as vacancies of oxygen (anionic vacancy) which are the donors capable to double ionizing event, and their concentration defines the volume of conductivity of aluminum oxide. In connection with what the size of a potential barrier for electron injection out of metal into the isolator conducting band goes down and makes up according to source [13] $\varphi_B = 1.23$ eV . If the concentration of traps is great, the distance between traps isn't enough, so electrons from a metal electrode have possibility to tunnel directly onto traps and to pass isolator – hopping conductivity that we observed within the limits of frequencies from 10^3 up to 10^5 Hz at constant voltage of 1 V.

Electron density n in conduction band is connected with electron density on the n_t -traps in the following relation :

$$n = \theta n_t , \tag{1}$$

where θ is the constant specifying the extent of trap filling, it depends on depth of trap occurrence traps E_t.

Existence of traps (vacancies) leads to considerable reduction of current since being originally empty, they entrap injected electrons not allowing them to move in electric field. Consequently, the entrapped electrons make up a volume characteristic, and don't participate in current transfer through isolator. Electron density on the traps [14]

$$n_t \approx \frac{\varepsilon\varepsilon_0 U}{ed^2} = \frac{\varepsilon\varepsilon_0 E}{ed} . \tag{2}$$

The purpose of work is research of current-voltage characteristic (*I-V* characteristic) of films of amorphous aluminum oxide, for identification of the mechanism of charge transportation within $Si/Al_2O_3/Al$.

2. Methods and Materials

Films used in work are received through method atomic layer deposition (ALD) on *p*-type silicon Si (100) surface-supported substrate with its resistivity of ρ = 10 Ohm • cm from a mixed gas of trimethylaluminum of $Al(CH_3)_3$ (TMA) and water vapour at substrate temperature of 300 K. Directly before the loading of the fin into the ALD reactor cell Si-fin were processed by one-percent solution of fluorohydrogen acid HF for the purpose of removal of natural SiO_2 oxide. Aluminum contact points in area of 0.24 mm^2 were applied on isolator with thermal evaporation in vacuo. The thickness of oxide layer measured with the help of ellipsometer LEF – 3 was 78 nm. Measurements of current-voltage characteristic are executed at room temperature on the spectrometer "Concept 41" (Novocontrol Technologies), measuring voltage on a sample was in limits from 0.01 till 1 V.

3. Results and Discussion

The most probable mechanisms of charge transportation are the following: overbarrier Schottky emission, Frenkel-Poole mechanism and the currents limited by a space charge.

I-V characteristic of the structure is measured at the room temperature and at an alternating current. By the shape of *I-V* characteristic it can be concluded that in these films the conductivity mechanism is monopolar (electronic) [15].

When analyzing the experimental data and assessing the received results we used the following measurements: dielectric capacitance of aluminum oxide at frequency of 10^{-1} Hz ε=94, high-frequency dielectric capacitance $\varepsilon_\infty = 9.2$.

I – V characteristic at logarithmic scale of lg*I* - lg*U* (Figure 1) shows existence of three marked sites corresponding to power characteristics of a current (*I*) from applied voltage (*U*).

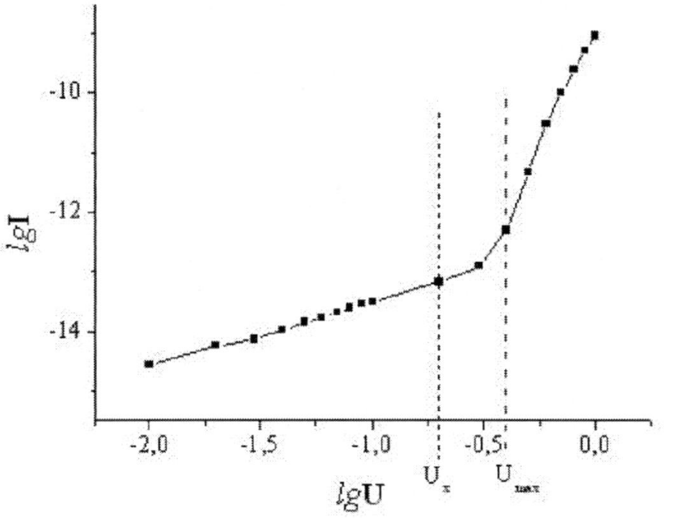

Figure 1. Sample *I-V* characteristic at the room temperature at the frequency of 10^{-1} Hz in coordinates of currents limited by a space charge

On the interval of voltage ranging from 0.01 V up to 0.2 V lgI ~ lgU, i.e. out the Ohm law is fulfilled

$$J = en_0\mu\frac{U}{d}, \tag{3}$$

where J – is electric current density in isolator, n_0 – is concentration of balanced thermally generated free carriers, μ- is current carrier mobility, U – the voltage applied to a sample, d – thickness of isolator.

The Ohm law is fulfilled at low voltages while the average concentration of n_i injected non-balanced free carriers doesn't become comparable with n_0 concentration of those carriers which are released thermally and are therefore neutralized. Existence of an ohmic site in loow fields allows to speak about existence of free electrons in isolator with n_0 concentration .

At higher voltages on the interval ranging from 0.2 V up to 0.4 V when n_i starts to prevail over n_0, current trapping quadratic dependence of voltage $\lg I \sim 2\lg U$ is observed, *I-V* characteristic submits to dependence:

$$J = \frac{9}{8}\theta\varepsilon\mu\frac{U^2}{d^3}. \tag{4}$$

Voltage of transfer from an ohmic site of *I-V* characteristic to quadratic $U_X = 0.2$ V, is given by the following:

$$U_x = \frac{4en_0d^2}{3\theta\varepsilon}, \tag{5}$$

where n_0 – is concentration of free carriers at thermal balance.

The equation (4) be fair until Fermi level F is lower than level of trap occurrence E_t, at some voltage Fermi's level crosses trapping level, this voltage is called voltage of maximum trap infill

$$U_{max} = \frac{eN_td^2}{2\varepsilon}, \tag{6}$$

where N_t - is concentration of total number of traps in a unit of isolator volume, d - is isolator thickness, ε - is dielectric capacitance of isolator, ε_0 - is dielectric constant.

At $U=U_{max}=$ 0.4 V current starts to increase sharply, the number of injected electrons will be equal to number of traps. The received result corresponds to a case when all defects are identical and their levels lie superficially (above Fermi's level, as $U_x < U_{max}$) (t.i. they are small). As the currents limited by a space charge are described with power like current-voltage characteristic of $I=C\cdot U^k$ type [16], growth of a current observed on chart at voltage higher than 0.4 V lg$I$ ~ 8.25 lg$U$, it allows us to assume that the main mechanism of current transportation is currents limited by a space charge. Dependence of a current on voltage is described by a formula

$$I = e\mu N_c\left(\frac{\varepsilon\varepsilon_0}{N_t}\right)^l \frac{U^{l+1}}{d^{2l+1}} = e\mu N_c\left(\frac{\varepsilon\varepsilon_0}{N_t}\right)^l\left(\frac{E}{d}\right)^l, \tag{7}$$

where N_c – is effective density of states in conduction band.

I-V characteristic in fields higher than 0.4 V proves existence of traps with exponential (continuous) distribution of density on energy in amorphous aluminum oxide.

In isolator with traps a current is in $1/\theta$ time less, than in isolator without traps. According to the schedule (Figure 1) we get $\theta = 0.85$.

According to (5) taking into account that $U_x = 0.2$ V, we get concentration of free balance current carriers of n_0=1,23 $\cdot 10^9$ cm^{-3}. According to (6) taking into account that U_{max}=0.4 V, we get concentration of levels of sticking (of traps) N_t=1.24$\cdot 10^{17}$ cm^{-3}. With the help of an ohmic site (3) we determine current carrier mobility in aluminum oxide: μ=2.473 $\cdot 10^{-4}$ m^2/V$\cdot$s = 247.3 cm^2/V$\cdot$s

At further increase of voltage new additional carriers of a charge are injected into isolator. These electrons already have no place to disappear, but remain in a conduct band. Therefore at voltages, higher than voltage of complete filling of traps, *I-V* characteristic will be the same, as well as for isolator without traps. That is the filled traps don't influence any more on *I-V* characteristic and they can be not taken into consideration.

Concentration of electrons in a conduct band, i.e. is higher than mobility edge

$$n = N_c\left(\frac{\varepsilon\varepsilon_0 U}{N_t d^2}\right)^l. \tag{8}$$

Characteristic energy of distribution of traps (depth of levels of sticking) E_t, is determined by a formula:

$$E_t = lkT. \tag{9}$$

The measurement l is determined by the schedule of lg I – by lg$U$ where *I-V* characteristic gives a straight line with l+1 tilt angle, and l= 7.25, E_0 = 0.19 eV.

Fermi level position is determined by a formula (will be changed accordingly to increase in voltage):

$$F = E_t \ln\frac{N_t d^2}{\varepsilon\varepsilon_0 U}, \tag{10}$$

we get F=(5.63÷6.50) eV.

4. Conclusions

Thus, the possible mechanism of charge transport in Si/Al_2O_3/Al structure is the currents limited by a space charge. Concentration of free balanced current carriers in Al_2O_3 makes up (1.23±0.01)$\cdot 10^9$ cm^{-3}, concentration of traps (1.24±0.01) $\cdot 10^{12}$ cm^{-3}, depth of their occurrence, counted from a bottom of conduct band, is 0.19 eV.

Литература

[1] Wilk G D Wallace R M Anthony J M 2001 High-k gate dielectrics: Current status and materials properties considerations *J. Appl. Phys.* **89** p 5243

[2] Lee M B Hahm S H Lee J H Song Y H 2005 Emission behavior of nm-thick Al_2O_3 film-based planar cold cathodes for electronic cooling *Appl. Phys. Lett.* **86** 12 pp 1235(11–13)

[3] Hickmott T W 2000 Polarization and Fowler—Nordheim Tunneling in Anodized Al-Al_2O_3–Au Diodes *J. Appl. Phys.* **87** 11 pp 79(03–12)

[4] Hickmott T W 2000 Voltage-dependent dielectric breakdown and voltage-controlled negative resistance in anodized Al-Al_2O_3-Au *J. Appl. Phys.* **88** 5 pp 28(05–12)

[5] Frenkel J I Exper J 1938 *Theoret. Phys.* **8** p 1292

[6] Ganichev S D Ziemann E Prettl W 2000 *Phys. Rev. B.* **61** p 10361

[7] Kolodzey J Ahmad E Adam T N Rau G Q I Olowolate J O Suehle J S Chen Y 2000 *IEEE Trans. Electron Device.* **47** p 121

[8] Lee C - H Park K - C Kim K 2005 *Appl. Phys. Lett.* **87** p 073510

[9] Choong-Ki Lcc Eunac Cho Hyo-Sug Lcc Kwang Soo Scol and Scungwu Han 2007 Comparative study of electronic structures and dielectric properties of alumina polymorphs by first-principles methods *Physical review.* **76** 245110 pp 1-7

[10] Gritsenko V A Tyschenko I E Popov V P Pass T V 2010 Isolators in nanoelectronics [V.A.Gritsenko, I.E.Tyschenko, V.P.Popov, T.V.Pass] exc. editor A.L.Aseev's *Russian Academy of Sciences, Sib. department, Inst. of physics of semiconductors named after Rzhanov's A.V of the Siberian Branch of the Russian Academy of Science* (Novosibirsk: Publishing house of the Siberian Branch of the Russian Academy of Science) p 258 pp 117-126

[11] Pass T V Gritsenko V A 2010 Application and electronic structure of isolators with high dielectric permeability *Successes of physical sciences* **180** 6 pp 213-217

[12] Perevalov T V Shaposhnikov A V Gritsenko V A Wong H Khan Zh Kh Kim Ch V 2009 Electronic structure α-Al_2O_3: ab initio modeling and comparison with experiment *Technic. Phys. Lett.* **85** 3 pp 197-201

[13] Nikiforov D K Korzhavy A P Nikiforov K G 2009 Influence of a dielectric nanolayer on issue properties of structures of Al – Al_2O_3 and Be – BeO *News of the Russian state pedagogical university of a name of A.I.Herzen* **11(79)** pp 153 - 159

[14] Lampert M 1973 Injection currents in firm bodies (Lampert M Mark P M World) p 416

[15] Novikov Yu N Gritsenko V A Nasyrov K A 2009 The Manyfon mechanism of trapionization Al_2O_3 in Al_2O_3: experiment and numerical modeling *Technic. Phys. Lett.* **89** 10 pp 599-602

[16] Lazarev V B 1979 Conductivity of oxide systems and film structures (Lazarev V B Krasov V G Shaplygin I S M Nauka) p 168

Investigation of the processes of accumulation and relaxation charge in composite films of polypropylene

E Fomicheva

Department of physics, Herzen State Pedagogical University, Saint-Petersburg, Russia

E-mail: e.e.fomicheva@gmail.com

Abstract. The work is devoted to the investigation of relaxation processes in composite films of polypropylene with aerosil. The investigation shows that relaxation of electret state is determined by the bulk conductivity. Two volume percentage of aerosil is optimal volume for electret stability of samples. The investigation of the sample by means of the method of thermostimulated depolarization reveals that particles of aerosil in composite samples play the role of macrodipoles.

1. Introduction

Range of application of polymeric electret materials is very wide: from household appliances to special equipment and medicine. This causes the need for a polymer with a high and stable electret characteristics. One of the most simple and accessible ways of creation similar material is added to the polymer matrix various dispersed fillers. However, despite the strong interest to the creation and use of composite materials, there are almost no models that explain the processes responsible for changing the electret properties of polymers after the introduction of these fillers. Therefore investigation of the influence of dispersed fillers on the electret properties of polymers and explain of relaxation processes in composite materials is an urgent problem of polymer physics.

2. Experimental results

The objects of study are films of polypropylene (PP), a thickness of 300 microns. The filler is aerosil (brand A-175) with a particle size of about 12 nm.

Time and temperature stability of the electrets state for polypropylene were done using isothermal and thermoactivated relaxation of the surface potential. For the study of the time stability of the electrets state the samples were charged at a temperature of 70°C to 1 kV surface potential by corona discharge and the dependence of the surface potential of the time was obtained at a temperature of polarization. For the study of the temperature stability of the electrets state, the samples were charged at a temperature of 20°C and the temperature dependence of the surface potential in a linear heating mode was investigated. The experimental results (figure 1, 2) showed that the addition of aerosil leads to the increase of the time and temperature stability of the electrets state in polypropylene. Optimal content of aerosil to the stability of the electrets state is a number of 2 vol.%. of aerosil.

The relaxation time (time of decreasing of the surface potential in e times) of electrets state was calculated (table 1).

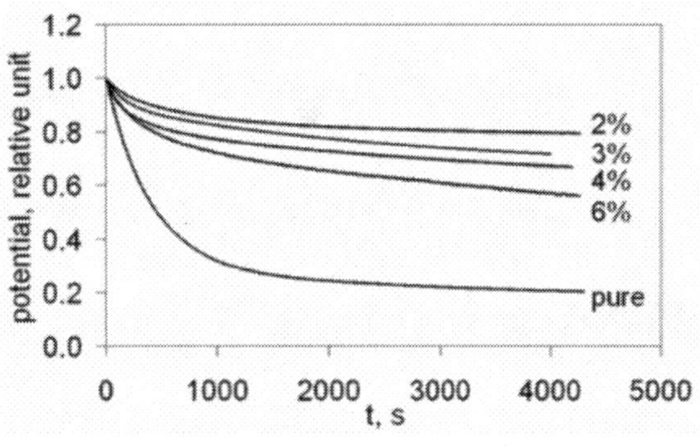

Figure 1. The time dependence of the surface potential for the samples charged in the negative corona discharge. T = 100°C. 1 – pure PP; 2 – PP + 2 vol.%; 3 – PP + 3 vol.%; 4 – PP + 4 vol.%; 5 – PP + 6 vol.% of aerosil.

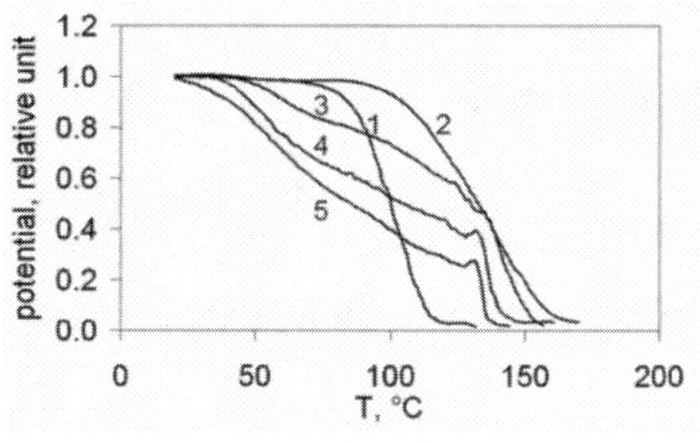

Figure 2. The temperature dependence of the surface potential for the samples charged in the negative corona discharge. Heating rate 5°C/min. 1 – pure PP; 2 – PP + 2 vol.%; 3 – PP + 3 vol.%; 4 – PP + 4 vol.%; 5 – PP + 6 vol.% of aerosil.

Table 1. Relaxation time of electret state and bulk conductivity of composite samples

Volume percentage of aerosil, %	Relaxation time (T = 100°C), s	Bulk conductivity (T = 140°C), 10^{-15} (Om·m)$^{-1}$
0	1994	192,5
2	13915	7,5
3	10000	12,2
4	8280	16,9
6	6050	53,7

Electrical conductivity of materials was studied using a method of thermostimulated polarization. The sample was heated from 20°C to 150°C in the applied electric field intensity of 500 V/mm and cooled in the same temperature range (without turning off the electric field) with the same speed. Cycles «heating-cooling» was held until the heating curve of the cycle does not coincide with the curve of the heating of the previous cycle. The results of measurements showed that the introduction of 2% vol. aerosil reduces the conductivity of polypropylene. During the increase of the percentage of filler conductivity is growing, but is still significantly less than the conductivity of the original polymer. Conductivity of the samples was calculated and the results for the temperature of 140°C given in table 1.

Comparison of the obtained values of the relaxation time and conductivity (figure 3) allows to make a conclusion that the relaxation of the electret state in the studied material is due to the conductivity.

This result can be explained by the interaction of particles of aerosil and water molecules in the polymer. There are hydroxyl groups in a polymer. The groups are formed from the molecules of water available in the volume of the polymer. The hydroxyl groups are attached to the surface atoms of silicon particles of aerosil. Arising out of the active centres of the surface absorb water, resulting in the characteristic shell, consisting of hydroxyl groups and of the adsorbed water, forms around of aerosil particles [1-4]. The presence on the surface of aerosol particles of hydroxyl groups attached to the silicon atoms, also confirmed by the existence of the IR spectra of composite polypropylene strip 3680 cm^{-1} (figure 4). Bands in this area of the spectrum are traditionally considered to be OH-fluctuations in the SiOH-groups [2, 5-7].

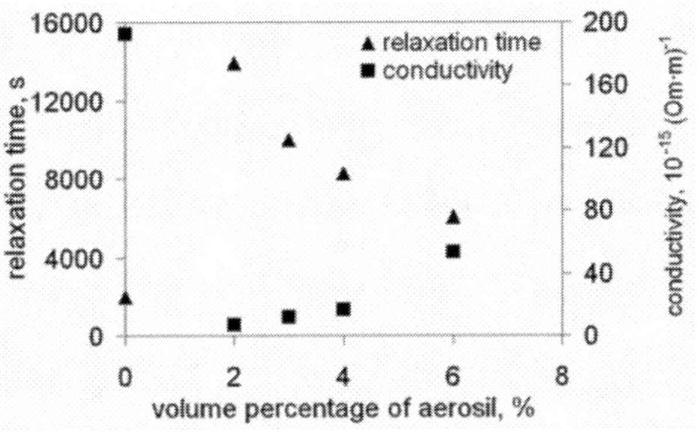

Figure 3. The dependence of the relaxation time and the conductivity from the volume percentage of aerosil.

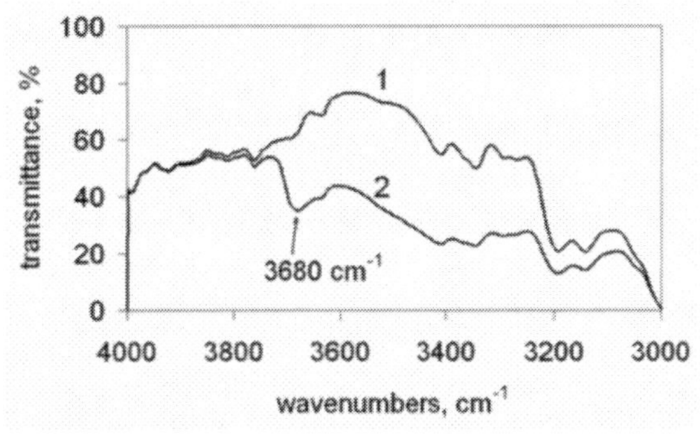

Figure 4. IR-spectra of pure polypropylene (1) and polypropylene with 6 vol.% aerosil (2).

According to the model of conductivity, developed by the authors [8-10], the water molecules are the catalyst for the conductivity. Therefore, decrease of their number in the volume of polymer leads to a decrease of conductivity of the material. This fact was observed in the curves of thermally stimulated polarization. However, the dependence of the conductivity of composite samples from the volume percentage of aerosil is not monotonous. Adding more than 2 vol.% of the filler leads to increase of conductivity of the material. To explain this fact were studied the composite samples by the method of small-angle X-ray scattering.

The experimental scattering intensity for composite films at different volume concentrations of aerosil and the experimental intensity of scattering for a pure polypropylene shows in figure 5a. The scattering indicatrix for the composite samples differ only a shift in the axis of ordinates (figure 5a). Moreover, normalized by the concentration curves coincide in the framework of the inaccuracy (figure 5b). This means that significant interference effects don't occur with increasing concentration of aerosil and, consequently, of aerosil particles don't form aggregates.

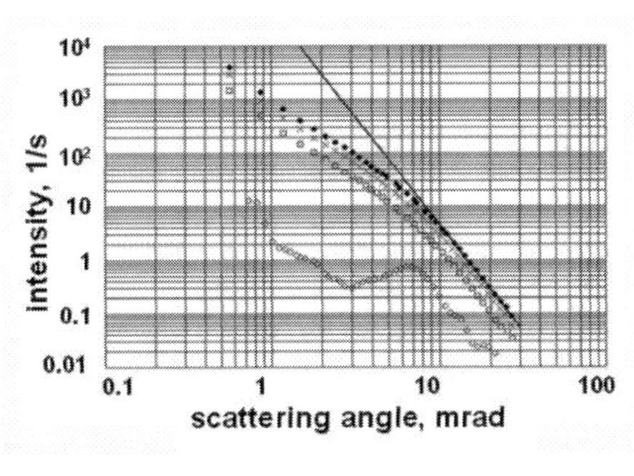

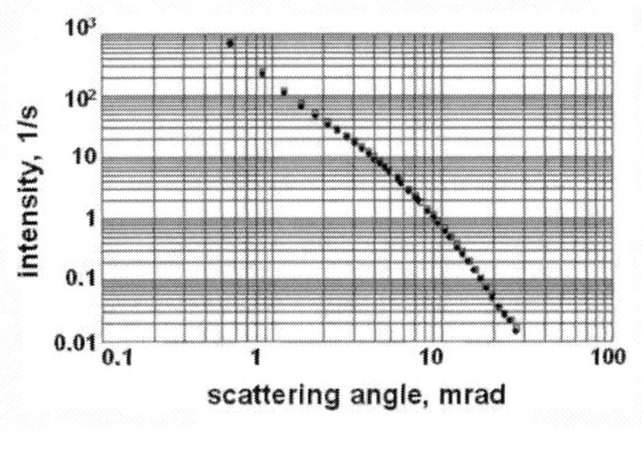

a b

Figure 5. Results of the experimental scattering intensity for composite films: a) pure polypropylene (○○○), polypropylene with 2 vol.% (□□□), 3 vol.% (×××), 6 vol.% (●●●) of aerosil; б) dependence normalized to the concentration for polypropylene with 2 vol.% (□□□), 6 vol.% (●●●).
—— line with $\alpha = 4$.

Journal of Physics: Conference Series **461** (2013) 012018 doi:10.1088/1742-6596/461/1/012018

All of the indicatrix of the scattering in the range of large angles have linear dependence with the slope $\alpha = 4$. According to the theory of X-ray small-angle scattering, extended linear dependence of the logarithm of the intensity of logarithm of the scattering angle indicates the fractal structure of the surface of the scattering object. The fractal dimension D_S of the surface can be calculated by the formula $D_S = 6 - \alpha$ [11]. Fractal dimension from the experimental curves is $D_S = 2$. According to the theory, the value of D_S for the fractal surface can vary from Ds = 2 (completely smooth) to $D_S = 3$ (full rough). Thus it can be argued that the surface of the aerosil particles is sufficiently smooth, based on the experimental data.

We can make the following conclusion from the results. Increase the volume of filler leads to a increase its free surface in the volume sample and, as a consequence, to increase numbers of OH-groups and adsorbed on them water molecules. The increase in the number of aerosil over 2 vol.%, apparently, leads to the emergence of conduction of the surface of the particles. The total conductivity of composite samples increases, as it is observed on the dependence of the conductance on the volume fraction of filler.

We investigated the role of the filler particles in processes of the relaxation of charge by thermally stimulated depolarization (TSD). The temperature dependence of the depolarization currents shows in figure 6. The position and shape of the peaks in the polypropylene and composites on its base are different. There is a new wide peak at a lower temperature on the spectra of composite materials. The position of the peak is independent of numbers of the filler (figure 6), and its height is proportional to the share of filler in the composite material (figure 7). The latter fact leads to the assumption that properties of the filler or to the properties of the phase boundary of polymer-filler responsible for formation the peak. The process observed in the pure polypropylene, are also present. It shows by the presence of a large current in the composites at 140-150°C. Investigation the dependence of the peak positions from the temperature polarization samples also showed differences in the spectra obtained for pure polypropylene and polypropylene with particles of aerosil. The position of maximum of the depolarization current of pure polypropylene is independent on the temperature of the polarization, but position of peak of the composite materials shifts to the region of high temperatures (table 2).

Table 2. The dependence of position of peak from temperature of polarization

T_{polar}, °C	T_{max}, °C pure PP	T_{max}, °C composite polymer
50	144	62
80		98
100	147	116
140	147	

These differences in the position and shape of the peaks, their behavior depending on the temperature of the polarization suggest the different nature of the processes taking place in composite and initial polymers. The process of the dipole relaxation appears in composite samples. The dipole moment induced in the particles in the applied of the electric field.

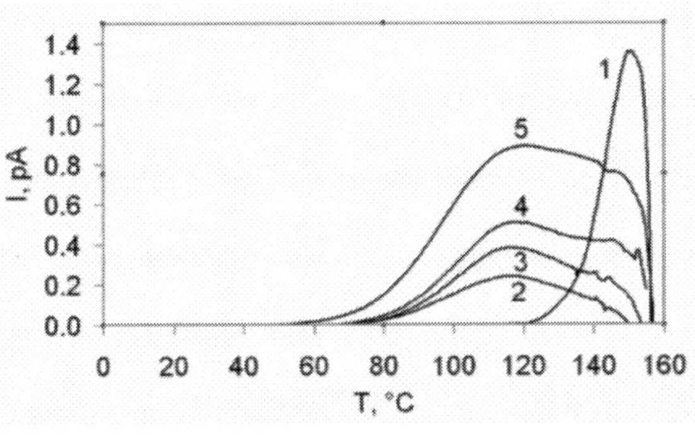

Figure 6. Currents of thermally stimulated depolarization. T_{polar} = 100°C, heating rate 5°C/min. 1 – PP; 2 – PP + 2 vol.%; 3 – PP + 3 vol.%; 4 – PP + 4 vol.%; 5 – PP + 6 vol.% of aerosil.

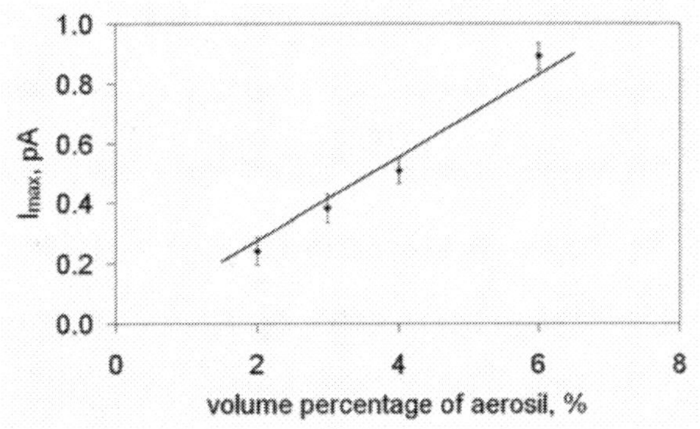

Figure 7. The dependence of maximum depolarization current from volume percentage of aerosil.

The mechanism of formation of the dipole moment can be explained by using the information about the structure of aerosil particles and polyelectrolyte model [2, 3, 8-10]. As mentioned above, there are hydroxyl groups on the surface of aerosil. If the group loses a proton, it becomes negatively charged $[O]^-$, and addition of a proton leads to the formation of a positively charged group $[OH_2]^+$. These groups move on the surface of the particle, when the particle is placed in an electric field. As a result of the particle becomes macrodipole.

The shift of the peak of depolarization current in the range of high temperatures with increasing temperature of the polarization means the presence of a dipole mechanism of relaxation (table 2). Dipoles that polarized at higher temperatures are better oriented along the field, and therefore require more energy for disordering.

The literature describes another way to distinguish the dipole peaks from the peak of the spatial charge: the dependence of the height of the peak from intensity of electric field is non-linear for the peak of the space charge [12]. Investigation of the composite polypropylene observed linear dependence of height of the peak from the intensity of the polarizing field (figure 8).

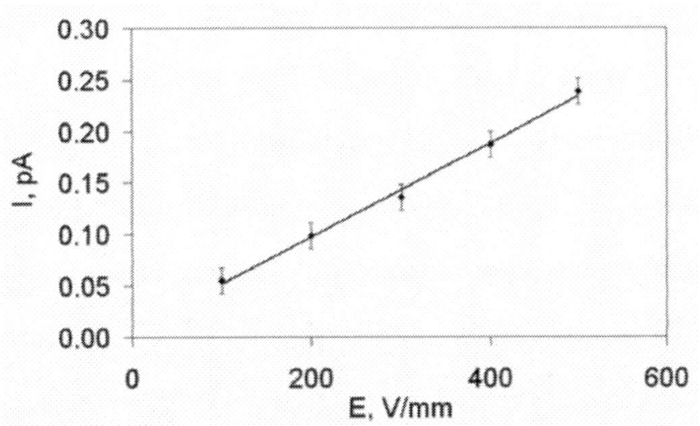

Figure 8. The dependence of maximum depolarization current from electrical field (PP + 2 vol.% of aerosil).

3. Conclusions

Thus, the main results of the work:

– relaxation of electret state in composite materials with aerosil is determined by the bulk conductivity;

– the addition of aerosil in the polypropylene leads to the increase of temperature and time stability of the electret state;

– optimal content of aerosil to the stability of the electrets state is a number of 2 vol.%. of aerosil;

– particles of aerosil in the composite material play the role of macrodipoles.

References

[1] Kac G.S., Milevski D.V. Napolniteli dlya polimernyh kompozicionnyh materialov: Spravochnoe posobie. // M.: Himiya, 1981. – 736 p.

[2] Sheka E.F., Markichev I.V., Natkanec I., Havryuchenko V.D. 1996 *Physics of Elementary Particles and Atomic Nucleis*, T. 27. **2**. 423-560.

[3] Parida S.K., Dash S., Patel S. and Mishra B.K. 2006 *Advances in Colloid and Interface Science*, Vol. **121**, Issues 1-3, 77-110.

[4] Roy M., Nelson J.K., MacCrone R.K., Schadler L.S., Reed C.W., Keefe R. and Zenger W. Polymer Nanocomposite Dielectrics – The Role of the Interface. // IEEE Transactions on Dielectrics and Electrical Insulation, Vol. 12, № **4**, 2005. P. 629-643.

[5] Vasil'ev V.V., Voicehovskii A.V. i dr. 2007 *Applied Physics*. **5**. 62-66.

[6] Katok K.V., Yanishpol'skii V.V., Tertyh V.A., Oranskaya E.I. 2008 *Nanostructured Materials Science*. **1**. 20-26.

[7] Silverstein R.M., Webster F.X. *Spectrometric Identification of Organic Compounds*, 1997, P. 119.

[8] Gorohovatskii Yu.A., Aniskina L.B., Viktorovich A.S., Gorohovatskii I.Yu., Karulina E.A., Tazenkov B.A., Temnov D.E., Chistyakova O.V. 2009 *Herzen University Journal of Humanities & Sciences*. **11(79)**. 47-61.

[9] Aniskina L.B. Viktorovich A.S., Galihanov M.F., Temnov D.E. 2010 *Herzen University Journal of Humanities & Sciences*. **135**. 24-36.

[10] Viktorovich A.S., Tazenkov B.A., Temnov D.E. 2008 *Herzen University Journal of Humanities & Sciences*. **10(64)**. 50-56

[11] Schmidt P.W. Use of scattering to determine the fractal dimension, In: Avnir D, editor. *The Fractal Approach to Heterogeneous Chemistry*, Wiley and Sons, New York (1989), 67–79.

[12] G. Sessler Electrets – M.: Mir, 1983. 487 p.

Journal of Physics: Conference Series **461** (2013) 012019 doi:10.1088/1742-6596/461/1/012019

Functional nanomaterials based on metal oxides with hierarchical structure.

E V Abrashova[1] I E Gracheva[2] V A Moshnikov[3]

[1] Postgraduate, St. Petersburg State Electrotechnical University "LETI", Pr. Popova 5, St. Petersburg, 197376 Russia
[2] Lecturer, PhD , St. Petersburg State Electrotechnical University "LETI", Pr. Popova 5, St. Petersburg, 197376 Russia
[3] Professor, Dr., St. Petersburg State Electrotechnical University "LETI", Pr. Popova 5, St. Petersburg, 197376 Russia

[1] E-mail: katabr@mail.ru
[2] E-mail:IEGrachova@mail.ru
[3] E-mail:VAMoshnikov@mail.ru

Abstract. The depending on evolution morphology of hierarchical porous nanostructured films based on metal oxide systems was investigated using atomic force microscopy. Gas sensitive properties of these structures were investigated. It was found that the characteristics of the degree of maturation of sols are gas-sensitive parameter to create functional porous layers.

1. Introduction

Sol–gel processes [1-3] are finding ever-increasing applications in various fields of science and ngineering. Models of various hierarchies formed during the sol–gel transition are of great scientific and practical interest. It is worth noting that these hierarchies [4-7] can be different in shape and origin.

In modern materials science, the principle of hierarchical self-assembly has acquired great importance. In essence, this principle lies in the synthesis of "building blocks" (modules) with specified sizes and shapes and in the further process of their integration. The hierarchical self-assembly can represent a multilevel process, when the integrated modules are the initial elements of larger sized aggregates (with larger characteristic interaction radii).

This work is aimed at investigating the mechanisms responsible for the formation of objects with a hierarchical porous structure by means of the self-assembly in sol–gel processes in which the initial fractal aggregates are formed over the entire volume of the colloid as a result of the diffusion-limited aggregation and the subsequent (more retarded) processes occur already with an increasing role of cluster–cluster interactions. The processes of formation, growth, and evolution of fractal aggregates can be controlled in the sol–gel synthesis.

2. Purposes

This work is aimed at investigating the mechanisms responsible for the formation of objects with a hierarchical porous structure by means of the selfassembly in sol–gel processes in which the initial fractal aggregates are formed over the entire volume of the colloid as a result of the diffusion_limited aggregation and the subsequent (more retarded) processes occur already with an increasing role of cluster–cluster interactions. The processes of formation, growth, and evolution of fractal aggregates can be controlled in the sol–gel synthesis.

Journal of Physics: Conference Series **461** (2013) 012019 doi:10.1088/1742-6596/461/1/012019

3. Experiment

Solutions of sols were obtained on the basis of binary heterochain inorganic polymers, where atoms form ion-covalent bonds between each other. The initial pre-courses used for the preparation of sol solutions are easily hydrolyzed components. Such components form polymolecules, or poly-solvated groups, due to their interaction with water. Tetraethoxysilane (TEOS, $Si(OC_2H_5)_4$) was chosen to obtain the nanostructured films of silicon dioxide. TEOS solutions helped to improve the film-forming qualities and have the ability to spread over the surface of porous silicon. To obtain the two-component oxide materials $10SiO_2 - 90SnO_2$ (mol. %), the hydrolysis and polycondensation of TEOS in the presence of inorganic salt of tin. Five types of film structures with a gradual increase in the exposure time of sol were obtained.

The optimum ratio of the basic components in an initial solution provides fast partial or full hydrolysis of alkoxides forming sol. The prepared sol solution is deposited on a rotating (3000 rev / min) glass substrate surface leading to a fuller hydrolysis of the alkoxides. The subsequent polycondensation and deposit of a sol thin layer on the substrate surface results in the formation of gel spatial at an annealing temperature of 600 °C.

Observations of nanostructured layer morphology of two component systems , using dynamic AFM (atomic force microscopy) methods, were carried out using an NTEGRA-Therma nanolaboratory (NT-MDT). Si-AFM probes NSG01 (NT-MDT) with a spring constant varying between 2.5 N/m and 10 N/m were used. The radius of curvature of the probes was ≥10 nm. We recorded AFM images at a resolution of 256 by 256 elements. Fractal dimension was calculated using Gwyddion.

Using impedance spectroscopy, we performed studies on the electrical properties of the nanomaterials. Measurements of the impedance spectra were carried out in the frequency range from 1 Hz to 0.5 MHz at 300 °C using an impedance meter «Z-500 P» (LTD «Elins»).

4. Results and Discussion

The sol–gel methods belong to technologies of molecular design. During the growth, fractal aggregates transform into percolation composites with controlled and reproducible sizes of pores in the range from several nanometers to several micrometers, or a porous structure is formed as a result of the spinodal decomposition and phase separation.

The main evolution stages of fractal aggregated systems were demonstrated using atomic force microscopy (figure 1–5).

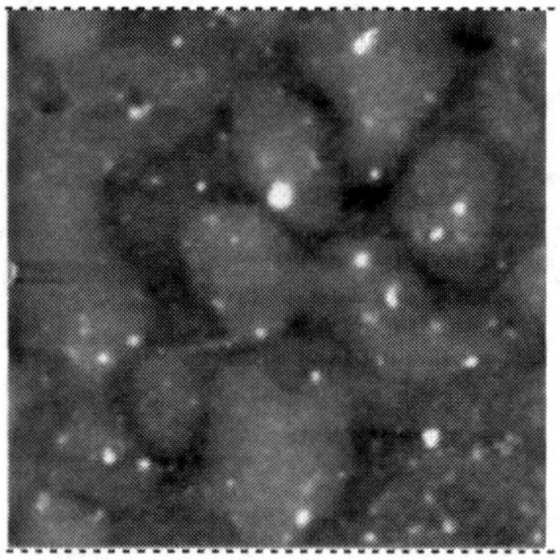

Figure 1. First type structures.

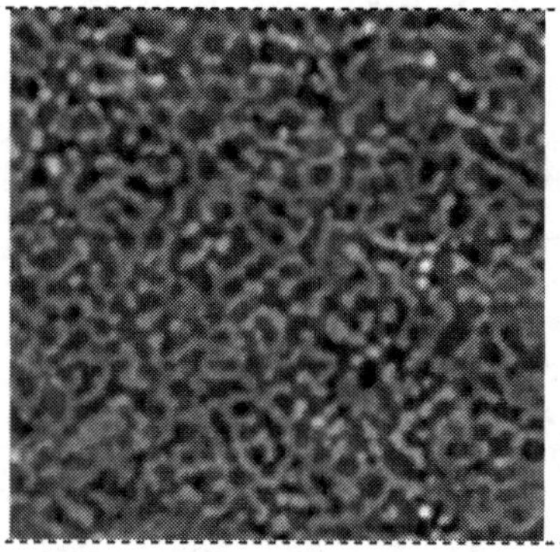

Figure 2. Second type structures.

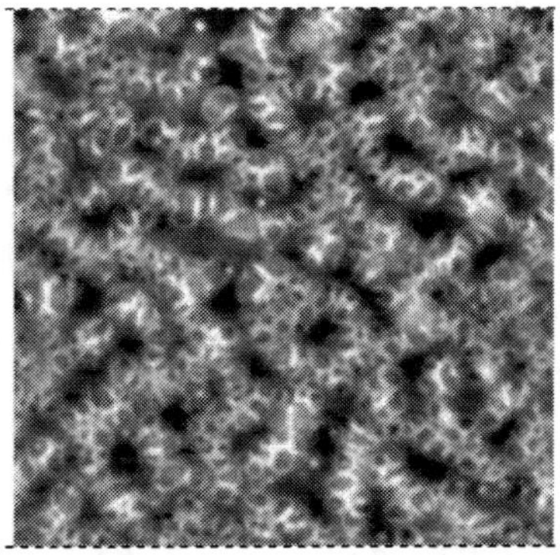

Figure 3. Third type structures.

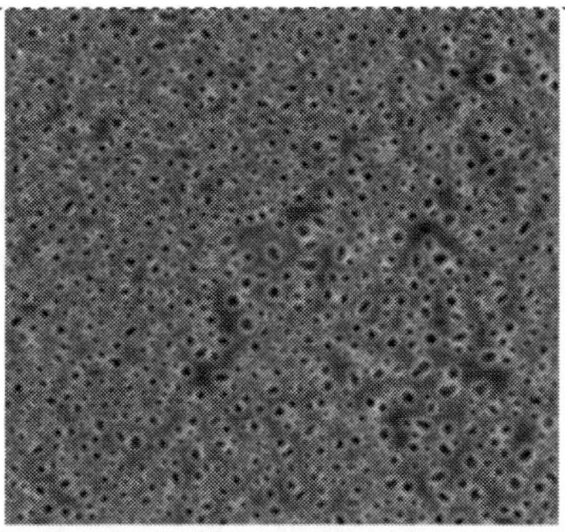

Figure 4. Forth type structures.

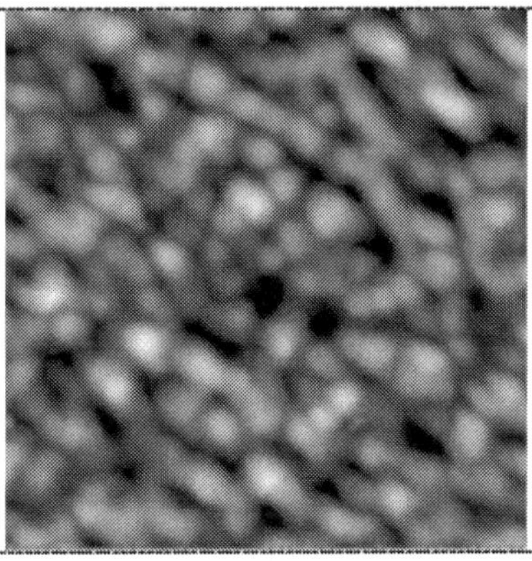

Figure 5. Fifth type structures

We revealed that local areas of irregular shape (figure 1: scanning area size is 5 μm × 5 μm; first type structures) in film nanocomposites formed in the early stages of ripening sols in nucleophilic growth. The fractal objects (figure 2: scanning area size is 10 μm × 10 μm; second type structures) are formed inside these areas with increasing exposure time of sols.

The formation of nanocomposites is determined by diffusion-limit aggregation and cluster–cluster aggregation conditions. It was shown that a further increase in the exposure time of sols leads to the formation of hierarchical structures with porous percolation isthmuses between the pores (figure 3: scanning area size is 10 μm × 10 μm; third type structures) in terms of spinodal decomposition. Revealed that hierarchical nanocomposites formed on glass substrates have pores with a size of two types 150-200 and 400-700 nm (figure 3), and on oxidized silicon substrates – 20 nm and 230-400 nm.

It was found that the fourth type structure corresponds to a porous structure with pore sizes of 10, 20 and 30 nm (figure 4: scanning area size is 10 μm × 10 μm), and the fifth type structures – corpuscular porous structure in which pores (with dimensions of about 10 nm) are the spaces between the particles (with size of about 40 nm) (figure 5: scanning area size is 1 μm × 1 μm).

It was found that the molecular weight distribution of the inorganic polymer chains in the sol strongly influences supramolecular structure of nanocomposites. It was shown that the longer the polymer chains in the sol, the more there is a strong tendency for spinodal decomposition. The dependence of the fractal dimension of the films on the degree of maturity of the sol is nonmonotonic (figure 6).

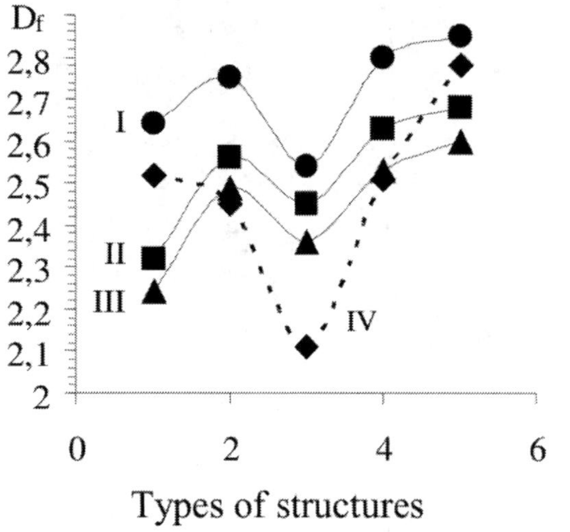

Figure 6. Dependence of the fractal dimension of the degree of maturation of the sol, calculated by:
I) decomposition;
II) triangulation;
III) counting cubes;
IV) of the power spectrum

The principle of operation of semiconductor gas sensors based on metal-oxide layers [8-13], which represent porous nanostructured composites, consists in changing the electrical conductivity of the material as a result of the chemisorption of gases. The first and necessary stage in the process of gas detection is the adsorption of oxygen on the surface of the semiconductor film. In the process, electrons are transferred from the bulk of the semiconductor to oxygen located on the surface of the film, which leads to a bending of the energy bands in the near-surface region to a depth determined by the Debye screening length. During the interaction of the reducing gas with negatively charged oxygen molecules on the surface of the semiconductor film, electron pass back into the bulk of the semiconductor and the reaction products leave the surface in the neutral form; in this case, the electrical resistance of the semiconductor material decreases. The measurements of the gas sensitivity were carried out using the reducing vapors of ethanol. The gas-sensitivity was calculated two ways:

$$S(Re) = \frac{Re(Z_{air}) - Re(Z_{gas})}{Re(Z_{gas})} \times 100\%$$

$$S(Im) = \frac{Re(Z_{air}) - Re(Z_{gas})}{Re(Z_{gas})} \times 100\%$$

where $Re(Z_{air})$ and $Im(Z_{air})$ – the value of active and reactive components of the complex impedance in air, $Re(Z_{gas})$ and $Im(Z_{gas})$ – the value of active and reactive components of the complex impedance in the presence of ethanol vapour (table 1). It was shown that gas-sensitive thin film layers decreases with increasing degree of maturation of the sol.

Table 1. The calculated sensitivity for the samples.

Sample type	first	second	third	forth
s(Re),%	400	130	90	14
s(Im), %	900	180	130	40

5. Conclusion

In this work, it has been shown that the control of the self_assembly in sol–gel processes provides ameans for controlling the structure of fractal clusters and their gas_sensitive properties. Five different stages of the evolution of fractal aggregated systems of tin and silicon dioxides have been determined, namely, , the formation of local areas of irregular shape in nucleophilic growth, diffusion_limited aggregation, cluster–cluster aggregation, formation of spherical species, the formation of hierarchical structures with porous percolation isthmuses between the pores in terms of spinodal decomposition, the formation of porous structure with pore sizes of 10, 20 and 30 nm corpuscular porous structure in which pores (with dimensions of about 10 nm) are the spaces between the particles. Investigated gas sensitive properties of these structures with respect to ethanol vapor. It was found that the characteristics of the degree of maturation of sols are gas-sensitive parameter to create functional porous layers.

The work is partially supported by Ministry of Education and Science of the Russian Federation, in the frameworks of Federal Task Program (FTP) «Scientific and pedagogic personnel of innovative Russia» for the period of 2009–2013.

References

[1] Brinker C. J., Scherer G. W. Sol-Gel Science. The Physics and Chemistry of Sol-Gel Processing. San Diego: Academic Press, 1990. 908 p.
[2] Handbook of sol-gel science and technology: processing, characterization, and applications / Ed. Sumio Sakka. New York, 2004.–V.1-3.
[3] Corriu R., Nguyên T. A. Molecular Chemistry of Sol-Gel Derived Nanomaterials. John Wiley & Sons, 2009
[4] One-step synthesis and gas sensing characteristics of hierarchical SnO_2 nanorods modified by Pd loading / P. Sun, Y. Yu, J. Xu et. al. // Sensors and Actuators B: Chemical. – 2011. – V. 160, № 1. – P. 244–250.
[5] Highly porous metal oxide polycrystalline nanowire films with superior performance in gas sensors / Jun Liu, Zaiping Guo, Kaixing Zhu et.al. // J. Mater. Chem. – 2011. – V.21. P. 11412–11417.
[6] Hierarchical nanostructured semiconductor porous materials for gas sensors /V.A. Moshnikov, I. E. Gracheva, V. V. Kuznezov et al. // Journal of Non-Crystalline Solids. – 2010. – V. 356, N. 37-40. P. 2020–2025.
[7] Nanostructured materials obtained under conditions of hierarchical self-assembly and modified by derivative forms of fullerenes / Irina E. Gracheva, Vyatcheslav A. Moshnikov, Evgeniya V. Maraeva et. al. // Journal of Non-Crystalline Solids. – 2012. – V. 358, N. 2. – P.433–439.
[8] Morrison R.S. Selectivity in Semiconductor Gas Sensor // Sensor and Actuators. –1987. –V.12. –P.425–441.
[9] Metal Oxide Nanostructures and Their Gas Sensing Properties: A Review / Yu-Feng Sun, Shao-Bo Liu, Fan-Li Meng et al. // *Sensors* 2012. – *12*(3). – P. 2610–2631
[10] A computer-aided setup for gas-sensing measurements of sensors based on semiconductor nanocomposites / I. E. Gracheva, A. I. Maksimov, V. A. Moshnikov, M. E. Plekh // Instruments and Experimental Techniques. – 2008. – V. 51, № 3. P. 462-465.
[11] Net-like structured materials for gas sensors / I. E. Gracheva, V. A. Moshnikov, S. S. Karpova, E. V. Maraeva // Journal of Physics: Conference Series. – 2011. – V. 291, № 1. – P. 012017.
[12] Ponomareva A. A., Moshnikov V. A., Suchaneck G. Mesoporous sol-gel deposited SiO_2-SnO_2 nanocomposite thin films // IOP Conf. Series: Materials Science and Engineering 30 (2012) 012003.
[13] Gracheva I. E., Spivak Yu. M., Moshnikov V. A. AFM techniques for nanostructures materials used in optoelectronic and gas sensors // IEEE Eurocon-2009. Eurocon-2009 – Saint-Petersburg, 2009. – P. 1250–1253.

Studying Ultradisperse Diamond Structure within Explosively Synthesized Samples via X-Ray Techniques

M D Sharkov[1], M E Boiko[1], S N Ivashevskaya[2] and N S Belyakova[3]

[1]Ioffe Physico-Technical Institute of the RAS, 26 Politekhnicheskaya Ul., St. Petersburg, Russian Federation

[2]Geology Institute, the Karelian Research Center of the RAS, 11 Pushkinskaya Ul., Petrozavodsk, Russian Federation

[3]St. Petersburg Polytechnical State University, 29 Politekhnicheskaya Ul., St. Petersburg, Russian Federation

E-mail: mischar@mail.ioffe.ru

Abstract. XRD (X-Ray Diffraction) and SAXS (Small-Angle X-Ray Scattering) data have been measured for a pair of samples produced with the help of explosives. XRD peaks have shown the both samples to contain crystal diamond components as well as graphite ones. Basing on SAXS analysis, possible presence of grains with radii up to 30–50 nm within all the samples has been shown. Structure components with fractal dimension between 1 and 2 in the sample have been detected, this fact being in agreement with the assumption of diamond grain coating similarity to onion shells. In order to broad rocking curves analysis, the standard SAXS treatment technique has been complemented by a Fourier filtering procedure. For the sample #1, rocking curve components corresponding to individual interplanar distances with magnitudes from 5 nm up to 15 nm have been separated. A hypothesis relating these values to the distances between concentric onion-like shells of diamond grains has been formulated.

1. Introduction

Ultradisperse diamonds (UDD), or diamond nanoparticles, belong to the set of the most frequently synthesized modern materials. Such strong interest is due to the fact that they have a series of special features, i.e. high adsorption capacity, high thermal conductivity, hydrophoby, large specific surface. These properties provide multiple prospects of UDD applicability.

The spatial size magnitude of a stable diamond nanoparticle is situated within a wide range of values, from 5 to 50 nm typically. The nanoparticle can be described as a diamond core surrounded by the shell consisting of two phases. These are, namely, the diamond one (constituted by sp^3-hybridized carbon atoms) and the amorphous graphite phase with carbon atoms in the sp^2 state [1]. According to published data, UDD particle shells differ in the both chemical composition and thickness [2]. The images of individual UDD clusters obtained by the method of transmission electron microscopy (TEM) showed the nanoparticles to have a well-faceted polyhedral shape. Digital modeling stated this polyhedral shape to influence the surface of small-size particles [3]. Diamond nanoparticles tend to form primary (up to 100 nm) and secondary (more than 100 nm) aggregations with fractal structure.

Within the number of materials containing big quantities of UDD there is the primary product of diamond detonation synthesis (the procedure of obtaining nanodiamonds using explosives). This primary product is represented by a black powder consisting of 30–60 per cent of UDD, 40–60 per cent of non-diamond carbon, and impurities containing metals (15% or less). Non-diamond carbon can include ultradisperse graphite, amorphous carbon, and also unstable carbon structures like carbines [4]. Presence of different carbon-based substances within the primary diamond detonation synthesis product confines its usefulness in chemically active media and compositions and, besides, results in enhanced adsorption of non-polarized compounds as well as low sensitivity to polarized and ionic ones. Therefore, the substance produced by diamond detonation synthesis is subject to diverse purification procedures, particularly, ones based on thermal treatment [5].

One of main tasks in studying finely grained materials (similar to UDD produced by detonation synthesis) is their domain structure investigation. A special role is played by the techniques that do not destroy the analyzed sample. Within the number of such methods there is the technique of small-angle x-ray scattering (SAXS).

SAXS is usually applied to tasks of determining spatial sizes for clusters, crystallites, powder and polycrystalline grains, layer thicknesses and superlattice periods, all these values being able to have order of magnitude from nanometers to hundreds of nanometers. The method of SAXS allows one to solve a triple problem constituted by defining sizes of large sample structure homogeneities (grains and pores), determining geometrical properties of the sample atomic matrix (so-called fractal dimension), and estimating lattice parameters for superstructures presented within the sample [6].

2. Experiments

The measurements have been performed for two samples of UDD obtained by detonation synthesis and received from the company JSC Sinta (Minsk, Belarus). The sample denoted #2 hereafter was subject to heat treatment unlikely another sample (#1). The both ones have been characterized by a set of physical and chemical techniques: powder x-ray diffraction (Thermo Scientific ARL X'TRA Powder X-ray Diffraction System, $\lambda_{Cu} = 1.54$ Å), scanning electron microscopy (the microscope VEGA II LSH, TESCAN Ltd., with a module for microprobe analysis), mass-spectrometry (the mass-spectrometer with inductively coupled plasma ICP-MS Xseries2, ThermoFisher Scientific). SAXS data have been obtained within the frames of several sequences of experiments at the laboratory set DRON-8 of the company Bourevestnik (St. Petersburg, Russia) at the $K_{\alpha 1}$ Cu beam wavelength.

The standard procedure of SAXS analysis [6] includes data treatment in three different data ranges. The angular scale is transformed initially into the scale of wave vector magnitudes $q = 4\pi \sin \theta/\lambda$ (with θ being one half of the scattering angle). In the vicinity of the incident beam direction the SAXS curve is broadened; the FWHM (full width at half maximum) of this SAXS peak allows one to estimate homogeneity sizes in the sample. At the slopes of the small-angle peak the SAXS curve can be described as a power function explained by the Porod's model [7]. The power magnitude is related to the so-called fractal dimension of the sample structure indicating the atomic matrix geometry. For 3D dense dispersed media, the power magnitude is to be 4 (the Porod law), whereas its value for a low-dimensional matrix is close to the dimension itself. Finally, when the sample contains some layering or superlattice the wide interplanar distances will be represented at the SAXS curve as Bragg peaks whose positions will then obey to the Bragg law $qd = 2\pi n$ (d is an interplanar distance here, and n is an integer).

3. Results and discussion

The both samples consist generally of carbon (about 80%). Analyses of micro-impurities in the samples showed them to contain enlarged quantity of copper. There were also detected unessential fractions of Zn, Pb, Ti, Cr, Ni, Mn, La, Fe, P, Mg, Al, Si, Ca, Na, K, Cl, S. Figure 1 represents the powder XRD data for both the samples #1 and #2.

On processing these data the two samples appeared to be constituted by nano-diamond (Num. 00-058-1638 in the ICDD database) and graphite (Num. 00-041-4187 in the ICDD database). A series of

Journal of Physics: Conference Series **461** (2013) 012020 doi:10.1088/1742-6596/461/1/012020

intensive peaks (2θ/deg: 43.7, 75.5, 91.4, 120.5, 141.1) corresponding to diamond lattice reflections of Cu K_α photons can be observed at figure 1, as well as a separate peak at 2θ=26.6^0 which must correspond to the graphite peak at 26.4^0. Also the XRD data of the both samples contain a set of five intensive peaks representing the copper spinel reflections (2θ/deg: 30.2; 35.5; 54; 56.5; 62), this fact being completely consistent with the results on the samples chemical composition. It can also be concluded from the data at figure 1 that the sample #1 thermal treatment did not give rise to substantial decay of non-diamond carbon forms. The sizes of coherent scattering regions calculated with the help of the Scherrer model (from the FWHM of the most intensive peak, for the (111) reflection here) happened to be about 4 nm. These results derived from modeling are consistent with the data obtained from direct SAXS measurements.

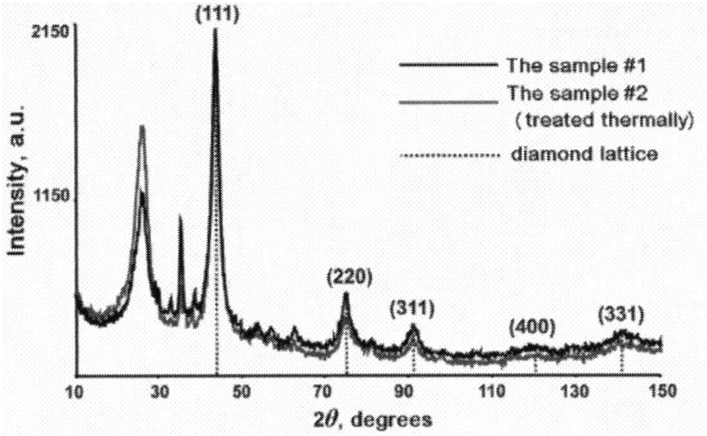

Figure 1. The XRD signals for the samples #1 and #2.

SAXS data analysis has led to following results. Different specimens have appeared to contain grains with spatial sizes up to 30–50 nm. Particularly, the sample #1 SAXS data (figure 2a) treatment has demonstrated the central peak to correspond to particles with the size of about 30 nm. The meaning of 50 nm has been obtained on processing the specimen #1 SAXS data recorded after the sample had been deformed (figure 2b). In turn, SAXS data for the sample #2 have resulted in the grain radius estimation of 40 nm approximately (figure 3).

Low-dimensional structural components have also been detected within the samples. This fact is consistent with the hypothesis of onion-like UDD grain coverage. Particularly, the deformed sample #1 SAXS data have turned out to contain fragments with power dependence whose power values are close to 1.5 and 1.8. Besides, the sample #2 SAXS function has been shown to include parts with power decrement characterized by power magnitudes of about 1.8 and 4. The value 4 is typical of 3D dense grained substance (that is the Porod law [7]) whereas power exponents about 2 must describe a 2D component presented within the sample under investigation.

For the sake of clarifying the specimen #1 SAXS data it was suggested to expand the standard SAXS processing technique by a procedure of Bragg peak Fourier filtering. This procedure consists of several steps. The data subject to filtering are Fourier transformed initially, then the image is divided by a typical Fourier transformed signal shape, and this result is treated using the inverse Fourier transform. Such processing has given rise to splitting the broad Bragg peak into a variety of separate contributions. Each of them is to represent an individual interplanar distance in the sample matrix. The distance values (situated in the current case within the interval from 5 nm to 15 nm) may describe differences between particular pairs of concentric onion-like shells.

SAXS analysis has thus yielded estimations for UDD grain sizes in two UDD samples produced via detonation synthesis and confirmed the assumption of onion-like concentric shells of UDD grains.

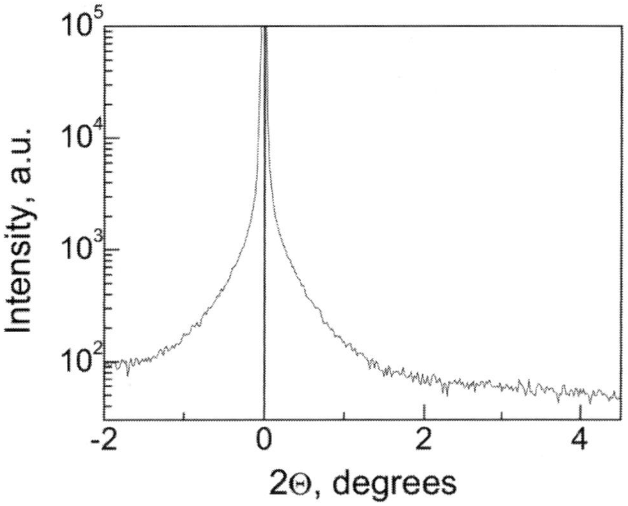

Figure 2a. Recorded SAXS data for the UDD sample #1 before its deformation.

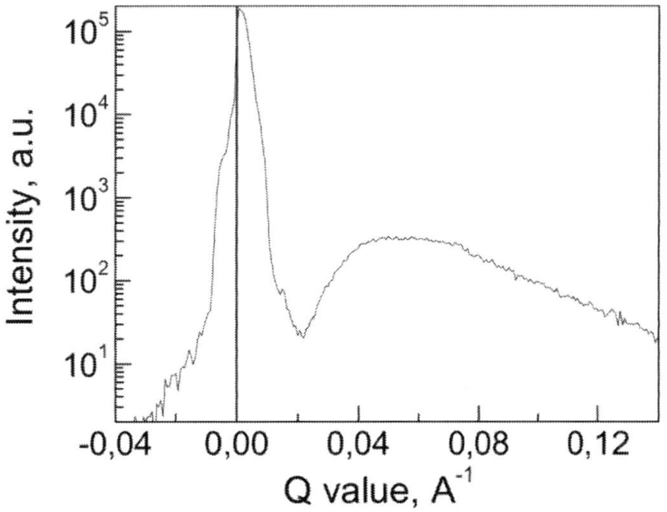

Figure 2b. Recorded SAXS data for the sample #1 before its deformation.

4. Conclusions

The SAXS data for two UDD samples produced by detonation synthesis have been obtained and treated. SAXS data processing for two DC samples has given appreciations for nanodiamond spatial sizes about 40–50 nm. The model of UDD grains surrounded by onion-like shells has been confirmed, with double Fourier transformation being the first time used for this purpose.

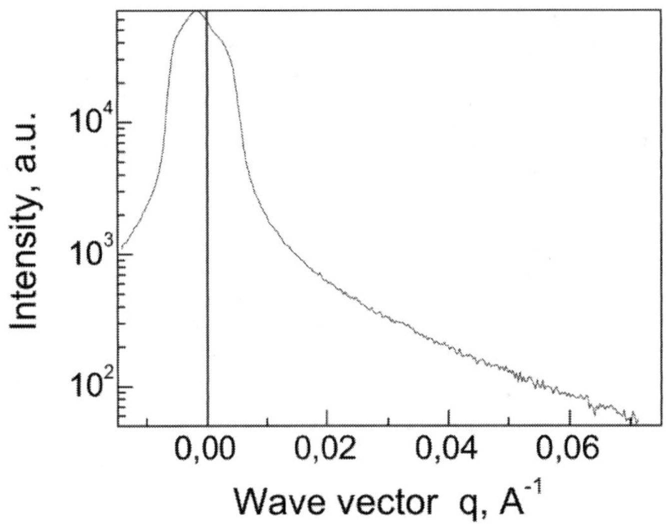

Figure 3. Measured SAXS data for the sample #2 (which was earlier subject to heat treatment).

Acknowledgements
The authors are grateful to their colleagues:

- Inina I.S., Ternovoy A.N., and Paramonov A.S. (Geology Institute of the RAS Karelian Scientific Center, Petrozavodsk, Russia);
- Ivanova T.I. and Kurskov A.V. (Investigation Development Manufacture "Bourevestnik", St. Petersburg, Russia).

References

[1] Redlich P, Banhart F, Lyutovich Y and Ajayan P M 1997 *Carbon* **36** 561
[2] Kulakova I I 2004 *Phys. Solid State* **46** 636
[3] Barnard A S and Sternberg M 2007 *J Mater Chem* **17** 4811
[4] Savvakin G I and Trefilov V I 1991 *Sov. Phys. Doklady* **36** 785
[5] Ivashevskaya S N 2011 *Conference FKS-2011: Abstracts* (Gatchina, Leningrad Region, Russia: Petersburg Nuclear Physics Institute) p 73
[6] Vasilevskaya T N and Zakharchenya R I 1996 *Phys. Solid State* **38** 1711
[7] 1982 *Small-Angle X-Ray Scattering* ed O Glatter and O Kratky (London: Academic Press)

Low temperature growth of ITO transparent conductive oxide layers in oxygen-free environment by RF magnetron sputtering

D Kudryashov, A Gudovskikh and K Zelentsov

Renewable Energy Laboratory, St. Petersburg Academic University-Nanotechnology Research and Education Centre of Russian Academy of Sciences, 8/3 Khlopina Str, St Petersburg, 194021

E-mail: kudryashovda@spbau.ru

Abstract. Indium Tin Oxide (ITO) thin films were grown at room temperature (RT) in oxygen-free environment by rf-magnetron sputtering on glass and Si(100)-substrates. The effects of argon pressure, sputtering power and film thickness on the electrical and optical properties of ITO films were investigated. For a 100 nm thick ITO films grown at RT in argon pressure $1.95\cdot10^{-3}$ mbar and sputtering power of 50 W, the transmittance was near 90% at 480 nm and resistance was $5.4\cdot10^{-4}$ Ohm·cm. It has been shown that the sputtering power plays an important role in electric properties of ITO films. SEM images of these samples show smooth surface with sharp substrate/ITO interface.

1. Introduction

Indium Tin Oxide is used in a wide range of important applications: LCD panels, solar cells, touch panels and electronic inks due to its high transparency in the visible part of the spectrum and low electrical resistivity ~10^{-4} Ohm·cm. It is a degenerate n-type semiconductor with a large bandgap of around 4 eV [1]. Among various sputtering techniques ITO films can be obtained by rf-magnetron sputtering resulting in thin films with good homogeneity and thickness uniformity.

Electro-optical properties of sputtered ITO films are sensitive to the growth conditions, such as substrate temperature, sputtering power or pressure [2]. Usually ITO films are grown under argon-oxygen environment and elevated temperatures (>350° C). This causes various restrictions on substrate material choice – only high thermal and oxygen resistance materials can be used. A few publications have reported high quality ITO films grown at low temperatures [3,4]. This work presents the optical and electrical properties of ITO films grown at room temperature by rf-magnetron sputtering in oxygen-free environment.

2. Experimental

ITO films were deposited at room temperature on Si(100)-wafers and microscopic glass slides by rf-magnetron sputtering techniques (BOC Edwards Auto 500 RF) using (90% In_2O_3-10%SnO_2) 3" ITO-targets supplied by LTS Chemicals. Substrates were mounted on 300 mm stainless steel rotating disk. Distance between target and substrate was 100 mm. The sputtering chamber was evacuated to less than $5\cdot10^{-6}$ mbar prior to deposition. High purity argon gas (99.999%) was introduced at rate of 2-10 sccm. Before deposition, the target was always presputtered in argon plasma for 10 min to remove contaminants. No oxygen was added during deposition. The working pressure was varied from $1\cdot10^{-3}$

to $4 \cdot 10^{-3}$ mbar and rf power was varied from 20 to 65W. All the ITO samples were with film thickness about 100 nm.

Film thickness was measured with a stylus profiler AMBiOS XP-1. The resistivity of ITO films was measured with a conventional four-probe technique. The Hall mobility and carrier density of the films were determined by Ecopia HMS-5000. The microstructure of the samples was analyzed using a scanning electron microscope SUPRA 25-30-63. The optical transmittance of the films was measured using spectrophotometer based on Solar Laser Systems M266 monochromator. Refractive and absorption indexes of ITO film were measured using HORIBA JOBIN YVON and Horiba UVISEL2 Spectroscopic ellipsometers.

3. Results and Discussion

Figure 1 shows the ITO growth rate as a function of argon pressure at rf-power of 50 W. The curve has a maximum at gas pressure of $1.95 \cdot 10^{-3}$ mbar. At higher pressures, there is an increase of the collisions number between sputtered and argon atoms, resulting in decreasing deposition rate. Low argon pressure reduces the sputtering efficiency due to low carrier's concentration near magnetron area [5]. All subsequent experiments were performed at argon pressure of $1.95 \cdot 10^{-3}$ mbar.

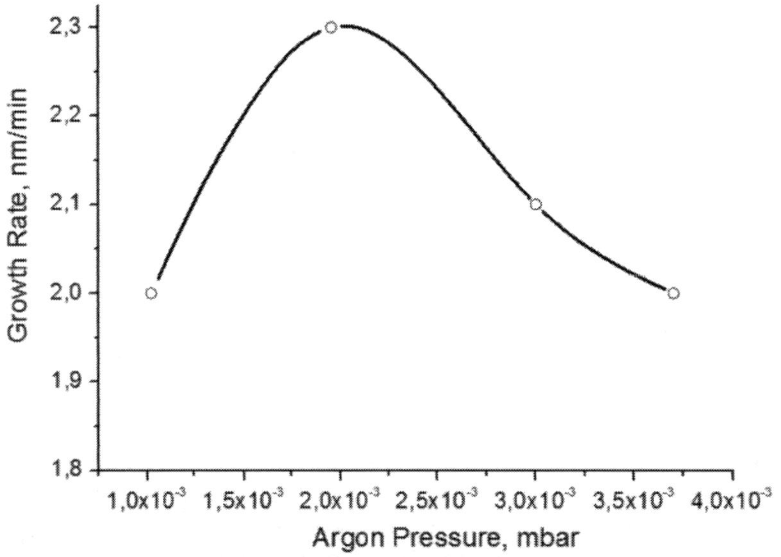

Figure1. ITO growth rate at rf-power of 50W as a function of the argon pressure

SEM image of ITO film sputtered at rf-power of 50 W is shown in figure 2. A smooth surface without visible defects is observed. Similar morphology was observed for thin ITO films grown at rf-power of 20-65 W.

The deposition rate is found to be linearly dependent on the rf-sputtering power. Experimentally determined proportional coefficient was 2.98 nm/(hour·W). Similar dependences were observed in [3,4]. At sputtering power above 100 W periodic arc flashes over the magnetron area were observed. Therefore the subsequent experiments were performed at rf-power of 20-65 W.

The resistivity (Rho) of an ITO films sputtered on Si-substrate as well as on glass slices decreases with increasing of sputtering power (figure 3) and reaches a minimal value of $5.4 \cdot 10^{-4}$ Ohm·cm at 50 W. Further increase of sputtering power results in increasing Rho. The measured Hall mobility of the majority carrier for the ITO film grown at 50 W was 17.4 $cm^2 \cdot V \cdot s^{-1}$ and the carrier concentration was $6.8 \cdot 10^{20}$ cm^{-3} (regardless of the substrate). Similar relations were obtained in [3] where authors shown changes in ITO structure (from a mixed amorphous/crystalline phase to polycrystalline) with increasing of rf-power. The results of the XPS analysis performed in [3] indicate that thin ITO films with minimal resistivity achieve the Sn/In ratio value of 0.099 as in sputtering target.

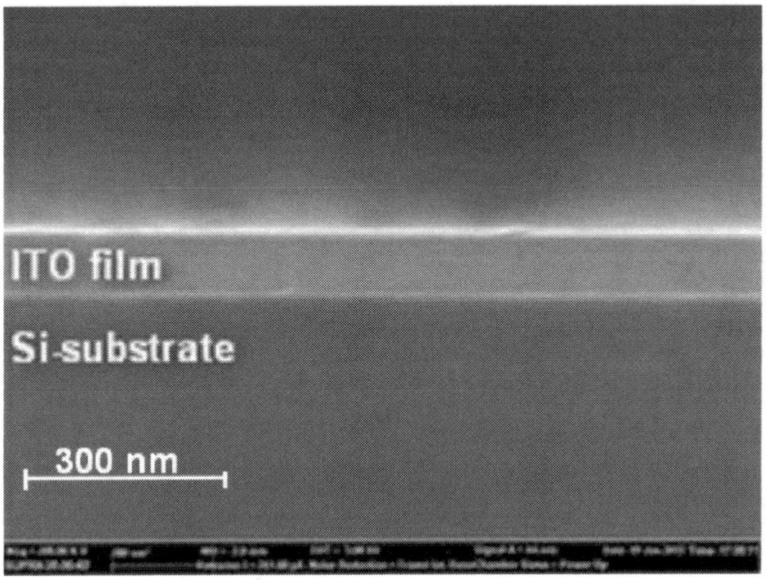

Figure 2. SEM image of rf-sputtered ITO film (100 nm) deposited at 50W and argon pressure of $1.95 \cdot 10^{-3}$ mbar

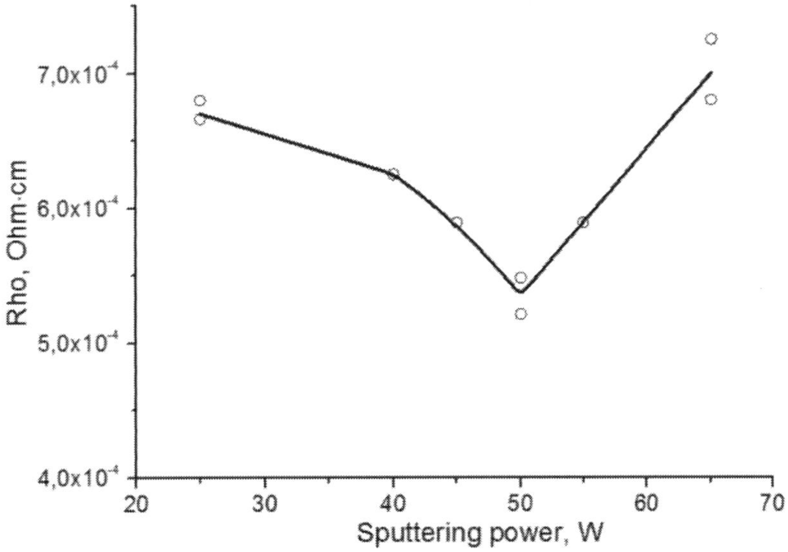

Figure 3. Resistivity of the ITO films as a function of the rf-power

Sputtering power as well as argon pressure did not have a distinct influence on the refractive index (n) of ITO films. It was equal to 1.97 ± 0.01 (at 632.8 nm) under rf-power of 20-65 W and working pressure of $1\text{-}4 \cdot 10^{-3}$ mbar. A measured plot of the refractive index as a function of the wavelength for the sputtered at 50 W ITO thin film can be seen in figure 4. It can be seen that in the main spectral transmission region the refractive index increases towards shorter wavelength. Additionally the squared line in figure 4 shows the extinction coefficient (k) as a function of the wavelength. The absorption coefficient, α, is related to the extinction coefficient, k, by the following formula:

$$\alpha = 4\pi k \lambda^{-1}. \tag{1}$$

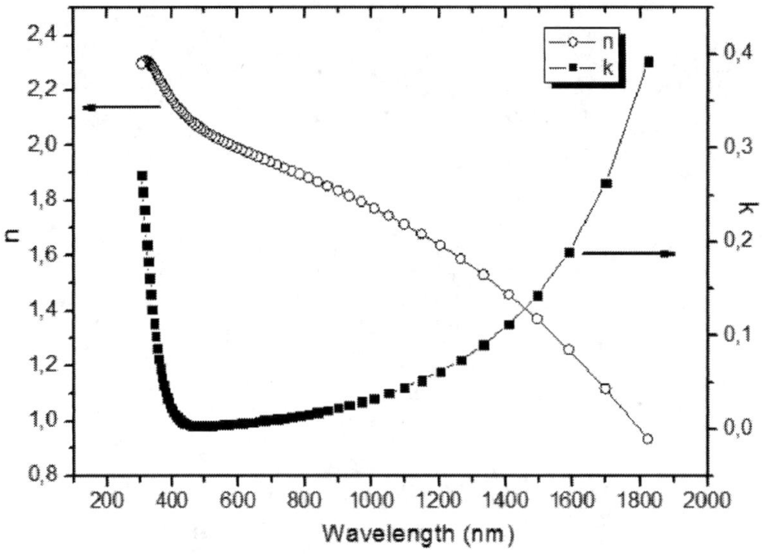

Figure 4. Measured optical constants of 100-nm ITO film sputtered at power of 50 W.

The transmittance spectra of the deposited ITO films at different rf-power are shown in figure 5. The transmittance of the glass slices was also measured for comparison (thick line).

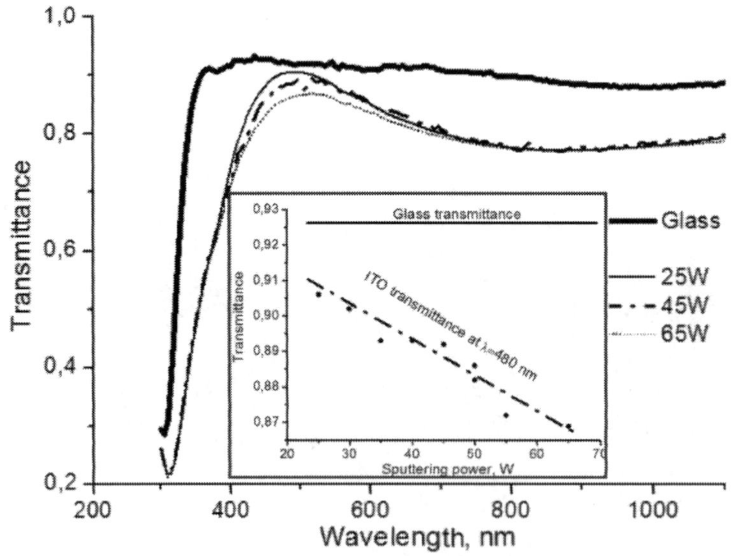

Figure 5. Optical transmittance spectra of ITO films vs rf-power. Inset shows a transmittance value of ITO films at λ=480 nm vs rf-power.

It can be seen that the 100 nm ITO films deposited on 1.0-mm microscopic glass slides were 80-90 % transparent in the visible region. Below 400 nm ITO films show strong absorption because of sub-band and band-to-band absorption. Also it was shown (figure 5 inset) that ITO transmittance λ=480 nm increases with decreasing of rf-power (at constant ITO thickness). It can be explained by changes in crystalline structure of ITO films with rf-power [3]. ITO films deposited at rf-power of 25 W and argon pressure of $1.95 \cdot 10^{-3}$ mbar had the best transmission with the resistivity of $6.7 \cdot 10^{-4}$ Ohm·cm. Lower rf-power values result in low deposition rate (< 74 nm/hour) and unstable plasma behavior

ITO films grown under rf-power of 50 W had the best conductance with transmittance of 85-90% in visible region. Also it was found no significant effect of argon pressure on the electrical and optical properties of the ITO films.

Conclusion

High quality 100 nm ITO films were grown on Si-substrate and glass slices by rf-magnetron sputtering at room temperature in oxygen-free environment. The sputtering power was shown to play a major role in ITO film resistivity and transmittance. For a 100 nm thick ITO films grown at RT in argon pressure of $1.95 \cdot 10^{-3}$ mbar and sputtering power of 50 W, the transmittance was near 90 % at 500 nm and resistivity was $5.4 \cdot 10^{-4}$ Ohm·cm.

4. Acknowledgment

This study was partially supported by the Russian Foundation for Basic Research (project № 12-08-01257a).

References

[1] Kim H and Gilmore C M 1999 Electrical, optical, and structural properties of indium–tin–oxide thin films for organic light-emitting devices *J. Appl. Phys.* **86** 6451
[2] Tuna O, Selamet Y, Aygun G and Ozyuzer L 2010 High quality ITO thin films grown by DC and RF sputtering without oxygen *J. Phys. D: Appl. Phys.* **43** 055402
[3] Chuang M J 2010 ITO films prepared by long-throw magnetron sputtering without oxygen partial pressure *J. Mater. Sci. Technol.* **26** 577
[4] Wu W F and Chiou B S 1994 Properties of radio-frequency magnetron sputtered ITO films without in-situ substrate heating and post-deposition annealing *Thin Solid Films* **247** 201
[5] Schiller S, Heisigand V, Goedicke K 1977 Use of the ring gap plasmotron for high rate sputtering *Thin Solid Films* **40** 327

Journal of Physics: Conference Series **461** (2013) 012022 doi:10.1088/1742-6596/461/1/012022

Synthesis optimization of photonic crystals based on silicon and vanadium dioxides

A A Akhmadeev, E V Sarandaev and M Kh Salakhov

Department of Physics, Kazan Federal University, 18 Kremlevskaya St, Kazan 420008, Russia

E-mail: Albert.Akhmadeev@kpfu.ru, Sarev@inbox.ru, mkhsalakhov@gmail.com

Abstract. The photonic crystal is the material which structure is characterized by periodic distribution of refraction index in the spatial directions, which have the photonic band gaps in a spectrum of own electromagnetic states. There are numerous approaches of the creation of photonic crystals. In the present the optimal conditions of synthesis of photonic crystals based on silicon dioxide as well as the inverse photonic crystals based on vanadium dioxide are investigated. It is known that the synthesis process is influenced by many different factors. We have studied the dependence of the particle size on the concentration of reagents, as well as on the duration of the reaction. These studies are important for the production of samples of photonic crystals with a definite structure.

1. Introduction

One of the most important areas of modern physics is the study of periodic optical nanostructures, known as photonic crystals (PC) in which the dielectric constant varies periodically in space with period allowing Bragg diffraction of light. It is known that crystals of all types can disperse some radiation, provided that the period of the crystal lattice is of the same order as the wavelength of the radiation. Similarly, photonic crystals do not transmit the light with a wavelength comparable to the period of the photonic crystal structure, at the same time PCs are transparent to a wide range of electromagnetic radiation spectrum. These spectral bands are called "photonic band gap" (PBG) [1]. The works on photonic crystals are motivated by a number of promising applications, such as high-performance light emitting diodes, low-threshold lasers, optical waveguides with sharp bends and optical microchips. Nowadays there are numerous approaches to the creation of PC by using the lithography [2], interferential holography [3] and self-assembly of colloidal particles [4]. One of the first materials that have been considered as photonic crystals were synthetic opals, consisting of close-packed spherical particles of silica. Self-assembly techniques are very promising, since they are quite simple in terms of hardware design and have no fundamental restrictions both on the sample size and the number of photonic crystals produced in one cycle of synthesis. However, the synthesis process is influenced by the factors like setting of the concentration ratio of the reactants, reaction temperature, sequence of blending and more. So our purpose is to investigate the effect of concentration ratio of reactant and duration of the reaction on the particle size, making possible creation of PC with predetermined characteristics. Our results are relevant for the experimental verification of the effect predicted by the authors in [5].

2. Self-assembly method for the preparation of photonic crystals based on silica and vanadium dioxide

The preparation of photonic crystals based on silica (SiO_2) particles is carried out in two phases. First, suspension with SiO_2 particles was synthesized, in the second phase the particles were deposited in the close-packed structure. Suspension of silica particles was synthesized by the method described in [6]. According to this method, spherical particles are prepared by hydrolysis of tetraethyl orthosilicate (tetraethoxysilane, TEOS, $Si(O\text{-}C_2H_5)_4$) in hydroalcoholic medium with ammonia as a catalyst. The amount of reactants was varied in different ranges (100 - 170 ml of ethanol; 0 - 45 ml of water; 11 - 14 ml of aqueous ammonia; 2.5 - 9 ml of TEOS).

Self-assembly of colloidal particles was performed using a vertical deposition [4]. Glass substrate was immersed vertically into the suspension of the microspheres. This forms a meniscus at the interface "liquid-air-substrate", where the particles are drawn by capillary action. With the evaporation of the liquid meniscus moves down the surface of the glass, leaving a thin film of close-packed microspheres.

Synthetic opals can be used for the producing of inverse opals using template method [7]. To perform this the voids between the particles in synthetic opals are filled with the desired compound, which is transformed into a solid structure as a result of chemical reactions. After that the colloidal particles are removed using the processes of dissolution or thermal decomposition. Thus we synthesized inverse PC based on vanadium dioxide (VO_2). In turn, vanadium dioxide is attractive due to the rapid semiconductor-metal phase transition [8], so that the material can be used to create ultra-fast switch.

3. Photonic crystals characterization techniques

Photonic crystals parameters were measured by atomic force microscopy (AFM) (Ntegra Prima, NT-MDT) and by optical spectrophotometry (Lambda-35, Perkin Elmer). Each method has a certain range of applicability.

3.1. Atomic-force microscopy method

AFM produces three-dimensional image of the sample with high resolution, but it makes possible to visualize only the surface layer of the PC. Figure 1 shows AFM image of synthetic opal based on SiO_2. As a result of AFM image processing the particle size was determined (560 ± 15 nm).

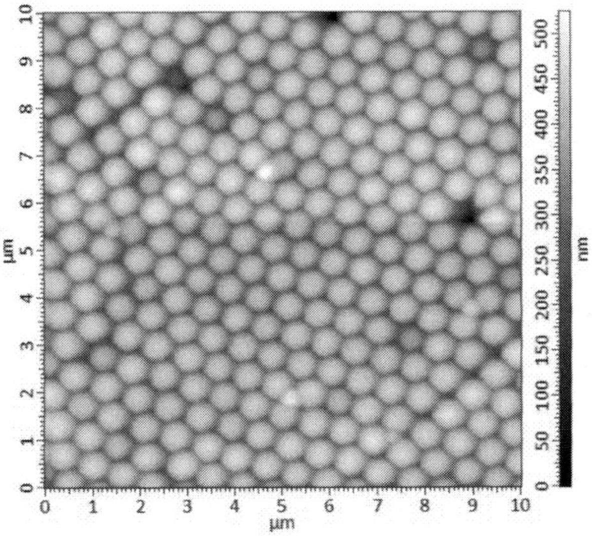

Figure 1. AFM image of a silica-based PC.

Journal of Physics: Conference Series **461** (2013) 012022 doi:10.1088/1742-6596/461/1/012022

3.2. Spectrophotometric method

Spectrophotometric method is based on the fact that the spectral position of the band gap obeys the Bragg's law:

$$\lambda = 2d_{(111)}\sqrt{n_{eff}^2 - \sin^2\Theta}, \quad (1)$$

where λ is the wavelength, $d_{(111)}$ is the period of the structure (the distance between the (111) planes), n_{eff} is the effective refractive index, Θ is the angle between the normal to the sample surface and the direction of the incident light. Equation 1 is written with the Snell's law. The effective refractive index n_{eff} of the photonic crystal depends on the refractive index of the SiO_2 particles ($n = 1.45$), the refractive index of air ($n_{air} = 1$) and the volume fraction of microspheres ($f = 0.74$):

$$n_{eff} = \sqrt{n^2 \cdot f + n_{air}^2 \cdot (1 - f)}. \quad (2)$$

Figure 2 shows the transmission and reflection spectra of PC samples based on silica. According to equation 1, the dependence of band gap spectral position (λ) on the incidence light angle (Θ) is linear in terms of λ^2 – $\sin^2\Theta$ (figure 3). Using a linear least-squares approximation, we have calculated the effective refractive index $n_{eff} = 1.35 \pm 0.02$ for sample A and $n_{eff} = 1.34 \pm 0.03$ for sample B (theoretical value calculated using the equation 2: $n_{eff} = 1.35$), the period of the structure $d_{(111)} = 460 \pm 20$ nm and 470 ± 20 for samples A and B respectively, the particle diameter $D_A = 560 \pm 20$ nm and $D_B = 570 \pm 20$ nm, which agrees well with the AFM. These parameters can be determined by the transmission spectra for all layers of photonic crystal unlike the using AFM method.

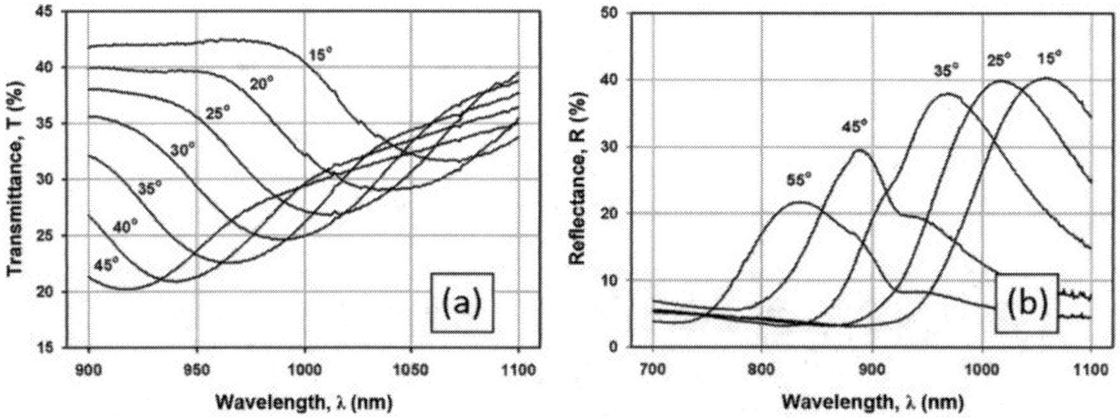

Figure 2. Transmission (a) and reflection spectra (b) of samples A and B respectively.

For the inverse photonic crystal based on vanadium dioxide reflection spectrum was recorded (figure 4 (a)). Contour width is about of 50 nm, which is 2-3 times less than in the spectra of similar samples from the reference [9] (figure 4 (b)). In this way, the phase transition of vanadium dioxide will be accompanied by an abrupt change of the band structure that is required to create quality femtosecond optical switches.

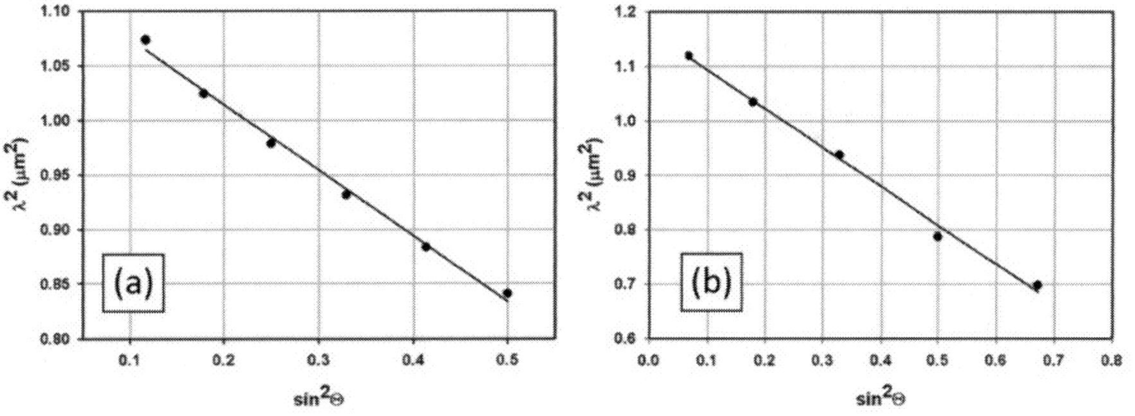

Figure 3. The spectral positions of the band gaps for different angles of incidence of light on the sample surface, (a) for sample A and (b) for sample B.

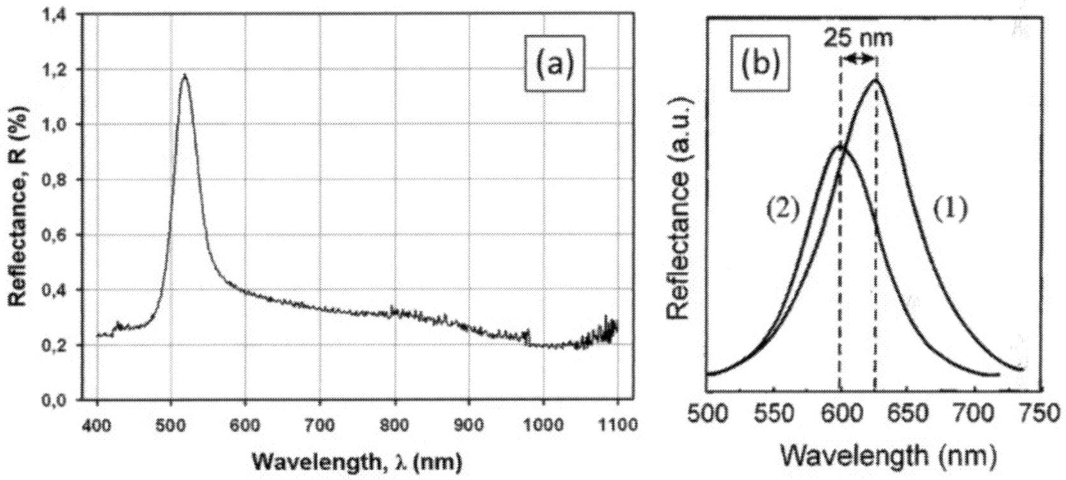

Figure 4. The reflection spectra of the inverse PCs based on VO_2, prepared in our work (a) and in the reference [9] (b).

4. Silica particle size dependence on the concentration of the reactants and the reaction duration

During the synthesis of silica microparticles the selection of the reactants concentration significantly affects on the size of the particles. Therefore it is necessary to know the size of the particles at a certain ratio of the reactants and how the diameter depends on the amount of a particular component.

4.1. The particle size dependence on the water and TEOS concentrations

The water concentration changes during the synthesis can lead to a fairly significant change in the size of the colloidal particles. Figure 5 (a) illustrates the influence of the molar concentration of water on the size of the synthesized microparticles in comparison with data from reference [10]. The concentrations of ammonia and TEOS was 2.3 M and 0.15 M, respectively. According to the data of [10], these concentrations were 1.0 M and 0.17 M. Similarly, the dependence of the silica particles diameter on the amount of TEOS was obtained (figure 5 (b)). As the authors of [11], an important advantage of the variation of tetraethoxysilane amount is the monotony of this dependence, which is convenient for predicting the size of the particles obtained under certain conditions.

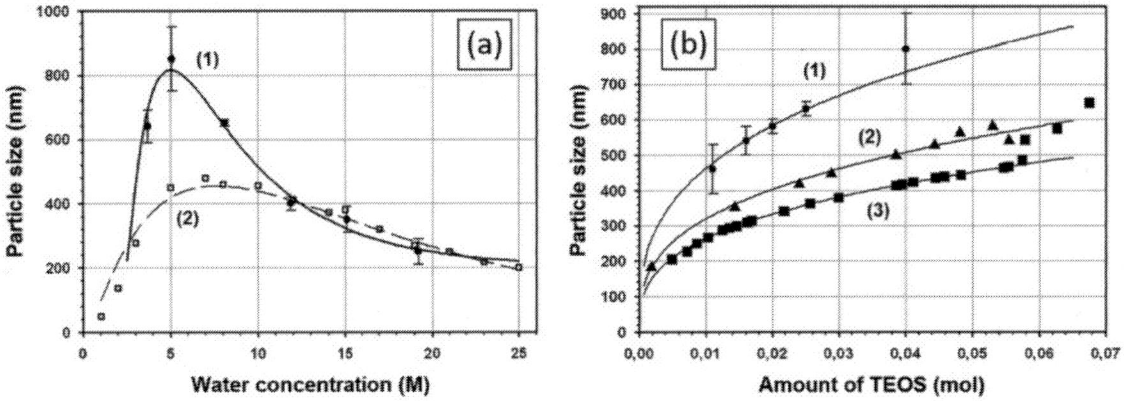

Figure 5. a) Dependence of silica particle size on the water concentration: 1) according to our data, 2) according to the reference [10]; b) Dependence of the particle diameter on the amount of TEOS: 1) according to our work, 2-3) according to data published in [11].

4.2. The particle size dependence on the reaction duration

It is also important to know how fast the particles grow during the reaction, or how soon their size reaches its maximum. To this end, the colloidal solution was prepared using the following amounts of reagents: ethanol - 162.8 ml, water - 4 ml, aqueous ammonia - 14 ml, TEOS - 8.9 ml. After reaction starts the particle sizes were measured at certain intervals of time. As can be seen from the figure 6, for 180 minutes (3 hours) the particles grow significantly and the average diameter increases to 800 nm. When the duration of the reaction $t > 180$ min size of microparticles remains virtually unchanged. Thus, the synthesis of SiO_2 microparticles with a size of about 800 nm can be completed within three hours.

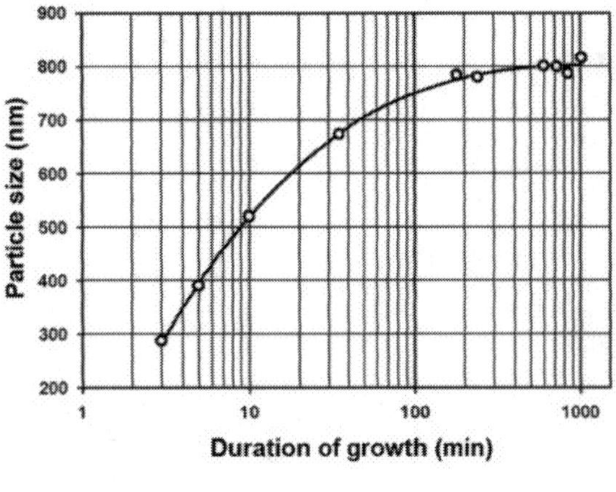

Figure 6. Dependence of the microparticle size on the reaction duration.

5. Conclusions

The dependence of silica particles diameter on the concentration of water and TEOS was investigated, as well as the influence of the reaction duration. We have shown that the experimental dependence is in agreement with the reference data, and allows to achieve a determined size of microparticles under appropriate conditions. Synthesized inverse photonic crystal based on vanadium dioxide has a pronounced band gap, which is important for the creating a high-quality ultra-fast optical switches.

Acknowledgments

The authors thank the staff from A.Butlerov Institute of Chemistry for collaborative participation in the scientific activities and the provision of chemical equipment.

References

[1] Yablonovitch E 1987 *Phys. Rev. Lett.* **58** 2059–62
[2] Yao P, Schneider G J, Prather D W, Wetzel E D and O'Brien D J 2005 *Optics Express* **13**(7) 2370–76
[3] Duneau M, Delyon F and Audier M 2004 *J. Appl. Phys.* **96**(5) 2428–36
[4] Vlasov Y A, Bo X-Z, Sturm J C and Norris D J 2001 *Nature* **414** 289–293
[5] Gainutdinov R K, Khamadeev M A and Salakhov M K 2012 *Phys. Rev. A* **85** 053836
[6] Stober W, Fink A and Bohn E 1968 *J. Colloid Interface Sci.* **26** 62–69
[7] Khokhlov P E, Sinitskii A S and Tretyakov Yu D 2006 *Dokl. Chem.* **408** 61–64
[8] Morin F J 1959 *Phys. Rev. Lett.* **3**(1) 34–36
[9] Golubev V G, Davydov V Y, Kartenko N F, Kurdyukov D A, Medvedev A V, Pevtsov A B, Scherbakov A V and Shadrin E B 2001 *Appl. Phys. Lett.* **79** 2127–29
[10] Bogush G H, Tracy M A and Zukoski C F 1988 *J. Non-Cryst. Solids* **104** 95–106
[11] Santamaria Razo D, Pallavidino L, Garrone E, Geobaldo F, Descrovi E, Chiodoni A and Giorgis F 2008 *J. Nanopart. Res.* **10** 1225–29

Journal of Physics: Conference Series **461** (2013) 012023 doi:10.1088/1742-6596/461/1/012023

Dielectric properties of sodium nitrite particles embedded into porous alumina

N O Alexeeva

Pskov State University, Lenin square, 2, 180000 Pskov, Russia

E-mail: nataliaal@yandex.ru

Abstract. Porous anodic alumina templates were fabricated using a two-step anodization in oxalic and phosphoric acid solutions. Dielectric studies were carried out for ferroelectric sodium nitrite embedded into porous alumina. Phase transition temperature in sodium nitrite within porous alumina is shifted compared to the bulk ferroelectric Curie point. The sign of this shift depends on the experimental conditions of sample preparation.

1. Introduction

Anodic aluminium oxide films formed by the electrochemical oxidation of aluminium have been widely studied to protect and decorate the surface of aluminium and to make inorganic membranes. In recent years, nanoporous anodic aluminium oxide with relatively regular pores, narrow distributions of pore sizes and interpore spacings has become a popular template system for the synthesis of various nanostructures. This trend originated from the discovery of Masuda and Fukuda a novel process so-called two-step anodization in 1995 [1]. They reported that the pore regularity can be improved by a long anodization time under appropriate conditions, known as 'self-ordering regimes'. Hexagonally arranged arrays with circular pores can be obtained under these regimes.

Porous alumina templates have advantages as a basis for nano-patterning. These are the fundamental nanometer scale of the pores, the control of their size over large areas achievable by varying the anodizing parameters, their high aspect ratio and highly regular structure of the pore arrays.

The arrangement and size of alumina pores are controlled by the applied voltage, temperature, type and concentration of the electrolyte. In this paper dielectric properties of $NaNO_2$ within porous alumina fabricated under different anodizing parameters were studied.

2. Experimental

Porous alumina was fabricated by the electrochemical method. Two-step anodization was used. Experimental setup was designed according to diagram described in [2]. It is shown in figure 1. The electrochemical cell consisted of two-electrode system, the lead electrode acting as the cathode and Al sheet acting as the anode. The Al sheet was inserted between the electrolyte container and heat-conducting substrate. During anodization, the surface of aluminium heated up, leading to a requirement for a cooling system. Cooling system was based on Peltier element. The temperature was measured by semiconductor temperature sensor with digital interface. Temperature sensor was integrated into the heat-conducting substrate and allowed to measure temperature with the accuracy of

0.5 ^{O} C. Temperature regulation was carried out with the help of microprocessor control system. The microprocessor control system was based on the microcontroller Atmega16 from Atmel. The microcontroller measured the temperature in the reaction zone and maintained the set temperature by varying the supply voltage of Peltier element. In this system PID regulation algorithm was used.

The anodized sample was cleaned and inserted into 40% solution of sodium nitrite for several days. Then the sample was wiped with moist cloth and dried.

AFM-images were obtained by the scanning probe microscope Nanoeducator from NT-MDT.

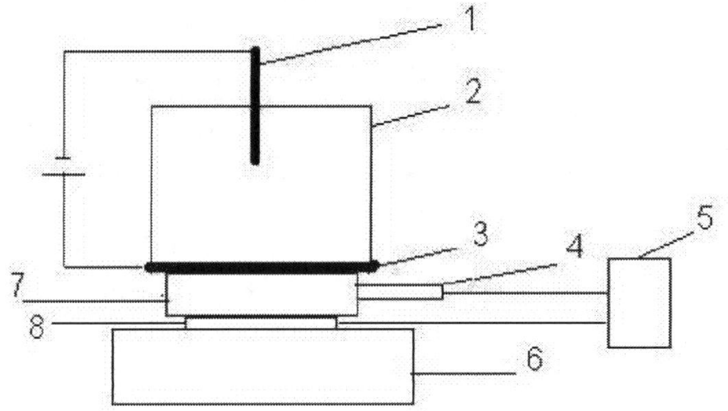

1 – lead electrode, 2- electrolyte container, 3- aluminium sheet, 4 – temperature sensor, 5- microprocessor control system, 6- cooling radiator, 7- heat-conducting substrate, 8- Peltier element.

Figure 1 Experimental setup

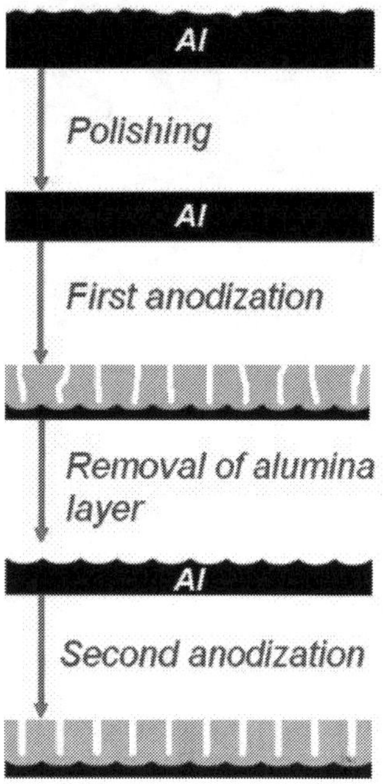

Figure 2 Two- step anodization

Figure 2 illustrates the procedure of fabricating alumina templates using two-step anodization.

The first anodization was carried out at a constant voltage of 40 V in 0.3 M oxalic acid at 8 ^{O} C or at a constant voltage of 160 V in 0.1 M phosphoric acid at 3 ^{O} C for about 10 hours.

In the first anodization step pores initiate at almost random positions and order thereafter by self-adjusting during the anodizing process of long duration, pore arrangements on the surface are disordered and have a broad size distribution. Ordered pore domains can only be obtained at the bottom of the layers. So, the pores produced in the first anodization step are not parallel to one another. To fabricate ordered nanopore arrays, the first porous oxide film, which contains the barrier layer at the bottom, is selectively dissolved in mixture 20 g/l CrO_3 and 35 ml/l H_3PO_4 at 80 ^{O} C. The barrier layer is not flat but consists of periodically arranged crests and troughs. These fluctuations result in a dimpled and undulating surface that has the same spatial ordering as the barrier layer. After the removal of the porous film, the periodic concave patterns that remain, as shown schematically in figure 2, act as a self-assembled template for the second anodization process. In the second anodization step the same parameters are used as in the first anodization step.

AFM-image of the aluminium surface after removal of the porous oxide is shown in figure 3. AFM-image of the porous alumina after the second anodization step is shown in figure 4. The average pore size is about 200 nm.

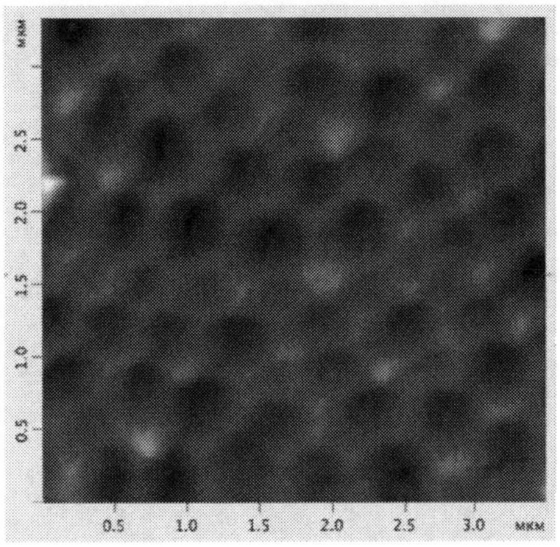

Figure 3 AFM-image of the aluminium surface after removal of the porous oxide (160 V, 0.1 M phosphoric acid, 3 ^{O} C)

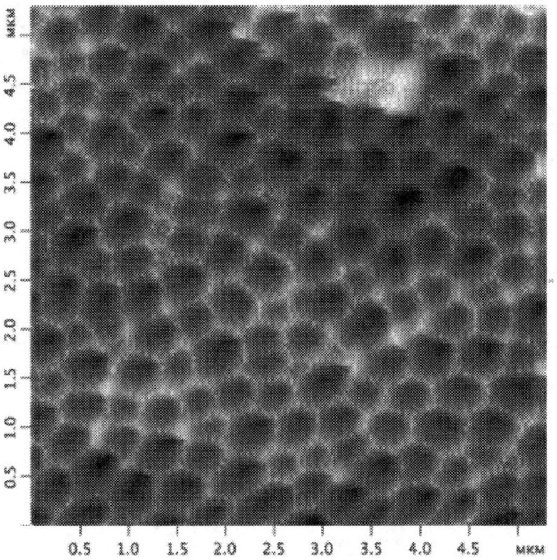

Figure 4 AFM-image of the porous alumina after the second anodization step (160 V, 0.1 M phosphoric acid, 3 ^{O} C)

3. Results

Dielectric measurements of the filled porous alumina were carried out at the frequency of 1 kHz by an E7-13 impedance meter. The temperature dependences of sample capacity C(T) and sample conductivity G(T) were measured upon continuous warming with a heating rate of 1 to 2 degrees per minute. Silver (sputtered on the surface of the porous alumina with ATC ORION SERIES SPUTTERING SYSTEM) or graphite were used as electrode materials.

The temperature dependences of capacity C(T) and conductivity G(T) of two samples under study are shown in figures 5 and 6. According to our preliminary results curves C(T) and G(T) have peculiarities at the temperatures near 425 K and 450 K.

Thus phase transition temperature in sodium nitrite within porous alumina is shifted compared to the bulk ferroelectric Curie point(437K) [3, 4].

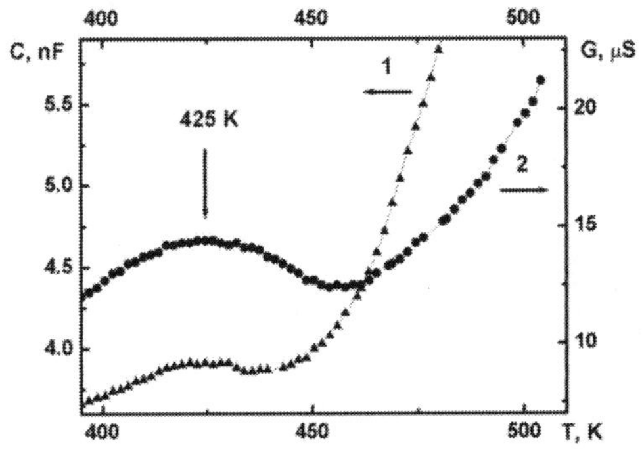

Figure 5 Temperature dependences of sample capacity (1) and sample conductivity (2). Conditions of sample preparation: 160 V, 0.1 M phosphoric acid, 3 ^{O} C.

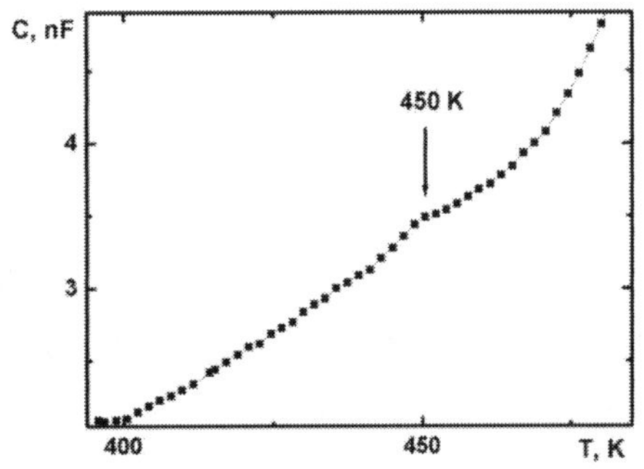

Figure 6 Temperature dependence of sample capacity. Conditions of sample preparation: 40 V, 0.3 M oxalic acid , 8 ^{O} C.

4. Conclusion

Porous alumina was fabricated by the electrochemical method using two-step anodization.

Our dielectric studies of $NaNO_2$ embedded into porous alumina showed that the phase transition temperature can be higher or lower than Curie point in bulk $NaNO_2$ depending on the experimental conditions of sample preparation.

These results are in accordance with modern theoretical models based not only on the phenomenological approach to the size effect on phase transitions in ferroelectric particles [5], but also taking into account an influence of pore walls on ferroelectric properties of nanoparticles [6] as well as long-wave dispersion interactions [7].

Acknowledgments

The author is grateful to V.G. Solovyev (Pskov State University, Pskov, Russia) for help in carrying out the experiments.

This work was supported by the Ministry of Education and Science of Russian Federation according to the program "Development of Scientific Potential of Higher Educational Institutions".

References

[1] Masuda H, Fukuda K 1995 *Science* **268** 1466
[2] Belov A N *et al* 2011 *Nanotechnics* **4 (28)** 58
[3] Gurevich V M 1969 *Electrical conductivity of ferroelectrics* (Moscow: Standards committee) pp 215-218
[4] Strukov B A, Levanyuk A P 1995 *Physical principles of the ferroelectric phenomena in crystals* (Moscow: Science) p 30
[5] Zhong W L, Wang Y G, Zhang P L *et al* 1994 *Phys. Rev.* B **50** 698
[6] Tien Ch, Charnaya E V, Podorozhkin D Yu *et al* 2009 *Phys. Status Solidi* B **246** 2346
[7] Alexeeva N O, Vanin A I, Pan'kova S V *et al* 2008 *Proc. Int. Conf. "Physics of Dielectrics (Dielectrics-2008)"* (St Petersburg: Herzen State Pedagogical University of Russia Press) vol 2 p 201

Journal of Physics: Conference Series **461** (2013) 012024 doi:10.1088/1742-6596/461/1/012024

Radiative characteristics of semiconductor injection lasers based on narrow asymmetric waveguides

S D Chervinskii[1,4], **B S Ryvkin**[2], **Yu M Shernyakov**[2,3], **A S Payusov**[2,3], **N Yu Gordeev**[2,3]

[1] St.Petersburg State Polytechnical University, Polytekhnicheskaya 29, 195251, St.Petersburg, Russia

[2] Ioffe Physical-Technical Institute, Polytekhnicheskaya 26, 194021, St.Petersburg, Russia

[3] St.Petersburg Academic University - Nanotechnology Research and Education Centre, Khlopina 8/3, 194021, St. Petersburg, Russia

E-mail: semen.chervinsky@gmail.com

Abstract. Quantum well InGaAs/GaAs lasers based on narrow asymmetric waveguide were created and investigated in pulse and CW modes. Broad-area 50μm stripe lasers showed internal quantum efficiency as high as 95%, threshold current density as low as 160 A/cm^2. Wall-plug efficiency in CW mode reached the value of 75%. The obtained parameters make the concept of narrow asymmetric waveguide promising for high-power laser diodes.

1. Introduction

High-power semiconductor lasers are widely used for pumping active medium, in material processing, in medical, and in a number of other applications. Combination of high optical power and good laser beam quality is essential for all these applications. In the recent years the most used concept for obtaining high optical power was one based on broadened waveguides [1-3]. However, at very high injection current densities these lasers faced significant limitations due to carrier accumulation in the waveguide layer. Recently E. Avrutin and B. Ryvkin proposed new laser waveguide design to overcome these limitations [4]. For these lasers based on narrow asymmetric waveguide (NAW) it was analytically shown that they can have smaller free-carrier absorption losses at very high injection current densities with no degradation of other main parameters [5].

Key feature of the lasers based on narrow asymmetric waveguide design is remarkably different refractive indices of *p*- and *n*-claddings. Thus, only fundamental mode exists in the waveguide, asymmetry of the waveguide caused asymmetric field penetration in *p*- and *n*-claddings (figure 1). Whereas free-carrier absorption coefficient in p-cladding sufficiently exceeds that one in n-cladding, the value of internal losses consequently decreased. Carrier accumulation in the NAW and corresponding optical losses are small. Moreover, noticeable field penetration in *n*-cladding results in a spot size comparable to those of broadened waveguides [6].

[4] To whom any correspondence should be addressed.

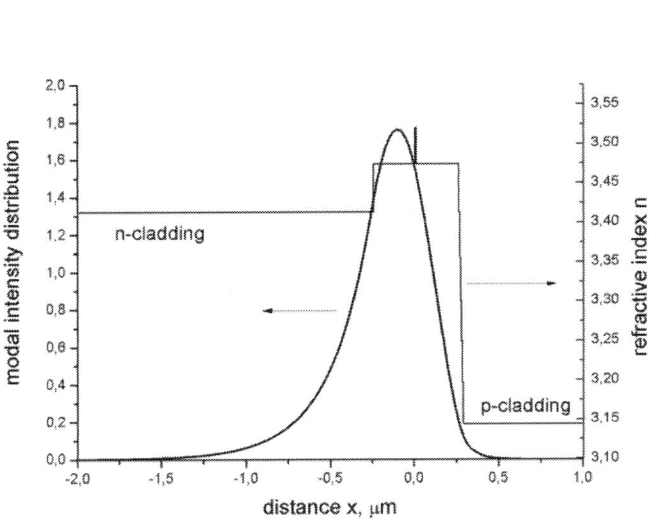

Figure 1. Refractive index and transverse optical mode profile for lasers based on narrow asymmetric waveguide.

2. Laser diode design

Laser waveguide profile and simulated optical mode intensity profile are shown in figure 1. The laser wafers were grown by metal-organic chemical vapour deposition (MOCVD). Asymmetric waveguide is formed by 5.0μm-thick $Al_{0.1}Ga_{0.9}As$ *n*-cladding (graded doping 0.2-2×10^{18}cm^{-3}) and 1.0μm-thick $Al_{0.6}Ga_{0.4}As$ p-cladding (graded doping 0.5-2×10^{18}cm^{-3}) with 0.5μm-thick GaAs in between. Active region contains one 90Å-thick InGaAs quantum well (emitting wavelength 1060nm). The laser wafer was processed into lasers with 50μm stripe width and cavity length of 0.5-2mm. No facet coatings were used. The laser samples were mounted *p*-side down on cooper heatsinks using indium solder.

3. Laser diode characteristics

The laser samples were measured both in pulsed and CW modes at 25°C. The lasers showed threshold current density as low as 160 A/cm^2. The lasing wavelength at room temperature varies in the range of 1050-1070 nm depending on the cavity length and the pump current. Resistivity of lasers based on narrow asymmetric waveguide was equal to 4.6-6.0×10^{-5}Ω·cm^2. Differential quantum efficiency reached the value of 92%. Internal losses and internal quantum efficiency were calculated in a common way from a set of light-current curves measured for lasers with different cavity lengths. The lasers showed internal quantum efficiency as high as 95% and intrinsic losses as low as 2.5 cm^{-1}. Laser beam divergence in vertical direction was as large as 35-40 deg. FWHM, which corresponds to the value obtained in the waveguide numerical simulation. The concept of narrow asymmetric waveguide is very promising for decreasing the beam divergence as it allows designing the optical mode profile. Our calculations shows that with 0.4 μm-thick GaAs waveguide layer laser beam divergence in vertical direction will be only ~25 deg. FWHM. Large internal quantum efficiency allowed obtaining very high values of CW wall-plug efficiency which is defined as the optical power out relative to the electrical power in. The last one is the product of the drive current and the total voltage across the laser terminals. Wall-plug efficiency for the best laser sample was as large as 75 % (figure 2) at the pumping current of 1.3 A (corresponding output optical power – 1.2W).

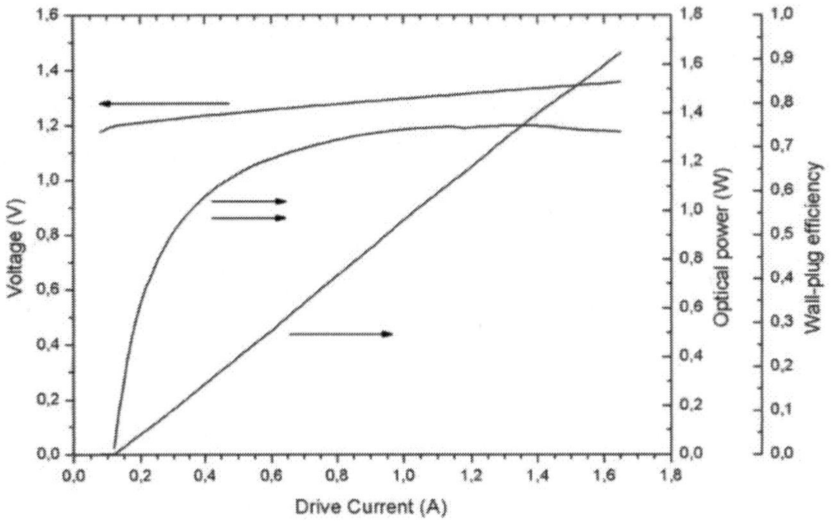

Figure 2. Optical power, voltage, and wall-plug efficiency vs. drive current for the laser based on narrow asymmetric waveguide with the cavity length of 1.4 mm.

4. Conclusion

Lasers based on narrow asymmetric waveguide were experimentally tested and showed good radiative characteristics corresponding with analytically predicted ones. The concept can be very effective to reduce the vertical beam divergence.

Acknowledgement

The authors are grateful to V.M.Lantratov and N.D.Il'inskaya for laser wafer growth and processing. The work was partly supported by Russia's Federal Program "Scientific and Educational Manpower for Innovative Russia" (grant no 12.740.11.1176).

References

[1] Al-Muhanna A, Mawst L J, Botez D, Garbuzov D Z, Martinelli R U and Connolly J C 1998 *Appl. Phys. Lett.* **73**, 1182

[2] Pikhtin N A, Slipchenko S O, Sokolova S N, Stankevich A L, Vinokurov D A, Tarasov I S and Alferov Zh I 2004 *Electron. Lett.* **40**, 1413

[3] Grump P et al 2009 *Proc. of SPIE* **7198**, 719814

[4] Avrutin E A and Ryvkin B S 2006 *Proc. of the Int. Conf. on Transparent Optical Networks (ICTON'2006)* (Piscataway, NJ: IEEE) Paper No. ThA2.4; Ryvkin B and Avrutin E 2009 *J. Appl. Phys.* **105**, 103107

[5] Ryvkin B S and Avrutin E A 2005 *J. Appl. Phys.* **97**, 123103

[6] Ryvkin B S and Avrutin E A 2005 *J. Appl. Phys.* **98**, 026107

Ultraviolet photodiodes based on AlGaN solid solutions

I Lamkin, S Tarasov

Saint Petersburg Electrotechnical University "LETI", Educational research laboratory of photonics of the department of micro-and nanoelectronics

E-mail: ialamkin@mail.ru

Abstract. We present the characteristics of ultraviolet photodiodes based on Schottky barriers to the Aluminum Gallium Nitride (AlGaN) solid solution. The paper is devoted to the creation of ohmic and rectifying contacts to $Al_xGa_{1-x}N$ solid solutions with a high ($x > 0.5$) proportion of AlN. The most important property of ($Al_xGa_{1-x}N$) materials is their solar blindness: the wide bandgap (Eg = 3.4 eV for GaN; 6.2 eV for AlN) makes these materials insensitive to the visual and IR light. The study of solid solutions with various mole fractions of Al has created solar-blind and see-blind ultraviolet photodetectors.

1. Ohmic contacts for ultraviolet photodiodes

In the recent years AlGaN-based devices are receiving a lot of interest. Nitrides are widely used for luminescent devices [1]. AlGaN solid solution semiconductors of the wurtzite crystal structure are increasingly used as ultraviolet photodetectors for applications such as data storage, combustion control, flame sensors, radiation dosimetry, atmospheric ozone monitoring, solar-blind detection, pollution detection, and polarization-sensitive detection [2, 3]. Optimum device performance, however, is still limited by the lack of low-resistance Ohmic contacts, both on light-emitting diode and photodiode structures. Contact formation on n-type AlGaN is understood best, and a good standard metallization scheme on n-type GaN does exist.

In this study, the properties of Ti and Ti/Al contacts, both on $Al_xGa_{1-x}N$, have been systematically studied as a function of the annealing temperature. The n-type $Al_xGa_{1-x}N$:Si ($x = 0.5 – 0.6$) and undoped $Al_xGa_{1-x}N$ ($x = 0.08$ - 0.7) epitaxial layers with a thickness of 1 μm were grown by PA MBE on c-sapphire substrates. The thermal vacuum evaporation of different metals (Ti, Au, Ni, Ag, In, Al) was used to deposit different single- and double-layer contacts onto the AlGaN layer surface cleaned preliminary with organic solvents. Thicknesses of metal layers were in the range of 15 – 75 nm. The best results were achieved for the Ti layer thicknesses of 15 nm, and the top layer Al thicknesses of 35 nm. Before the deposition of the metal contacts, the structures' surface was cleared using various chemicals, in particular, H_2O_2, CCl_4, HCl, followed by structures washing in distilled water. For better metal adhesion structures were heated to a temperature of 300 °C during the deposition. The samples were annealed after the contact deposition in the different vacuum conditions (10^{-2} - 10^{-5} Torr) of a metal deposition chamber and without the use of nitrogen or argon gas (figure 1). The n-type AlGaN layers with different contacts were characterized by specific contact resistance determined through the transmission line measurements (TLM). Study of influence of the residual gas pressure in the vacuum chamber during the annealing of ohmic contacts on their resistance is shown in figure 1. It is seen that a decrease in pressure leads to a decrease in the resistance of contact. Most likely, this effect can be explained by the influence of oxygen in the residual gas, causes oxidation of the metal layers of titanium and aluminum, followed by the formation of composition structure on the surface of metal oxide. Figure 2 shows the I-V characteristics of the Ti/Al contacts to n-type $Al_xGa_{1-x}N$:Si with aluminum mole fraction of x = 0.5.

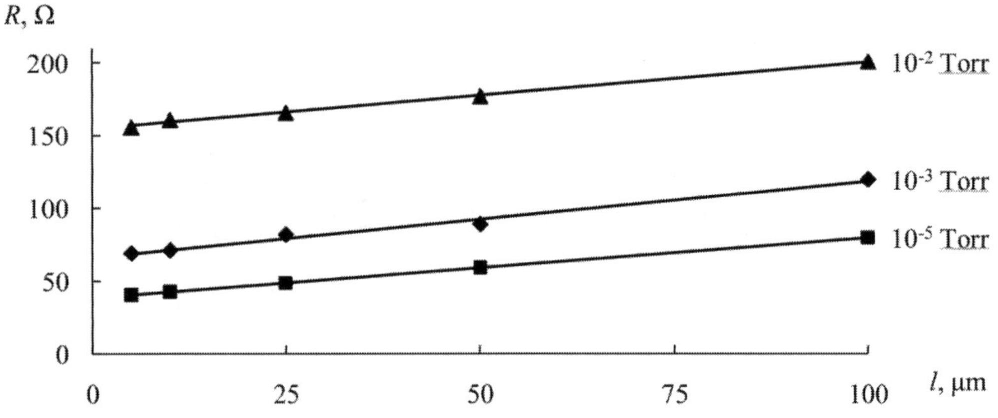

Figure 1. Resistance vs distance between the contacts deposited at the different residual pressures in the vacuum deposition setup.

As one can see, the I-V characteristics of the contacts are non-linear for untreated structures as well as for structures annealed at relatively low temperatures. Nonlinearity decreases with the temperature increase and at 750 °C the contacts become ohmic. Further increase of the annealing temperature leads to a slight increase of slope, but this increase should be recognized as insignificant.

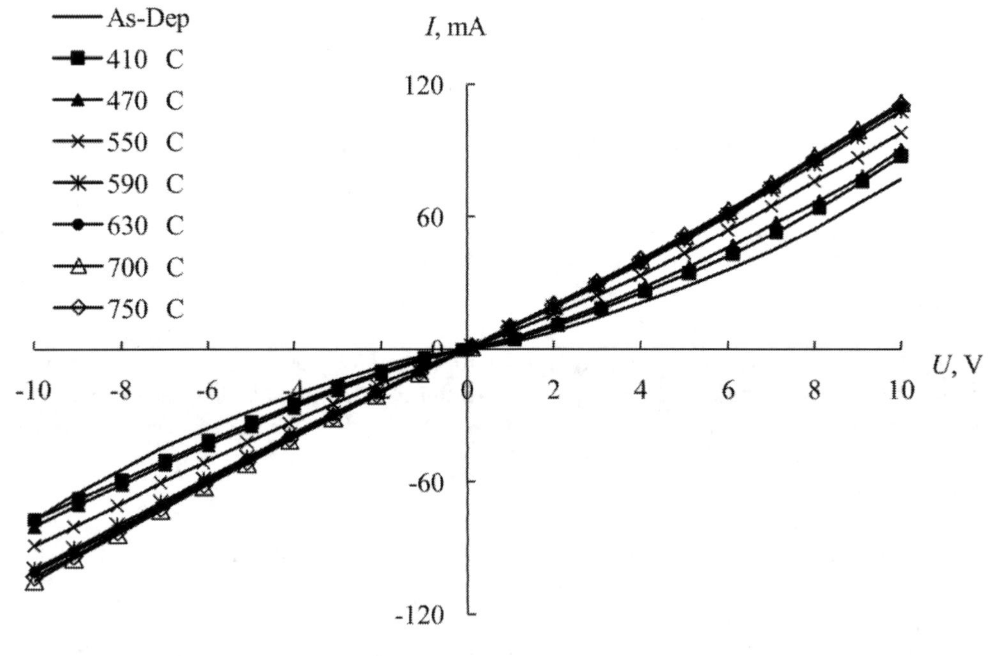

Figure 2. Current-voltage characteristics of the ohmic Ti / Al contacts to the $Al_{0.5}Ga_{0.5}N$ layer annealed at the different temperatures for 10 minutes.

To calculate the contact resistance with the TLM and to determine the optimal technological parameters of their growing, a number of structures has been created on the basis of the double-layer Ti/Al contacts with different thicknesses of the metal layers. I-V characteristics for different distances *l* between pads in these structures are shown in figure 3. The structures were annealed in vacuum at a temperature of 750 °C for 10 minutes. It is obvious that the measured resistance reduces with the decreasing of l. The resistivity was calculated by approximating the dependence of the resistance on the distance l to the intersection with the y-axis (when $l = 0$). The lowest TLM-resistance of $8 \cdot 10^{-5}$ $\Omega \cdot cm^2$ was achieved for the Ti/Al (15/35 nm) contact annealed at the temperature as high as 750 °C for sufficiently long time (above 10 min). The use of other relationships between thicknesses of metals in contact has led to a significant increase of the resistance.

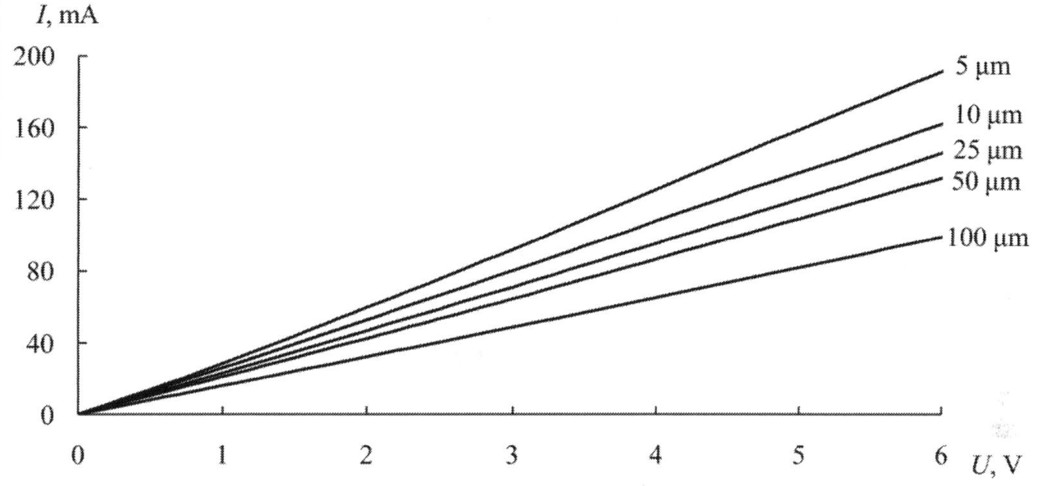

Figure 3. Current-voltage characteristics at different distances between the contacts.

2. Ultraviolet photodiodes based on Schottky barriers

$Al_xGa_{1-x}N$ is an attractive candidate for many UV detector applications, including flame and heat sensors, missile plume detection, secure-from-earth inter-satellite communications, UV calibration and monitoring devices for medical and biological applications, and other commercial applications such as plasma diagnostics and engine monitoring. These roles have primarily been filled in the past by photomultiplier tubes and Si detectors. The photomultiplier tubes are inefficient and inferior to semiconductor devices. Si has a narrow bandgap (1.1eV) and therefore requires filtering to operate as a UV detector. GaP photodetectors can be used for UV spectral range; their peak response is at about 410 nm. Moreover Ag-GaP Schottky photodiodes can be used as see-blind photodiodes [4]. However, they have an indirect band structure, which affects their performance. In contrast, the bandgap of $Al_xGa_{1-x}N$ can be tailored between 3.4 and 6.2 eV. This provides an obvious advantage over Si, as filtering can be bulky and/or expensive. The robustness of $Al_xGa_{1-x}N$ also offers the potential for stable device operation in hostile environments. SiC, although similar in robustness to GaN, has a bandgap of 2.9 eV, which is narrower than GaN, and no potential to tailor the bandgap through alloying. Therefore SiC detectors would also require filtering for many UV applications.

The Schottky contacts consist of semitransparent 15 nm thick Au disks. Au, Ni, Ag, Sn were investigated as materials of rectifying contacts. Diode diameters ranged from 1.5 mm to 3 mm. Figure 4 also plots the spectral response of $Al_xGa_{1-x}N$ Schottky diodes with different Al contents (x = 0.08… 0.9). The spectral response measurements were carried out in the 200 - 400 nm range, using a xenon lamp, a diffraction grating monochromator and a lock-in amplifier. The cut-off wavelength shifts from 360 nm to 240 nm. The characteristics were studied without application of voltage. It is shown that an increase in surface states greatly reduces the sensitivity to short wavelengths. The height of the barrier metal-semiconductor effect on long-wavelength edge. The $Al_{0.08}Ga_{0.92}N$ solid solution can be used to produce see-blind photodiodes. This UV sensor has a sensitivity in the 240 – 360 nm wavelength range. The $Al_xGa_{1-x}N$ solid solution with a mole fraction $x > 0.38$ of Al can be used as a solar-blind photodetector in the atmospheric conditions. For example, it makes it easy to detect ozone holes and harmful UV radiation for people. Increasing the share of Al to 0.9 has resulted in the separately of a UV photodetector sensitive to the vacuum ultraviolet, which has a long-wavelength edge of the photoeffect at 200 nm. The short-wave edge is determined primarily by the state of the metal - semiconductor and the corresponding value of the surface recombination velocity. The most sensitive short-wavelength region showed a structure with gold contacts, which can be attributed to a lower surface recombination velocity in such samples.

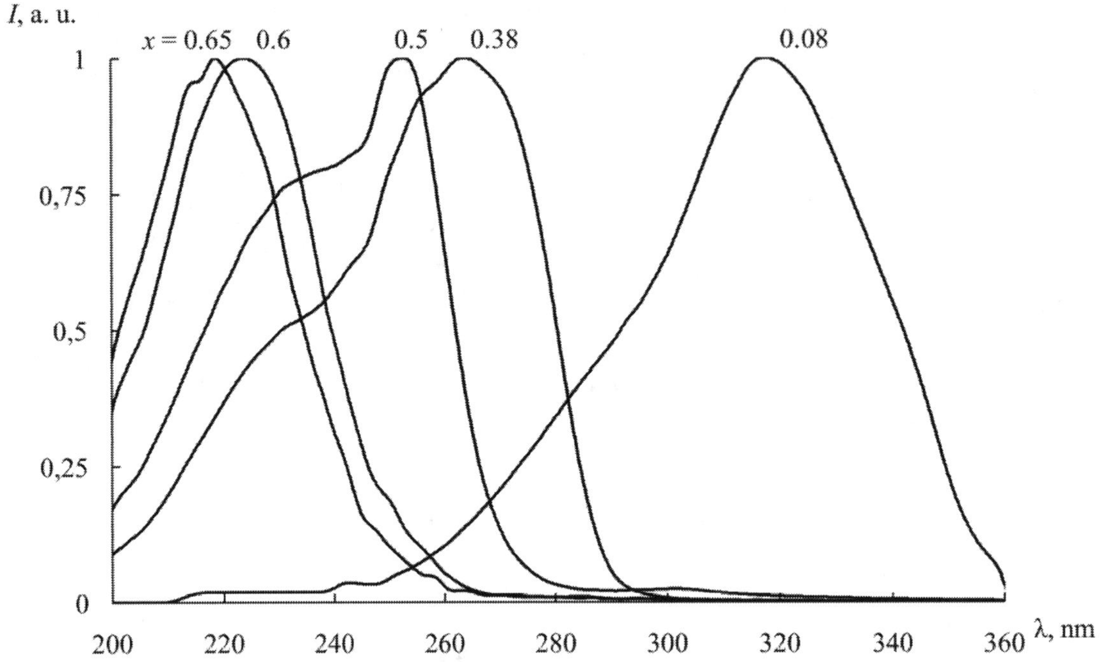

Figure 4. The spectral characteristics of photodetectors based on the sensitivity of the contact metal - $Al_xGa_{1-x}N$ with different mole fractions Al.

On the basis of contacts metal - AlGaN solid solutions, a range of photodetectors were created with the following responses in the ultraviolet spectrum:

- $Al_{0.08}Ga_{0.92}N$ → "see-blind", $\lambda < 360$ nm
- $Al_{0.42}Ga_{0.58}N$ → "solar-blind", $\lambda < 290$ nm
- $Al_{0.90}Ga_{0.30}N$ → "vacuum" UV photodetector, $\lambda < 200$ nm

The studies have lead to the conclusion of a good prospective of photosensitive structures based on metal - semiconductor nitrides contacts for usage in the ultraviolet region of the spectrum. The photodiodes with gold rectifying contacts have shown the best performance. These contacts provide not only a higher photosensitivity, but also several other advantages, such as good resistance to oxidation and high conductivity.

References

[1] Menkovich E A, Tarasov S A, Lamkin I A Luminescence of nanostructures based on semiconductor nitrides *Functional Materials* vol. 19, № 2, p. 233, 2012.

[2] Monroy E, Calle F, Pau J L, E. Munoz, F. Omnes, B. Beaumont B, Gibart P AlGaN-based UV photodetectors *Journal of Crystal Growth* vol. 230, p. 537, 2001.

[3] Omnes F, Monroy E, Calle F, Munoz, Beaumont B, Gibart P AlxGa1-xN based UV visible-blind photodetector device applications *Opto-Electronics Review* vol. 8(1), p. 43, 2000.

[4] Pikhtin A N, Tarasov S A, Kloth B Ag-GaP Schottky photodiodes for UV sensors *IEEE Transactions on Electron Devices* vol. 50, № 1, p. 215, 2003.

Journal of Physics: Conference Series **461** (2013) 012026 doi:10.1088/1742-6596/461/1/012026

Photovoltaic Detector Based on Type II Heterostructure with Deep AlSb/InAsSb/AlSb Quantum Well in the Active Region for the Mid-Infrared Spectral Range

G. G. Konovalov, M. P. Mikhailova, I. A. Andreev, K. D. Moiseev, E. V. Ivanov, M. Yu. Mikhailov and Yu. P. Yakovlev

Ioffe Physical Technical Institute, Russian Academy of Sciences, Saint Petersburg, 194021 Russia

E-mail: glebkonovalov@ya.ru

Abstract. Photodetectors for the spectral range 2-4 μm, based on an asymmetric type-II heterostructure p-InAs/AlSb/InAsSb/AlSb/(p, n)-GaSb with a single deep quantum well (QW) or three deep QWs at the heterointerface, have been grown by metal-organic vapor phase epitaxy and analysed. The transport, luminescent, photoelectric, current-voltage, and capacitance-voltage characteristics of these structures have been examined. A high-intensity positive and negative luminescence was observed in the spectral range 3-4 μm at high temperatures (300–400 K). The photosensitivity spectra were in the range 1.2-3.6 μm (T = 77 K). Large values of quantum efficiency (η = 0.6-0.7), responsivity (S_λ = 0.9-1.4 $A \cdot W^{-1}$), and detectivity $D^*_\lambda = 3.5 \cdot 10^{11}$ to 10^{10} $cm \cdot Hz^{1/2} \cdot W^{-1}$) were obtained at T = 77-200 K. The small capacitance of the structures (C = 1.5 pF at V = −1 V and T = 300 K) enabled an estimate of the response time of the photodetector at τ = 75 ps, which corresponds to a bandwidth of about 6 GHz. Photodetectors of this kind are promising for heterodyne detection of the emission of quantum-cascade lasers and IR spectroscopy.

1. Introduction

In recent years, considerable attention has been given to development of new types of photodetectors for the mid-IR spectral range on the basis of heterostructures with quantum wells (QWs) in order to improve their parameters (make lower dark currents, improve the operation speed, etc. [1]). Fabrication of quantum-well photodiodes based on GaAs/AlGaAs and InGaAs/SbP double-barrier heterostructures for the wavelength range 3-5 μm has been reported [2, 3]. Photodiodes of this kind are promising for heterodyne detection and free-space communication systems. The spectral range 2-5 μm is also important for gas analysis, ecological monitoring, and medical diagnostics. One advantage of QW IR photodiodes is their fast response (on the order of several picoseconds [1]).

Free-space detection requires a bandwidth exceeding 10-20 GHz. In this case, information is transmitted by quantum-cascade lasers for which the recording time is limited by the slow response of the existing photodetectors. Another advantage of QW photodiodes is their operation under small biases with low dark currents and, accordingly, with a low noise level. Research and development work on such photodetectors has been underway for more than 15 years. The InAs(Sb)-AlSb system has a unique band diagram because of the large conduction band offset $\Delta E_C > 1.35$ eV, $\Delta E_V = 0.15$ eV and the possibility of fabrication of deep QWs [4]. This is a new promising material for optoelectronic devices, quantum-cascade lasers, field-effect transistors, and resonant-tunneling diodes [5-7].

This communication reports on the fabrication and study of a photovoltaic detector operating in the spectral range 2-4 μm based on an asymmetric type-II heterostructure p-InAs/AlSb/InAsSb/AlSb/(p, n)-GaSb with a single deep QW or several QWs at the interface.

2. Procedures for fabrication and study of the photodiodes

Nanoheterostructures containing one or three QWs, 20-nm AlSb/5-nm $InAs_{0.84}Sb_{0.16}$/20 nm AlSb, and 0.5-μm-thick (p, n)-GaSb capping layers were grown on p-type InAs:Mn (100) substrates by low-pressure metal-organic vapor-phase epitaxy (LP-MOVPE) in an AXTRON-200 installation at a temperature of 500°C in the atmosphere of hydrogen at the MOVPE laboratory of the Institute of Physics, Czech Academy of Sciences [8]. The band diagram of a structure with three QWs is shown in Fig. 1. The capping GaSb layer was nominally undoped, whereas the InAs substrate was doped with a Mn acceptor impurity to a concentration $p = 10^{17}$ cm^{-3} at T = 300 K. In structures with one or three QWs and 20-nm-thick AlSb barriers, the width of the InAsSb QWs was 5 nm. At T < 100 K, the substrate became a semi-insulator, which enabled magnetotransport measurements.

The magnetotransport properties of the structures were studied with rectangular samples with the Hall configuration and six ohmic contacts in fields of up to 5 T at a low temperature T = 77.4 K. The Hall mobility in the sample with a single 5-nm-wide $AlSb/InAs_{0.84}Sb_{0.16}/AlSb$ QW grown on a p-InAs:Mn (100) substrate was $\mu = 5000\ cm^2 \cdot V^{-1} \cdot s^{-1}$. The results of these measurements demonstrated that an n-type channel similar to that we observed previously [9] exists at the type-II p-InAs/AlSb interface. The transport broadening [10] $2\Gamma = 2\hbar/\tau_{tr}$, where $\tau_{tr} = \mu m^*/e = 10^{-13}$ s, was estimated at an effective electron mass in the InAsSb solid solution $m^* = 0.016m_0$. A value $2\Gamma = 16$ meV was obtained by taking into account the fact that the mobility is governed by scattering at interfacial irregularities [11]. As noted in [10], the value of 2Γ for the double InAs/AlSb heterostructure is related to the linewidth of intersubband transitions and well agrees with the data for our structure with a QW calculated from the half-width of the luminescence spectrum of a structure of this kind, $\Delta h\nu = 21$ meV at T = 77.4 K, found from the data of [12]. These results are indicative of the good quality of the heterointerface in the MOVPE-grown QW structures.

To study the electroluminescent, electrical, and photoelectric properties, we fabricated structures by means of standard photolithography and wet etching as mesa diodes with a sensitive area diameter of 300 μm. Electroluminescence (EL) spectra were recorded with a Digikrom-480 monochromator, Stanford S-580 phase-lock detector, and cooled InSb photodiode (Judson Ltd.). To stabilize measurements at temperatures above 300 K, the diodes were mounted on a special holder with a thermoelectric cooler.

The photosensitivity spectra were studied in the temperature range 77-300 K with an SPM-2 monochromator and a globar as a source of light. The quantum efficiency was evaluated by comparison with the sensitivity of a calibrated Carl Zeiss thermopile.

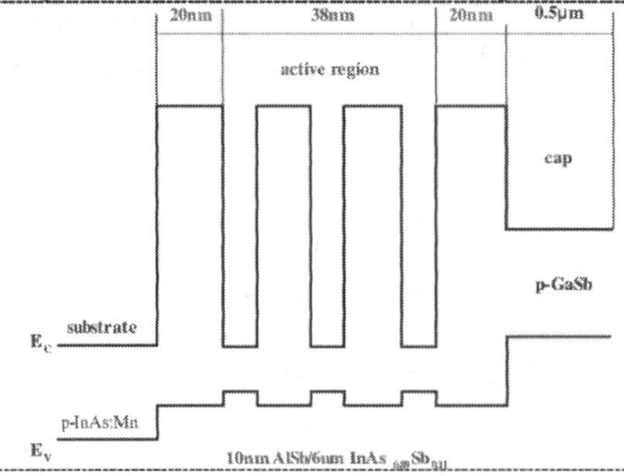

Figure 1. Band diagram of an asymmetric heterostructure with 3 deep AlSb/InAsSb/AlSb QWs at the interface.

3. Results and discussion

A study of the EL spectra in the temperature range 77–300 K revealed a high intensity of both positive and negative EL under a forward or reverse bias (minus at the p-substrate), respectively, at photon energies of 0.3-0.4 eV [12].

Figure 2 shows spectra of positive and negative EL for a structure p-InAs/AlSb/InAsSb/AlSb/p-GaSb at T = 28 and 106 °C. The temperature dependence of the optical power of the positive and negative EL at high temperatures (300-380 K) at a drive current i = 200 mA is shown in Fig. 3. It can be seen that the power of the negative EL increases and that of the positive EL decreases with increasing temperature. The high efficiency of the negative EL is due to the decrease in the intensity of nonradiative Auger recombination with increasing temperature. Moreover, as shown in [13], the Auger recombination can be suppressed at the type-II heterointerface. This makes it possible to use this structure by converting it to the light-emitting diode or photodiode mode at high temperatures. The current-voltage (I-V) characteristics of the structures under study, shown in Fig. 4, were of rectifying type and corresponded to an abrupt heterojunction. The dark currents for a triple-well structure were lower than those for the structure with a single well (see table 1).

The differential resistance was calculated from the I-V characteristics at small near-zero biases for structures with single and triple QWs at three temperatures of 77, 250, and 300 K. The measurement results are listed in the table, together with the R_0A products, where A is the size of the sensitive area of the mesa diode. Normalized spectral photosensitivity characteristics at T = 77 and 295 K are shown in Fig. 5 for structures with a single QW. The photosensitivity spectra were localized in the wavelength range 1.0-3.4 µm at T = 77 K and 1.2-3.8 µm at 295 K. The nature of the spectrum corresponds to that of the p-InAs/p(n)-GaSb heterojunction. The single-well structure shows an additional weak peak at 4.0-4.5 µm. It is noteworthy that a similar long-wavelength peak has also been observed previously in negative-EL spectra and its spectral position remained the same on changing the bias polarity [12]. We believe that this peak is due to the transition from the surface state of a Mn acceptor located close to the interface [13]. The absolute sensitivity at the peak of the spectrum for a sample with three QWs at the heterointerface was 1.5 times that of single-well structures in the photovoltaic mode. On applying a small positive bias, the photoresponse signal somewhat increased. It can be seen in the table that, as the temperature was raised from 77 to 300 K, the photosensitivity decreased by three orders of magnitude in accordance with changes in the differential resistance. The dark currents in the triple-well structures were lower than those in structures with a single QW. For example, the reverse current density at T = 77 K and V = −0.4 V was $J_1 = 5.5 \cdot 10^{-2}$ and $J_3 = 3.4 \cdot 10^{-2}$ $A \cdot cm^{-2}$ for single- and triple-well heterostructures, respectively.

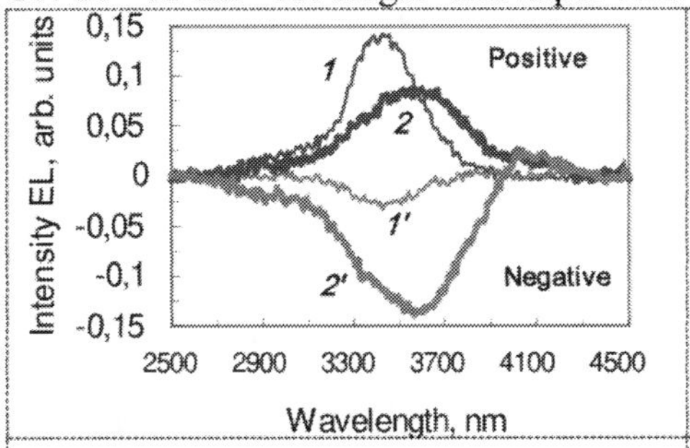

Figure 2. Spectra of the positive and negative electroluminescence from a p-InAs/AlSb/InAsSb/AlSb/p-GaSb heterostructure under forward ("+" at p-InAs) and reverse ("−" at p-InAs) biases at a drive current i = 50 mA and two temperatures T: (1, 1') +28°C and (2, 2') +106°C.

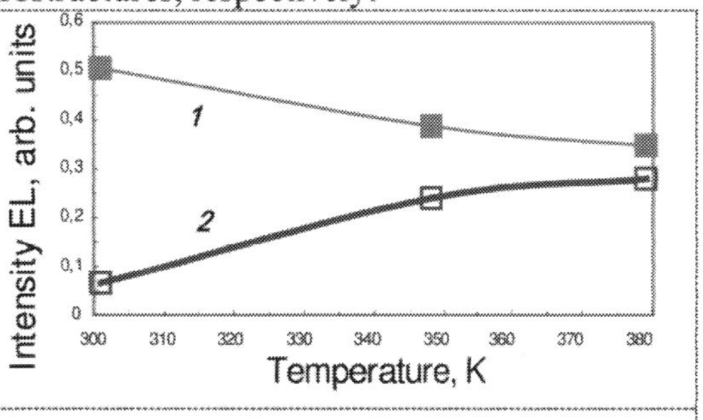

Figure 3. Temperature dependences of the EL intensity for a p-InAs/AlSb/InAsSb/AlSb/p-GaSb heterostructure with a single QW. (1) Positive EL (forward bias) and (2) negative EL (reverse bias). Drive current i = 200 mA.

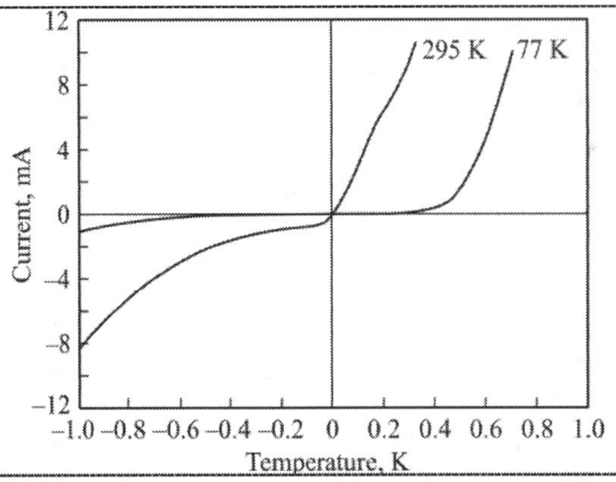

Figure 4. I-V characteristics of a heterophotodiode with 3 QWs in the active region at small biases. T = 77 and 295 K.

In the photovoltaic mode, the photosensitivity and quantum efficiency were estimated as $S_\lambda = 0.9\text{-}1.4\ A\cdot W^{-1}$ and $\eta = 0.6\text{-}0.7$. The equivalent noise power (with only the dark noise taken into account) and detectivity of the single-well photodiode were calculated at the peak of the spectrum at $\lambda = 3\ \mu m$ and $T = 77$ K by using the known relations. We obtained $NEP = 6\cdot 10^{-14}\ W\cdot Hz^{-1/2}$ and $D^*_\lambda = 3.5\cdot 10^{11}\ cm\cdot Hz^{1/2}\cdot W^{-1}$. The detectivity varied within the temperature range 77-200 K from $3.5\cdot 10^{11}$ to $10^{10}\ cm\cdot Hz^{1/2}\cdot W^{-1}$.

The large value of the detectivity of the photovoltaic detector based on the MOVPE-grown asymmetric type-II heterostructure p-InAs/AlSb/InAsSb/AlSb/(p, n)-GaSb with a deep quantum well at the heterointerface is comparable with the parameters of MBE-grown photodiodes based on p–n-InAs superlattices and QW detectors based on GaAlAs/GaAs and InGaAs/InP double heterostructures [1] operating in the spectral range 2-3 μm, as well as with the parameters of commercial InAs photodiodes.

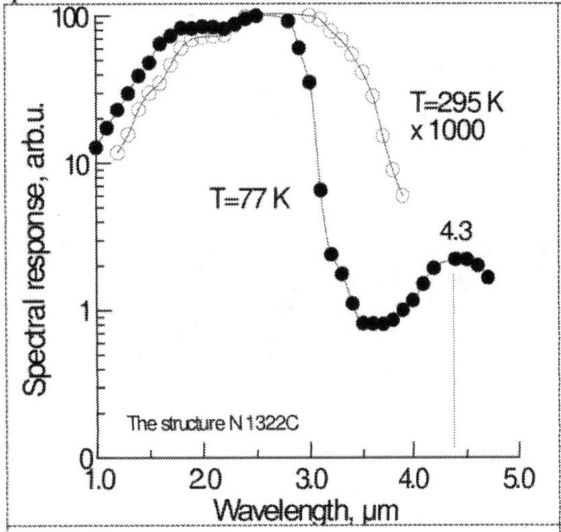

Figure 5. Normalized photosensitivity spectra for a photodiode with a single QW at T = 77 and 295 K.

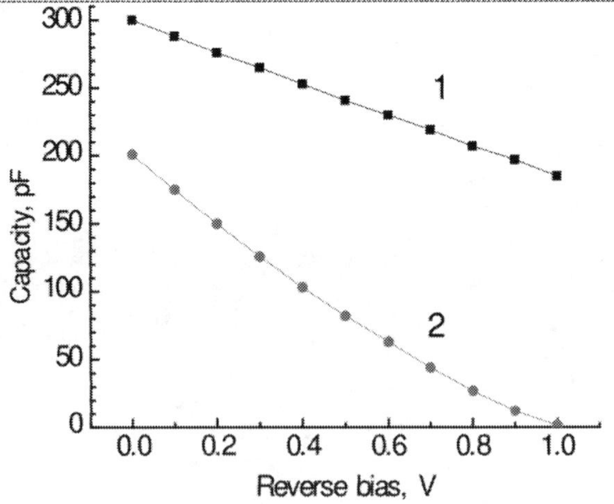

Figure 6. Capacitance vs. the reverse bias for a heterophotodiode with (1) single QW and (2) 3 QWs at T = 300 K.

Analysis of the capacitance–voltage characteristics demonstrated a marked difference between the parameters of structures with single and triple QWs at the heterointerface (Fig. 6). In the single-well structure, the capacitance weakly changed within the range C = 200-300 pF as the reverse bias was varied from 0 to 1 V. An interesting feature was that the capacitance sharply decreased from C = 200 pF at zero bias to 1.5 pF at T = 300 K for the triple-well photodiodes. These values

Journal of Physics: Conference Series **461** (2013) 012026 doi:10.1088/1742-6596/461/1/012026

correspond to a response time $R_LC = 75$ ps at a load $R_L = 50\ \Omega$. Such a behavior of the capacitance can be accounted for by the fact that, in the case of a series connection of several capacitors, the total capacitance is always smaller than the capacitance of any capacitor in the circuit. Indeed, we have in the triple-well sample several series connected p–n junctions containing four AlSb barriers. The operation speed must be independent of the sample area. Thus, we have in our case a fast photodiode with a bandwidth of ~10 GHz. Previously, we have described fast-response p–i–n photodiodes based on bulk GaInAsSb/AlGaAsSb heterostructures for the spectral range 1.6-2.4 µm, with a bandwidth of about 2 GHz [14], and avalanche photodiodes for the spectral range 2-4 µm [15]. The photodiodes with QWs in the active region, examined in this study, are promising for application in heterodyne detection of the emission of quantum-cascade lasers, information networks, medical diagnostics, and ecological monitoring.

Table 1. Dark currents and differential resistance at zero bias voltage for structures with one and 3 QWs at the heterointerface.

Parameter	Single-well structure N 1322		Triple-well structure N 1323	
	T = 300 K	T = 77 K	T = 300 K	T = 77 K
Differential resistance				
R_0, Ω, at 10 mV	28	$26\cdot10^3$	42	$3.3\cdot10^3$
$R_0\cdot A$, $\Omega\cdot cm^2$	$2.0\cdot10^{-2}$	1.8	$3.0\cdot10^{-2}$	2.4
Dark current I_d, A, at				
V = −0.2 V	$2.4\cdot10^{-3}$	$1.5\cdot10^{-5}$	$1.0\cdot10^{-3}$	$1.2\cdot10^{-5}$
V = −0.6 V	$2.8\cdot10^{-3}$	$7.2\cdot10^{-5}$	$1.6\cdot10^{-3}$	$5.0\cdot10^{-5}$

4. Conclusions

Asymmetric type-II p-InAs/AlSb/InAsSb/AlSb/(p, n)-GaSb heterostructures with one and 3 quantum wells at the heterointerface were grown by metal-organic vapor-phase epitaxy and their luminescent, transport, and photoelectric properties were studied. A high-intensity positive and negative electroluminescence was observed in the spectral range 3-4 µm at T = 300-380 K. A study of the temperature dependence of the positive and negative luminescence demonstrated that the light-emitting structures described in the communication can operate at high temperatures in the light-emitting diode/photodiode mode upon switching the bias voltage.

The spectral, current-voltage, and capacitance-voltage characteristics of mesa photodiodes with AlSb/InAsSb/AlSb quantum wells in the active region were for the first time studied in detail in the temperature range 77-300 K in the wavelength range 1-4 µm. Large values of the monochromatic responsivity (S_λ = 0.9-1.4 $A\cdot W^{-1}$) and quantum efficiency (η = 0.6-0.7) were obtained, and the detectivity was estimated at $3.5\cdot10^{11}$ to 10^{10} $cm\cdot Hz^{1/2}\cdot W^{-1}$ at T = 77-200 K. A sharp decrease in the diode capacitance with the reverse bias (C = 1.5 pF under a reverse bias V = −1 V, T = 300 K) was observed for the detector with three quantum wells in the active region, which corresponds to a response time of 75 ps and a bandwidth exceeding 6 GHz. The parameters of the quantum-well photodiodes in the active region examined in this study are comparable with characteristics of quantum-well photodiodes based on InGaAs/InP and GaAlAs/GaAs heterostructures grown by molecular beam epitaxy, as well as with parameters of commercial p-n InAs/InAsSbP photodiodes, but surpass these in operation speed.

References

[1] S. Haywood and M. Missons, in *Mid-Infrared Semiconductor Optoelectronics*, Ed. by A. Krier, Springer Series in Optical Sciences (Springer, 2006), p. 429
[2] B. F. Levine, J. Appl. Phys. **74**, R-01 (1996)
[3] E. Lina, A. Gusman, J. Sanchez-Rocha, et al., Infrared Phys. Technol. **44**, 383 (2003)
[4] W. Kruppa, M. J. Yang, B. R. Bennet and J. B. Boss, Appl. Phys. Lett. **85**, 774 (2004)
[5] J. Devenson, R. Teissier, O. Cathabard and A. N. Baranov, Appl. Phys. Lett. **90**, 111 (2007)
[6] D. H. Chow, E. T. Yu, J. R. Sonderson, et al., J. Appl. Phys. **68**, 3744 (1990)
[7] M. P. Mikhailova, K. D. Moiseev and Yu. P. Yakovlev, Semicond. Sci. Technol. **19**, R109 (2004)
[8] E. Hulicius, A. Hospodkova, J. Pangrac, T. Simehek, K. D. Moiceev, E. V. Ivanov, M. P. Mikhailova and Yu. P. Yakovlev, in *Proceedings of the 8th Internat. Conference on MIOMD*, Bad Ischl, Austria, 4–16 May, 2007, p. 184; A. Nakagawa, H. Kroemer and J. H. English, Appl. Phys. Lett. **54**, 1893 (1989)
[9] S. Sasa, M. Nakai, M. Turukawa, and M. Inoe, Conf. Ser. **187**, 363 (2005)
[10] M. P. Mikhailova, K. D. Moiseev, T. I. Voronina, et al., J. Appl. Phys. **102**, 13710 (2007)
[11] M. P. Mikhailov, E. V. Ivanov, K. D. Moiseev, Yu. P. Yakovlev, E. Hulicius, A. Hospodkova, J. Pangrac, and T. Simecek, Fiz. Tekh. Poluprovodn. **44**, 69 (2010) [Semiconductors **44**, 66 (2010)]
[12] 12. M. P. Mikhailova, T. N. Danilova, G. G. Zegrya, A. N. Imenkov, K. D. Moiseev, V. V. Sherstnev, and Yu. P. Yakovlev, Fiz. Tekh. Poluprovodn. **33**, 351 (1999) [Semiconductors **33**, 350 (1999)]
[13] K. Moiseev, E. Ivanov, V. Romanov, M. Mikhailova, Yu. Yakovlev, E. Hulicius, A. Hospodkova, J. Pangrac, and T. Simecek, Phys. Proc. **3**, 1189 (2010)
[14] I. A. Andreev, O. Yu. Serebrennikova, G. S. Sokolovskii, E. V. Kunitsyna, V. V. Dyudelev, I. M. Gadzhiev, A. G. Deryagin, E. A. Grebenshchikova, G. G. Konovalov, M. P. Mikhailova, N. D. Il'inskaya, V. I. Kuchinskii, and Yu. P. Yakovlev, Pis'ma v Zhurnal Tekhnicheskoi Fiziki, Vol. **36**, No. 9, pp. 43–49, (2010) [Tech. Phys. Lett., Vol. **36**, No. 5, pp. 412–414 (2010)]
[15] Yu. P. Yakovlev, I. A. Andreev, S. S. Kizhaev, E. V. Kunitsyna, and M. P. Mikhailova, Proc. SPIE **6636**,66360D-1 (2000)
[16] M. P. Mikhailova and I. A. Andreev, in *Mid-Infrared Semiconductor Optoelectronics*, Ed. by A. Krier, Springer Series in Optical Sciences (Springer, 2006), p. 542
[17] M. P. Mikhailova, I. A. Andreev, K. D. Moiseev, E. V. Ivanov, G. G. Konovalov, M. Yu. Mikhailov and Yu. P. Yakovlev, Fiz. Tekh. Poluprovodn., Vol. **45**, No. 2, pp. 251–255 (2011) [Semiconductors, Vol. **45**, No. 2, pp. 248-252 (2011)]

15th Russian Youth Conference on Physics and Astronomy (PhysicA.SPb) IOP Publishing
Journal of Physics: Conference Series **461** (2013) 012027 doi:10.1088/1742-6596/461/1/012027

Self-heating process influence on efficiency of luminescence of nitride-based heterostructures

E A Menkovich[1], S A Tarasov[1] I A Lamkin[1], S Yu Kurin[2], A A Antipov[2], A D Roenkov[2], I S Barash[2], H I Helava[2, 3], Yu N Makarov[2, 3]

[1]Saint-Petersburg Electrotechnical University "LETI", Prof. Popova 5, St. Petersburg 197376, Russia
[2]GaN-Crystals Ltd., 27 Engels Ave, St. Petersburg 194156, Russia
[3]Nitride Crystals Inc., 181 E Industry Ct Ste B, Deer Park, NY 11729, USA

E-mail: ankat.kate@gmail.com

Abstract. In this paper we report on dependence of the temperature of active layers (ALs) of heterostructures of light-emitting diodes (LEDs) based on AlGaN (UV LEDs) and InGaN (blue LEDs) on various current values (up to 150 mA). It is shown that the heating of the heterostructures is directly related to the concentration of defects. UV LEDs are characterized by a higher temperature than blue LEDs, they also demonstrate a lower wall-plug efficiency (WPE) (about 1.5% at 20 mA). The WPE of blue LEDs with and without the superlattice are 15% and 18%, respectively. To verify the accuracy of the performed measurements the theoretical calculation of the AL temperature according to Van Roosbroeck-Shockley theory and the model of 2D-combined density of states is carried out.

1. Introduction

Due to the unique properties of LEDs based on InGaN alloy they will soon replace conventional lighting. Their operation does not require high voltage, they are safe and low-cost. Their luminous efficacy is significantly higher than that of incandescent lamps. LEDs based on AlGaN alloy can be used as the sources of UV radiation that is also promising. These emitters can be applied in electrical engineering, medicine, science, and industry. They do not contain mercury and other poisonous substances, their recycling do not require special conditions and equipment. Therefore, large-scale application of LEDs based on nitride semiconductors allows reducing energy consumption costs as well as enhancing the state of the environment. Hence, the development of the production technology of nitride-based LEDs is an important goal.

Unfortunately, nowadays there are a number of limiting technological and other factors which decrease efficiency of luminescence of nitride-based LEDs. Self-heating of the LED AL which occurs at high current density is the most critical factor. When the temperature rises the amount of phonons increases that results in high probability of non-radiative recombination and low output power. Moreover, it leads to impurity cluster delocalization that decreases the lifetime of device. These limiting factors are especially relevant in AlGaN-based UV LED heterostructures.

The study of the influence of self-heating processes in AL on efficiency of nitride-based LED heterostructures allows us to find the reasons of this decrease. Moreover, p-n-junction temperature monitoring gives an opportunity to identify a structure defectiveness.

2. Sample and experimental technique

The paper analyzes the process of self-heating and its influence on luminescence efficiency in LED heterostructures based on solid solutions of AlGaN (peak wavelength is 362 nm) and InGaN (peak wavelength is 460 nm). The blue LED heterostructures were grown by metalorganic vapour phase epitaxy (MOVPE) (hereinafter referred to as type 1 crystals), the UV LED heterostructures (type 2 crystals) were obtained by chloride-hydride vapour phase epitaxy (CHVPE). Sapphire was used as a substrate for both types.

Type 1 crystals had a 2 μm thick GaN buffer layer grown on substrate to protect the AL from threading dislocations (TDs) which are formed because of lattice mismatch of substrate and epitaxial layers. Then a 2 μm thick n-GaN contact layer (CL) was grown. The LED AL contained a set of 5-10 InGaN quantum wells having a width of 4-5 nm. The wells were separated from each other by 25-50 nm thick GaN barriers. Then a 100 nm thick p-GaN top layer (TL) was grown. Part of LED heterostructures contained AlGaN/GaN superlattice with the lattice period of 20 nm between the p-GaN TL and the AL. The crystals containing the superlattice are called type 1A thereafter, those that do not contain – type 1B.

The design of the heterostructures of type 2 was worked out to take into account the requirement for acheiving of internal quantum efficiency maximum at high current densities. To meet this requirement a planar heterostructure (n+-n-p) was grown and a 100 nm thick single layer was used as the AL. In order to reduce the density of TDs stress relief layers were grown. Then n-type AlGaN hole-blocking layer, AlGaN, AL, p-type AlGaN electron-blocking layer, p-type GaN CL were subsequently grown. The total thickness of all epitaxial layers grown on substrate was about 6 μm.

For characterization of LED dies we used our invention – a state-of-the-art diagnostic complex which allows investigation of optical and electrical characteristics of both LEDs and laser diodes according to the international standard CIE 127:2007 and the Russian GOST 17616-82. The diagnostic complex employs a new method for determination of LED AL temperature which is based on the detailed analysis of electroluminescence spectra. The effects of noise including those induced by interference are considerably suppressed. The diagnostic complex is designed to investigate both separate and built-in LEDs and other devices based on them.

The developed diagnostic complex includes the following components: power supply system for measurements in constant- and pulsed-mode operation; measuring block which includes thermostat (cryostat) and probe station; optical subsystem; PC. Measurements can be carried out in automatized regime under control of the specialized software worked out on the basis of LabView. The diagnostic complex allows accurate measurements of any LEDs and devices based on them in the following ranges: emission wavelength: 200-1100 nm; direct current: 10^{-8}-10 A; voltage: 0-250 V; temperature: 10-500 K.

3. Experimental results and discussions

In order to evaluate self-heating processes influence on efficiency of luminescence of nitride-based LED heterostructures the main attention was paid to the study of spectral, power-current (P-I) and current-voltage (I-V) characteristics.

To calculate the dependence of self-heating on current density the investigation of the dependence of luminescence spectra on temperature was conducted (figures 1, 2). The temperature of the AL was determined by analysis of external heating influence on luminescence spectra and the impact of direct current on the shape and location of luminescence spectra (figure 1, 2).

Figure 2 shows that self-heating in type 2 LEDs (45°C at operating current (I_f) of 20 mA) is more intensive than in type 1 LEDs (30°C). This difference can be explained by a higher dislocation density in epitaxial heterostructures of type 2. When the direct current rised to 50 mA the temperature was equal to 90°C. Even at lower current than I_f the crystal temperature was higher than room temperature.

Theoretical calculation of the AL temperature was implemented to assess the experimental results. For this goal the theory of Van Roosbroeck-Shockley was applied with account of the model of 2D-combined density of states [1]. Amendments were made to the equations which allowed us to take into account potential fluctuations caused by the Coulomb field of impurities, variations of the composition of solid solutions, and rough heterointerfaces. Quasi-band-to-band transitions were considered in calculations. The model of 2D-combined density of states was used to take into account effects of dimensional quantization.

The results of calculation were compared with normalized experimental spectra. In this case the value of the band-gap used for the calculation was obtained from experimental data. Evaluation of the AL temperature was performed by using mathematical programs based on spectral properties obtained experimentally. Figure 2 shows theoretically calculated and experimentally obtained dependencies of the AL temperature. One can see that the discrepancy of experimental and theoretical data does not exceed 5% that proves the correctness of the proposed methods for determination of the AL temperature and reliable work of the diagnostic test-system.

Optical and electrical properties of the samples were studied. The following dependencies were obtained (figures 3, 4):

- Dependence of emission wavelength on current and temperature;
- Dependence of output power on current;
- Dependence of WPE on current;
- I-V characteristics.

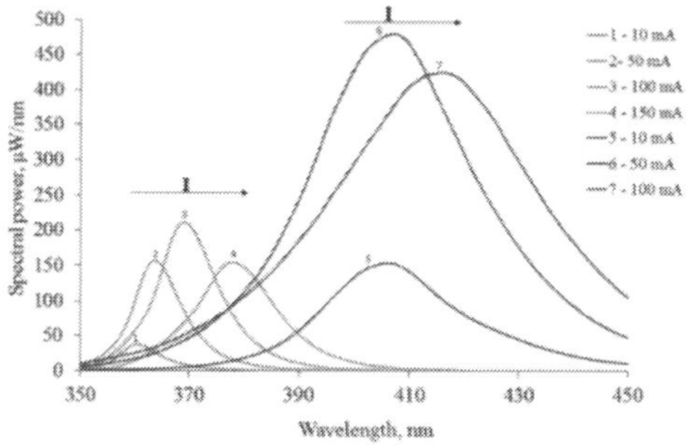

Figure 1. Dependence of spectral power on wavelength at different currents.

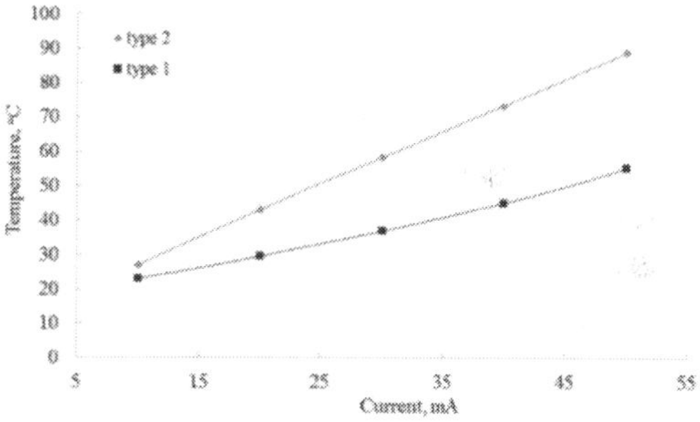

Figure 2. Dependence of temperature of active layer on current.

The analysis of these experimental data showed that the wavelengths of the type 1 LEDs were in the range of 460-462 nm, spectral full width at half-maximum (FWHM) – 30-40 nm at I_f and room temperature, while the LEDs of type 2 revealed 360-365 nm and 10-13 nm, respectively. The output power at I_f was equal to 9.5 and 1.15 mW for type 1 and type 2 LEDs, respectively. P-I characteristics deviation from linearity was observed in the samples of both type at currents higher than 30 mA. The maximum value of 23.1 mW was reached at the current of 90 mA for type 1 LEDs and 4.2 mW at 120 mA for type 2 LEDs. Output power reduction of LEDs of type 1 was negligible as compared with LEDs of type 2. Deviation of P-I characteristics from the linear dependence was caused by the structure heating that led to a higher non-radiative recombination rate and decrease of the emission intensity. When the ambient temperature increases there is a uniform shift of luminescence spectrum towards long-wave region by 0.5 nm for LEDs of type 1 and 1 nm for LEDs of type 2. Generally, wavelength shift of emission maximum caused by the current was 7 nm for the blue LEDs when the current changed from 10 mA to 100 mA, caused by the temperature (30-90°C) – 3 nm, for UV LEDs – 9.7 nm (10-100 mA) and 3.5 nm, respectively.

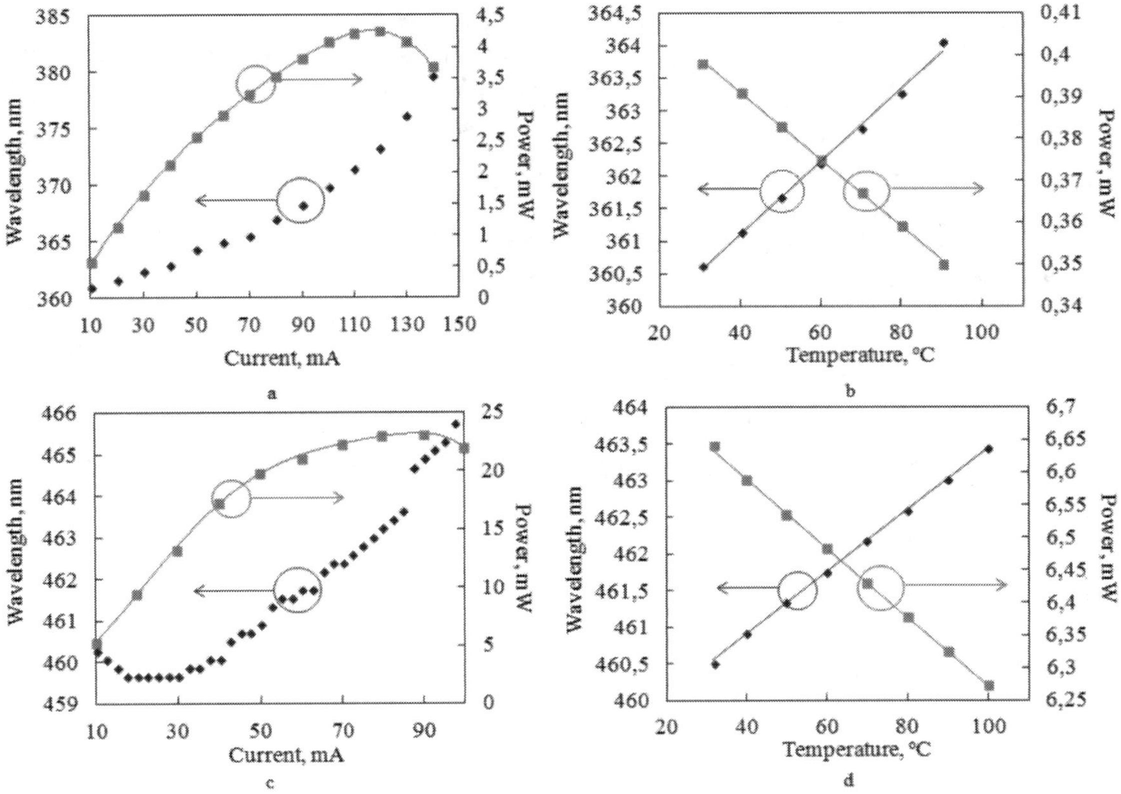

Figure 3. (a) Dependence of emission wavelength of UV LEDs on current and temperature. (b) Dependence of output power of UV LEDs on current. (c) Dependence of emission wavelength of blue LEDs on current and temperature. (d) Dependence of output power of blue LEDs on current.

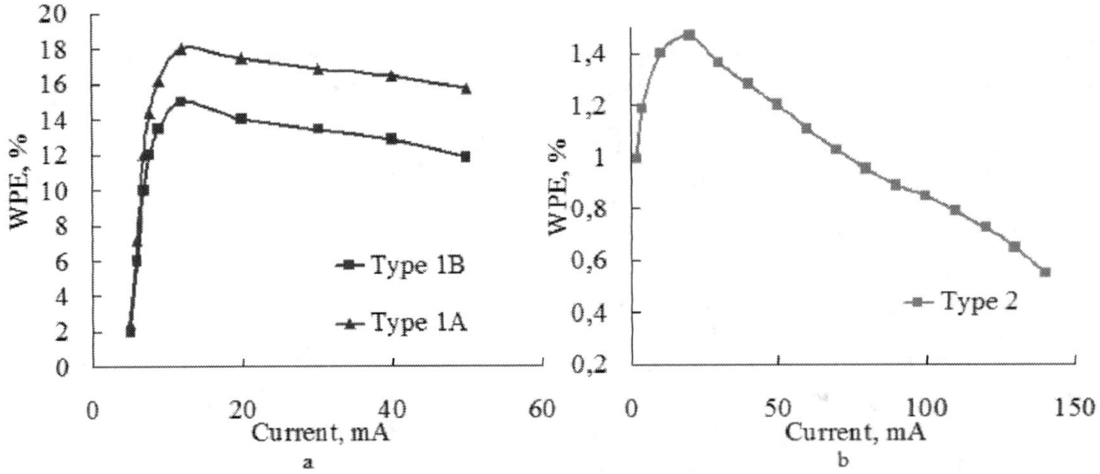

Figure 4. (a) Dependence of WPE of LED of type 1 on current. (b) Dependence of wall-plug efficiency of LED of type 2 on current.

The calculation showed that the LEDs of type 2 revealed the WPE as low as 1.5% at low currents. It can be caused by relatively high density of defects. The WPE of the LEDs of type 1B reached 15% and decreased to 12% at 50 mA, the LEDs of type 1A had the WPE as high as 18%, their maximum achievable power was 23.1 mW (figure 5).

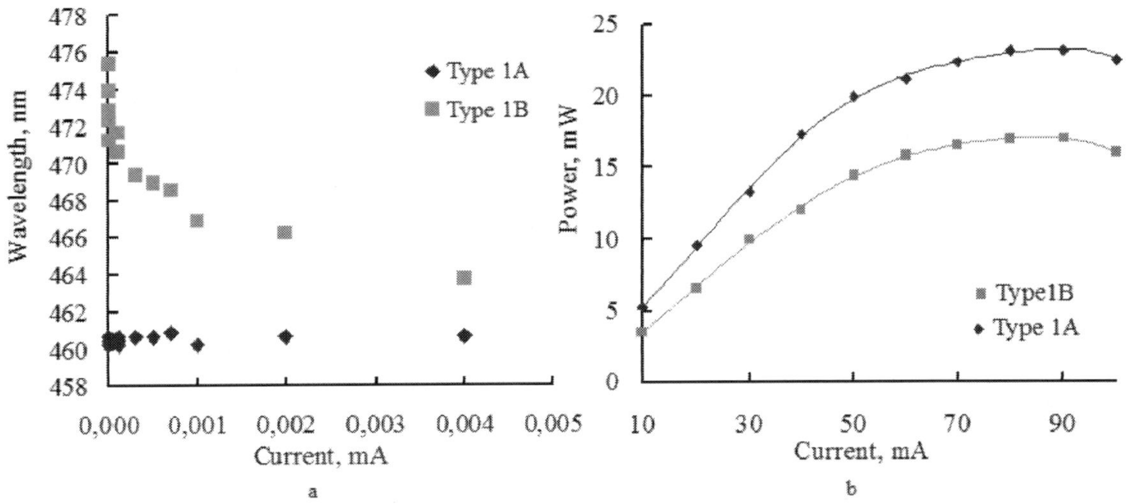

Figure 5. (a) Dependence of wavelength on low currents. (b) Dependence of output power on high currents.

The analysis demonstrated the considerable decline in density of defects in ALs of such heterostructures. It can be caused by lower values of elastic stresses. The analysis of emission wavelength maximum evolution at low currents showed that there was no wavelength shift in the structures of type 1A while it was considerable in the structures of type B. We assume that the shift of electroluminescence spectrum at low currents in the structures of type B is caused by the impact of elastic stresses and piezoelectric fields at the heterointerfaces. At low currents in the background of small stresses applied to the AL of the structures the contribution of piezoelectric fields is significant. Therefore, the well shape is considerably distorted under their influence. It results in a significant decrease of the distance between the layers constituting the heterostructure. When the external stress increases piezoelectric fields impact is compensated and the shift declines. Superlattice heterostructures demonstrate the absence of spectral maximum shift. We suppose it is caused by compensation of tensile and compressive elastic stresses due to the superlattice. Thus, the structure of type A demonstrates considerably higher stability of emission wavelength. Moreover, elastic stresses may relax during the LED operation by forming dislocations in the AL. Stress compensation caused by the superlattice should reduce formation of dislocations in type A LEDs. Therefore, this structure shows a better performance and has a longer lifetime.

4. Conclusions

The study determined the values of the LED AL temperatures for the structures based on AlGaN and InGaN alloys at different currents (up to 150 mA). It was shown that in case of structures with high density of defects the temperature can reach the value of 200°C and higher, i.e. the heating of the structures is directly related to the defect concentration.

To verify the accuracy of the performed measurement theoretical calculation of the AL temperature according to Van Roosbroeck-Shockley theory and the model of 2D-combined density of states was carried out. The calculation was made with account of potential fluctuations caused by Coulomb field of impurities, variations of the composition of solid solutions and rough heterointerfaces. The discrepancy of temperature values obtained by theoretical and experimental approaches was less than 5%. It proves the correctness of the developed method.

The direct interconnection of self-heating processes and luminescence efficiency was shown for the the above-mentioned structures. UV LED heterostructures were characterized by a higher temperature than the visible LEDs, they also demonstrated a lower efficiency (up to 1.5% at low currents). The WPE of blue LEDs without the superlattice reached 15% and decreased only to 12% at 50 mA. The structures with the superlattice revealed a better performance and showed the WPE of 18% and 1.5-fold higher maximum achievable output power. It proves a significant reduction of the density of defects in the ALs of studied structures. Moreover, considerably smaller influence of current values on emission wavelength was observed. It can be attributed to the compensation of elastic stresses.

References

[1] Schubert E F 2006 *Light Emitting Diodes* (New York: Cambridge University Press).

Acknowledgments

The work was partly supported by the Ministry of Education and Science of the Russian Federation (contract number 12.527.12.3001).

Journal of Physics: Conference Series **461** (2013) 012028 doi:10.1088/1742-6596/461/1/012028

UV LEDs for high-current operation

S Yu Kurin[1], A A Antipov[1], A D Roenkov[1], I S Barash[1], H I Helava[1, 2], E A Menkovich[3], S A Tarasov[3], I A Lamkin[3], N M Shmidt[4], and Yu N Makarov[1, 2]

[1]GaN-Crystals Ltd., 27 Engels Ave, St. Petersburg 194156, Russia,
[2]Nitride Crystals Inc., 181 E Industry Court, Suite B, Deer Park, NY 11729, USA,
[3]Saint-Petersburg Electrotechnical University "LETI", Prof. Popova 5, St. Petersburg 197376, Russia,
[4]Ioffe Physical Technical Institute of RAS, Polytekhnicheskaya 26, St. Petersburg 194021, Russia

E-mail: sergey.kurin@nitride-crystals.com

Abstract. In this paper we report on results of development of ultraviolet light-emitting diodes (UV LEDs) based on GaN/AlGaN heterostructures grown on Al_2O_3 (0001) substrates by chloride-hydride vapour phase epitaxy (CHVPE). Both UV LED heterostructures and packaged dies are investigated. UV LEDs proved performance capability at current density up to 125 A/cm^2 and revealed wall-plug efficiency (WPE) of 1.5% at operating current of 20 mA.

1. Introduction

UV radiation is largely used in a number of industries, medicine and measurement equipment. In the near UV area (320-390 nm) it is highly relevant to use UV radiation for photocatalytic purification of water and air as well as for photopolymerization processes. Nowadays mercury lamps generally serve as a source of UV radiation. Mercury lamps replacement by UV LEDs will provide ten-fold increase of lifetime of UV devices and will lead to considerable decrease of their power consumption as well as reduce mercury pollution of the environment.

Today the main problems preventing manufacturing of UV LED with high technical properties (low power consumption, long lifetime and high WPE) are posed by low quality of epitaxial heterostructures which are characterized by high density of structural defects as well as high prime cost of their obtaining by traditional epitaxial methods.

In order to solve some of the above-mentioned problems this paper considers a technological approach based on application of the CHVPE for the growth of UV LED heterostructures. It is a well-known method for fabrication of thick low-defect quasi-bulk or free-standing GaN materials and GaN-based templates served as substrates for device structures growth by other techniques such as molecular beam epitaxy (MBE) and metalorganic vapour phase epitaxy (MOVPE). It was also demonstrated that CHVPE can be successfully used to grow the whole device structures for application in visible and UV LEDs and power electronics. CHVPE method is convenient for mass production of semiconductor devices due to its low cost, flexibility of growth conditions, and good reproducibility.

2. Experimental technique

Nitride-based UV LEDs have some particular features compared to nitride-based LEDs of visible range which are to be taken into account for optimization of technological process of UV LED heterostructures growth.

Active region of UV LEDs with the wavelength less than 365 nm can be produced of materials not containing indium. In this case lattice constant mismatch of semiconductor materials constituting the heterostructure is relatively low that allows growing of thick active regions without stress relaxation accompanied by formation of dislocations [1].

The choice of a single layer of more than 100 nm thickness as an active region is motivated by the following assumption: the use of above-mentioned active region is promising to obtain an internal quantum efficiency (IQE) maximum at high current density (more than 100 A/cm^2). The use of multiple-quantum-well (MQW) structures as active regions of UV LED heterostructures is not suitable for the work at high current density because these structures are still facing problems of inhomogeneity of electron and hole injection in various wells. Another problem of LED heterostructures with MQW is IQE–I dependence drop that is observed even at low current density as a result of heating of the active region, carrier leakage, and Auger recombination [2].

CHVPE method is considered to be optimum for growing of thick stress relief layers (SRL) and thin active region. Compared to the traditional MOVPE method it gives an opportunity to control the growth rate over a large range: from micrometer to dozens and hundreds of micrometers per hour, thus, it is possible to grow both thick (several micrometers thick) buffer layers and thin (units-dozens of nanometers) active regions within one process. In MOVPE growth rates are low (as a rule, up to 3-4 μm/hour), so it is quite a difficult task to grow thick buffer layers, and in many cases templates ($AlGaN/Al_2O_3$ heterostructures in which AlGaN epitaxial layer has a thickness of several micrometers) are used as a buffer. In this case the process of LED heterostructure growth is divided into two steps: a) growth of template (for instance, using CHVPE method); b) growth of heterostructure on the template by using MOVPE method. Moreover, CHVPE method provides epitaxial layers of high crystal quality with high doping level which is limited only by inversion of the type of conductivity for Mg [3]. This technique offers advantages in growth rate, simplicity, and cost as well as the ability to grow III-V compound layer directly onto a substrate without the inclusion of low temperature buffer layer. Additional advantages potentially leading to high IQE of radiative recombination in CHVPE-grown UV LED heterostructures are due to the low impurity contamination [3, 4].

Table 1 shows comparison of CHVPE and MOVPE approaches which are used for UV LEDs production.

Table 1. Comparison of CHVPE and MOVPE approaches.

Parameter	CHVPE	MOVPE
Growth rate	Wide range (1-150 μm/hour)	Low (a few μm/hour)
NH_3 consumption	Low (< 10 slm)	Very high (> 30 slm)
Costs of raw materials	Low	High
Process pressure	Atmospheric	100-250
Use of H_2	No	20-50 slm

Epitaxial processes of the growth of UV LED heterostructures were performed in atmospheric pressure quartz hot-wall CHVPE reactor which is designed for deposition of the III group metals including GaN, AlN, InN, and their solid solutions on 2-inch diameter wafers. Ammonia (NH_3) and hydrogen chloride (HCl) were used as precursors. For GaN growth, HCl was passed over Ga source. To grow the AlGaN alloy, HCl was passed separately over the Ga and Al sources. Monosilane (SiH_4) and metallic Mg were used for n- and p-type doping, respectively. The reactor is equipped with 6-zone resistance heater providing the required temperature profile in the reaction area. The reactor includes both a growth zone and a dwell zone. The growth temperature was 1040°C. Design of the quartz reactor allows growing multilayer structures by either variation of precursor gas flows or transferring substrate between a growth zone and a dwell zone keeping at the same temperature.

UV LED dies of 0.31×0.31 mm^2 were subsequently fabricated. Ti/Al and Ni/Au metal layers were used as n- and p-type electrodes, respectively. Processing and packaging were performed using outside service.

3. Experimental results and discussions

UV LED heterostructures grown on Al_2O_3 wafers were tested by X-ray diffractometry (XRD), scanning electron microscopy (SEM), secondary ion mass spectrometry (SIMS), atomic force microscopy (AFM), electroluminescence (EL), photoluminescence (PL) methods.

Al_2O_3 (0001) wafers were used for epitaxial processes of UV LED heterostructures growth. UV LED heterostructures consisted of AlN/AlGaN (x~0.6) composed buffer layer, 6-8 pairs of AlGaN (x~0.1-0.15)/AlGaN (x~0.03-0.08) SRL, Si-doped n-AlGaN (x~0.05-0.08) barrier, (Al)GaN active region, and Mg-doped p-AlGaN (x~0.12) barrier. Mg-doped p-AlGaN (x~0.05) cover layer completed the structure. Here x is a molar fraction of AlN. All these layers were clearly distinguished in cross-sectional SEM image (figure 1) and SIMS-profile (figure 2).

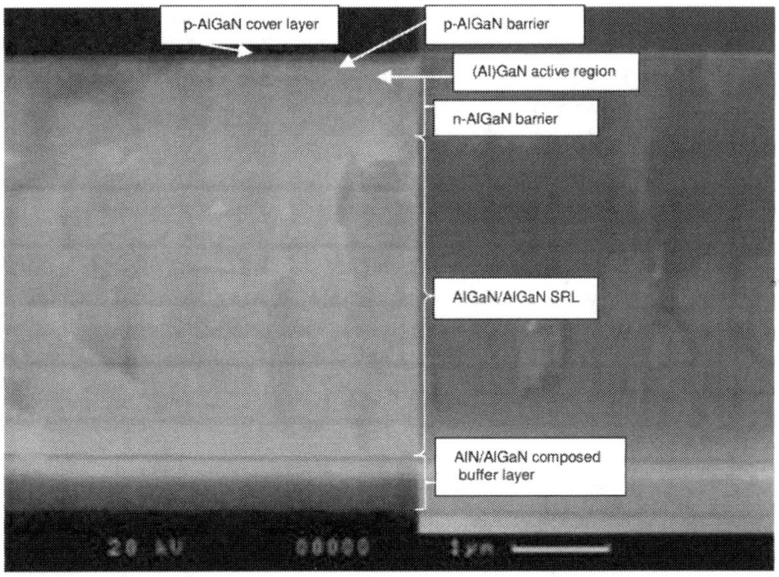

Figure 1. Scanning micrograph of cross-sectional view of UV LED heterostructure.

Journal of Physics: Conference Series **461** (2013) 012028 doi:10.1088/1742-6596/461/1/012028

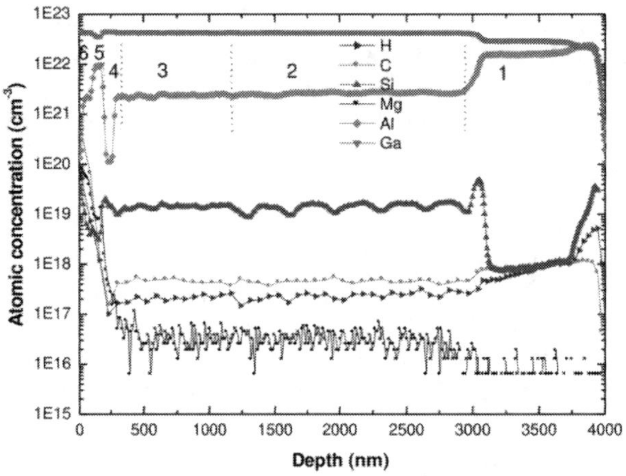

Figure 2. SIMS-profile of UV LED heterostructure: atomic concentrations of Ga, Al, doping impurities (Si, Mg) and background impurities (H, C) are shown. The sample surface is to the left. Points 1-6 show the layers of the heterostructure: 1 – AlN/AlGaN composed buffer layer; 2 – AlGaN/AlGaN SRL; 3 – n-AlGaN barrier; 4 – (Al)GaN active region; 5 – p-AlGaN barrier; 6 – p-AlGaN cover layer.

The full width at half-maximum (FWHM) of XRD peak (scan type: θ-2θ) in symmetrical reflection (0002) taken from AL of UV LED heterostructure was less than 400 arcsec. The dislocation density in the best samples measured by AFM technique varied from $9 \cdot 10^7$ cm^{-2} to $8 \cdot 10^8$ cm^{-2}.

The results of PL measurements are shown in figures 3, 4. The wavelength of PL maximum was in the range of 359.5 – 362 nm, the average PL FWHM was 11 nm. PL measurements revealed a uniform distribution of peak wavelength and FWHM across the wafer. The variation in peak wavelength from the center of the wafer to its edge, evaluated in terms of a standard deviation, was less than 1%, the variation in FWHM was about 2.5%.

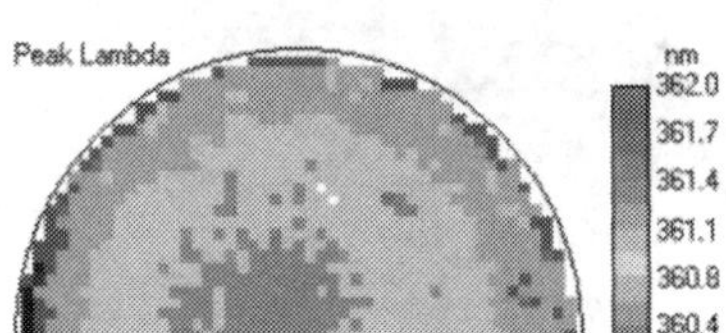

Figure 3. Peak wavelength distribution across the wafer.

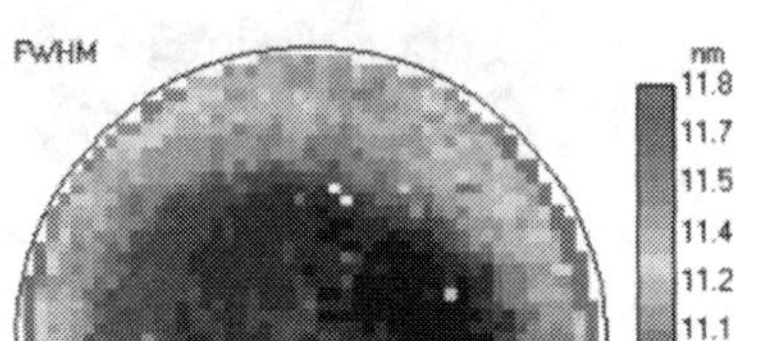

Figure 4. FWHM distribution across the wafer.

The peak wavelengths of the packaged dies were 360-365 nm at the operating current of 20 mA and the room temperature (figure 5), FWHM parameter was in the range of 10-13 nm. The output optical power at the operating current was equal to 1.14 mW. The deviation of power-current (P-I) characteristics from linearity was observed at the current values of more than 30 mA. It was caused by a structure heating. The maximum value of output optical power – 4.2 mW – was reached at the current of 120 mA. The peak wavelength shift at the current change from 10 mA to 140 mA was about 18.5 nm.

When the ambient temperature increased there was a uniform shift of the luminescence spectrum towards the long-wave region (figure 6). At the temperature change from 30°C to 90°C the peak wavelength shift was 3.4 nm.

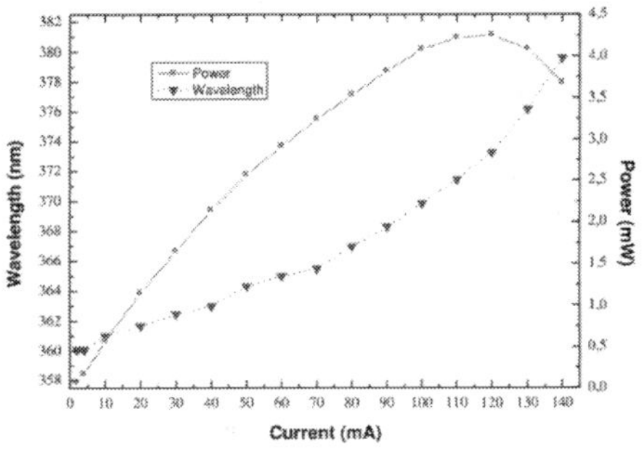

Figure 5. P-I characteristics of UV LED packaged die and dependence of its peak wavelength on the direct current.

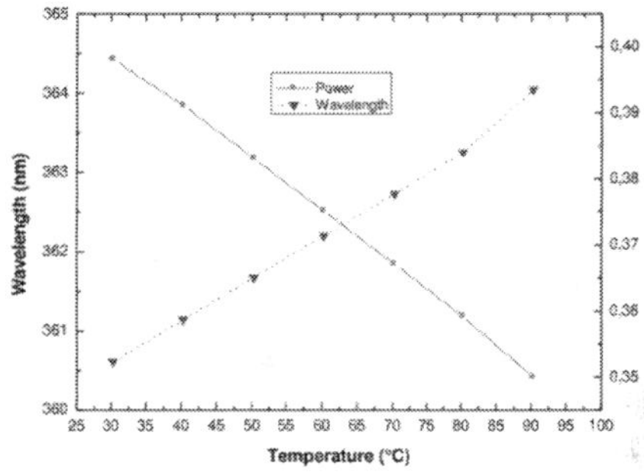

Figure 6. Peak wavelength of UV LED packaged die and its power dependence on the ambient temperature.

Figures 7, 8 show the EL spectra and WPE of UV LED die, respectively. At operating current of 20 mA peak wavelength was in the range of 360-365 nm and WPE was about 1.5%.

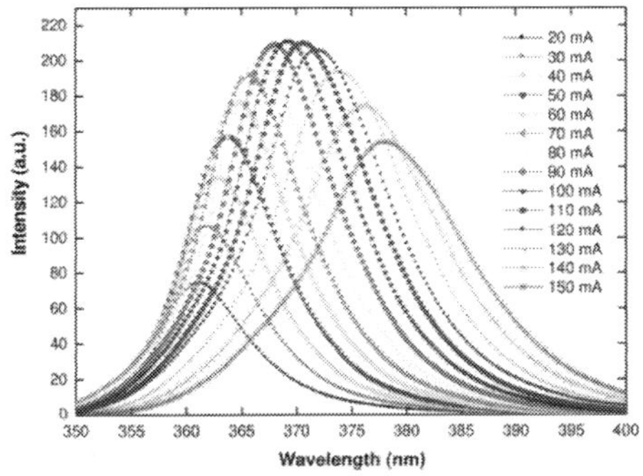

Figure 7. EL spectra of UV LED packaged die at different values of the direct current.

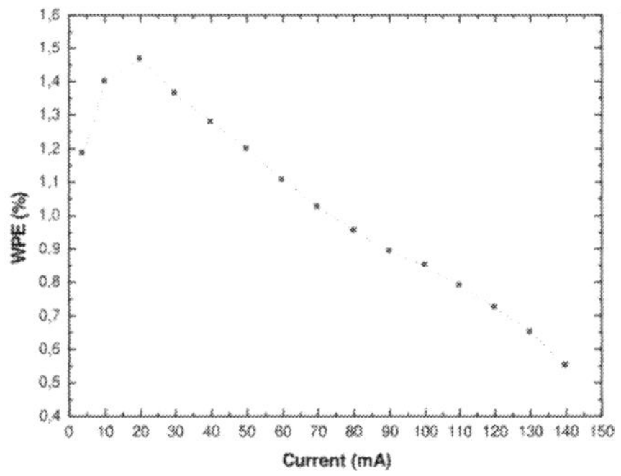

Figure 8. Dependence of WPE of UV LED packaged die on the direct current.

The output power results of our packaged dies compare favorably with the results reported by G. Smith *et al.* [5] on their CHVPE-grown UV LEDs at virtually the same wavelength. As for comparison with commercially available UV LEDs, the output power of our packaged dies has the same order of magnitude as Nichia's NSSU100C UV LEDs (spectral characteristics, e.g. FWHM and spectral curve shape, are identical).

4. Conclusions

Thus, CHVPE method can be used for the growth of GaN/AlGaN heterostructures of submicron thicknesses for the UV LEDs, which proved its workability at current densities up to 125 A/cm^2.

UV LED heterostructures design was worked out to meet the requirement for reaching IQE maximum at high current density. For this purpose, a single layer of more than 100 nm thickness was suggested as an active layer.

The analysis of UV LEDs showed that their FWHM parameter was in the range of 10-13 nm. At operating current 20 mA the active layer temperature was 45°C, the output optical power was equal to 1.14 mW, the WPE was 1.5% (dimensions of dies: 0.31×0.31 mm^2).

The performed study suggests the prospects of CHVPE technology for production of nitride-based semiconductor devices such as LEDs, solar cells, photodetectors etc.

References

[1] A. Antipov, I. Barash, V. Bublik, S. Kurin, Yu. Makarov, E. Mokhov, S. Nagalyuk, A. Roenkov, T. Chemekova, K. Scherbachev, and H. Helava, Mater. Electron. Tech. **1**, 52 (2012).

[2] K. Bulashevich, M. Ramm, and S. Karpov, Phys. Status Solidi C **6**, S804 (2009).

[3] A. Usikov, O. Kovalenkov, V. Soukhoveev, V. Ivantsov, A. Syrkin, V. Dmitriev, and A. Davydov, Phys. Status Solidi C **5**, 1829 (2008).

[4] V. Dmitriev and A. Usikov, Hydride Vapor Phase Epitaxy of Group III Nitride Materials, III-Nitride Semiconductor Materials, (Imperial College Press, London, 2006), chap. 1.

[5] G. Smith, T. Dang, T. Nelson, J. Brown, D. Tsvetkov, A. Usikov, and V. Dmitriev, J. Appl. Phys. **95**, 8247 (2004).

Acknowledgments

The work was partly supported by the Ministry of Education and Science of the Russian Federation (contract number 12.527.12.3001). We wish to thank Academic University – Nanotechnology Research and Education Centre of RAS for providing PL measurements.

Journal of Physics: Conference Series **461** (2013) 012029 doi:10.1088/1742-6596/461/1/012029

Laser performance investigation of a new UV active media $LiY_{0,3}Lu_{0,7}F_4:Ce^{3+}$ and $LiY_{0,3}Lu_{0,7}F_4:Ce^{3+}+Yb^{3+}$

E K Kuchaev, A S Nizamutdinov, V V Semashko, L A Nurtdinova and S L Korableva

Kazan (Volga Region) Federal University, 18 Kremljovskaja str., Kazan 420008, Russian Federation

kutchaev@mail.ru

Abstract. We show that the $LiY_{0,3}Lu_{0,7}F_4:Ce^{3+}$ and $LiY_{0,3}Lu_{0,7}F_4:Ce^{3+}+Yb^{3+}$ crystals are promising active media of UV spectral range with low-threshold lasing (30-90 mJ/cm^2). Due to crystal-chemical approach (additional doping by ions Yb^{3+}) we obtained the effect of suppression of spurious photodynamic processes. The active medium $LiY_{0,3}Lu_{0,7}F_4$: Ce^{3+}+ Yb^{3+} combines properties of saturable amplifier and oscillator with small saturation energy, and promising to generate pulses of ultrashort duration in the UV spectrum.

1. Introduction

In comparison with a set of the solid-state lasers of infra-red and visible ranges, the number of the solid-state lasers radiating in ultraviolet (UV) and vacuum ultra-violet (VUV) ranges is limited. There are problems [1] on a way of creation of these lasers, such as material solarization, excited state absorption (ESA) [2]. However, compact solid-state lasers of UV and VUV spectrum ranges development will allow solving variety of problems in various areas of science and technology as medicine, biology, microelectronics, nanotechnology, laser spectroscopy, materials precision processing, etc.

One of the approaches to solve the material solarization problem is in increasing crystal photochemical stability of solid-state materials with respect to UV radiation.

Objects of the investigation $LiY_{0,3}Lu_{0,7}F_4:Ce^{3+}$ and $LiY_{0,3}Lu_{0,7}F_4:Ce^{3+}+Yb^{3+}$ crystals can be attributed to the materials suitable for the construction of lasers operating in the UV region of the spectrum.

The purpose of this study is to investigate the lasing characteristics of active media $LiY_{0,3}Lu_{0,7}F_4:Ce^{3+}$ and $LiY_{0,3}Lu_{0,7}F_4:Ce^{3+}+Yb^{3+}$ and intracavity losses connected to these active media.

2. Objects of study and experimental setup

Double fluoride mixed crystals of scheelite structure, doped by ions Ce^{3+}, which are the objects of our investigation are homologous to known $Ce:LiYF_4$ and $CeLiLuF_4$ active media. Previous studies [3] have shown that the ratio of yttrium ions (30%), and lutetium (70%) provides the maximum ratio of cerium ions entering the matrix. Ce^{3+} content was 1 at. % in both samples and Yb^{3+} ions content was 1 at. %.

These crystals were synthesized in the crystal growth laboratory, the Department of Quantum Electronics and radiospectroscopy of Institute of Physics at the Kazan Federal University. These crystals are the new active media, generating in the ultraviolet range of the spectrum.

The samples were prepared for side pumping scheme. They were a half-cylinders of length 7.1 mm with polished parallel ends, which were side-polished plane angle, the value of which is chosen from Brewster fall of reflected from the mirrors of the lasing beam with a wavelength of 309 nm. Such a choice of the angles between the faces and the side planes is due to the need to obtain π-polarized light laser generation with minimal losses.

The experimental setup for measuring the lasing characteristics is shown in Figure 1. It comprises side pumping by 1st Stokes component of 266 nm laser radiation obtained from Raman shifter cell with H_2 and nonselective cavity for investigated crystals.

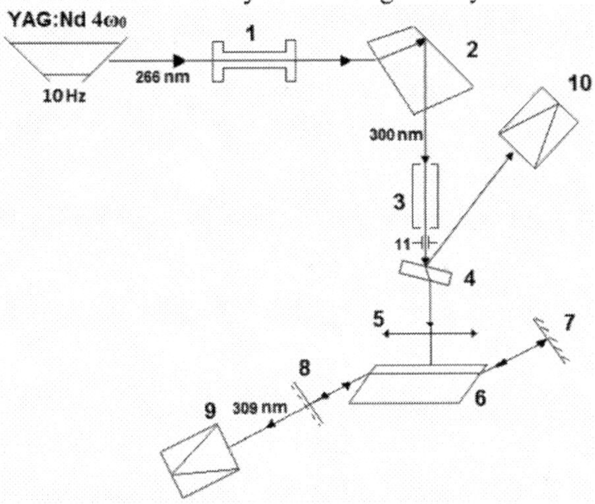

Figure 1. Experimental configuration scheme 1 – gaseous hydrogen Raman shifter; 2 – Pellin–Broca prism; 3 - optical attenuator; 4 - quartz plate; 5 - a cylindrical lens (f=35mm); 6 - active medium, 7 - totally reflecting mirror; 8 - output mirror; 9.10 – photodiodes; 11 - diaphragm.

3. Results. Lasing characteristics

Experimental plots of the output energy depending on the pump energy for these crystals for output mirrors with different reflection coefficients R are shown in Figures 2a and 2b.

As it is evident from the nature of the energy dependence of lasing on the pump energy (Fig.2a), there is a hysteresis in the generation characteristics: the course of the value of energy of laser oscillation with increasing pump energy is not the same as the corresponding dependence with decreasing pump energy. When reducing the pumping energy, the energy of the lasing is higher than the value when we increase pumping energy. Hysteresis confirms that dynamic processes induced in the active media which cause additional intracavity losses. Additionally, these losses are changing their value during the experiment.

For the active medium without ytterbium doping the maximum slope efficiency was $\eta = 14\%$ (R=70%). For the corresponding output mirror in the cavity there is a large part of the laser radiation, as compared to other mirrors. The fact that color centers are destroyed due to the absorption of laser radiation provides relatively high slope efficiency. This also explains the hysteresis of the lasing characteristics of Fig. 2a. An increasing number of color centers are being destroyed by the end of the experiment.

Also the graph (Figure 2a) shows that the area of the hysteresis loop is minimal when the output mirror with a reflection coefficient R=70% is used, which confirms the fact that the formed color centers are destroyed more efficiently.

For a sample additionally doped by ytterbium ions $LiY_{0,3}Lu_{0,7}F_4{:}Ce^{3+}{+}Yb^{3+}$ the hysteresis dependence was not observed. This suggests that photodynamic processes were suppressed by the addition of an effective electron traps (Yb^{3+}). However, the lasing saturation was observed. The maximum slope efficiency was found to be $\Delta\eta = 11\%$ with the output mirror reflecting R = 55% of the laser generation.

It is known that the dynamic saturation of the amplifying medium can be used for mode locking in schemes with slow saturable absorber. Thus, the active medium with ytterbium ions is promising for ultrashort pulses lasing.

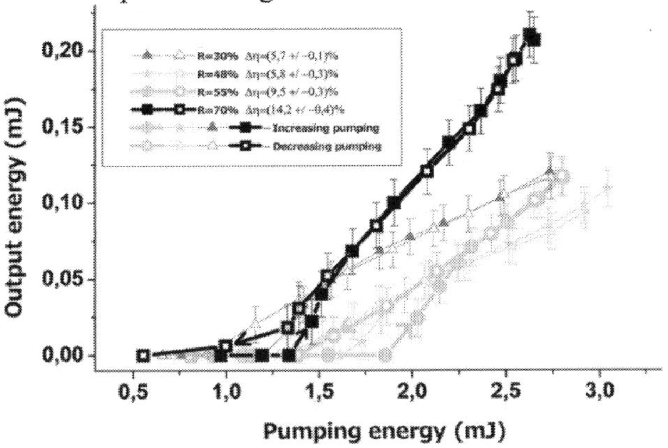

Fig 2a. Output lasing energy depending on the pump energy for $LiY_{0,3}Lu_{0,7}F_4:Ce^{3+}$ crystal for output mirrors with different reflection coefficients R

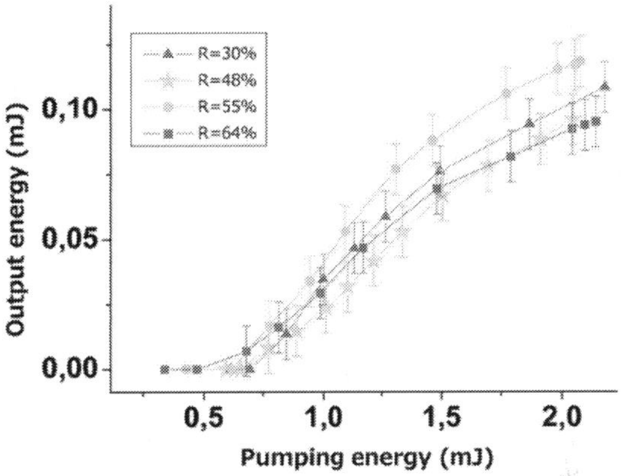

Fig 2b. Output lasing energy depending on the pump energy for $LiY_{0,3}Lu_{0,7}F_4:Ce^{3+}+Yb^{3+}$ crystal for output mirrors with different reflection coefficients R

Table 1. Slope efficiency for mirrors with different reflection coefficients

R, reflection of mirror, %	Slope efficiency $\Delta\eta$, %	
	$LiY_{0,3}Lu_{0,7}F_4:Ce^{3+}$	$LiY_{0,3}Lu_{0,7}F_4:Ce^{3+}+Yb^{3+}$
30	5,7 ±0,1	10,7±0,8
48	9,5 ±0,3	8,2±0,4
55	5,8 ±0,3	11,5±0,3
64	-	8,0±0,2
70	14,2 ±0,4	-

4. Intracavity losses calculation

Producing a further analysis of the data, we have calculated the intracavity losses. From [4] the energy balance equation is known:

$$E_g = \eta \cdot \Delta \cdot \frac{K_r}{K_r + \rho} \cdot (E_p - E_{th}), \quad (1)$$

Where E_g – lasing energy, η – quantum efficiency of luminescence, $\Delta = h\nu_g/h\nu_p$ – quantum efficiency of laser generation, K_r – "useful losses" coefficient corresponding to couplers reflectance, E_p – pumping energy, E_{th} – threshold pump energy and constant ρ – coefficient of the intracavity losses. However, in our case, the coefficient ρ should be a function $\rho = \rho_0 + \rho_{ph}(E_p, E_g^{cav}, R, t)$ [5], where

ρ_0 – passive intracavity losses, R – output mirror reflection coefficient, t – time. Functional dependence $\rho_{ph}=f(E_p,E_g^{cav},R,t)$ takes into account both the processes of formation of color centers under the pump, and their destruction, spontaneous (due to thermal vibrations of the lattice) and due to ionization of color center by laser radiation circulating inside the cavity (this process is called photobleaching).

After some simple mathematical transformations we can receive the final formula for the calculation of intracavity losses [5]:

$$\rho = \mathrm{K_r}\cdot\frac{(\eta\cdot\Delta\cdot E_p-\eta\cdot\Delta\cdot K_r\cdot a-E_g)}{\eta\cdot\Delta\cdot K_r\cdot a+E_g}\,, \qquad (2)$$

$$a=\frac{h\nu_p\cdot V}{\sigma_g\cdot\eta}\,, \qquad (3)$$

Where $h\nu_p$ – pump photon energy, V – pumping volume, σ_g – laser transition cross section.

Figure 3 shows intracavity losses depending on the pump energy $\rho(E_p)$ for output mirrors with different reflection coefficients R for the sample without Yb (Fig 3a) and with ytterbium (Fig 3b).

From the curves shown in Figures 3a and 3b we can see that by increasing of pumping energy total intracavity losses are reduced. This dependence is explained by the formation of color centers by pumping radiation and their subsequent destruction by laser radiation. The more lasing remains in the cavity, the lower amount of the color centers.

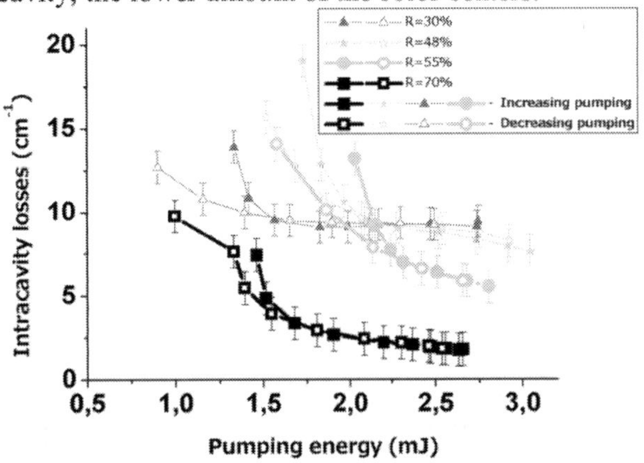

Figure 3a. Intracavity losses $\rho(E_{\text{н}})$ dependence on pumping energy for **$LiY_{0,3}Lu_{0,7}F_4$:Ce^{3+}** active medium with different reflection coefficients R of output mirror

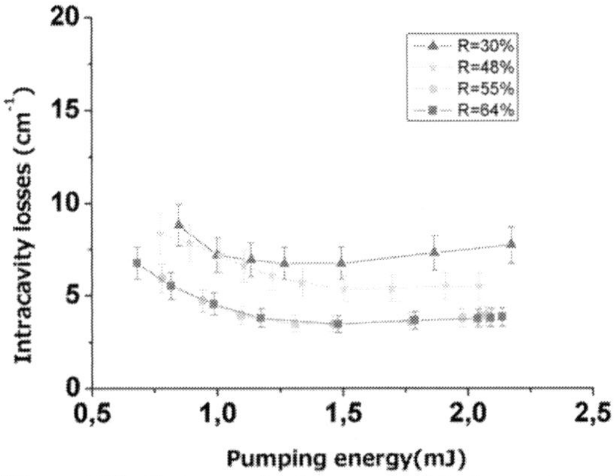

Figure 3b. Intracavity losses $\rho(E_{\text{н}})$ dependence on pumping energy for **$LiY_{0,3}Lu_{0,7}F_4$:Ce^{3+}+Yb^{3+}** active medium with different reflection coefficients R of output mirror

Curves for the sample without ytterbium shows that increasing the pump energy curve (Figure 3a) is not the same as pump reduction curve, which also illustrates the existence of color centers creation determined dynamic processes in the sample. If sample was additionally activated by ions of Yb^{3+} dependence is not so pronounced, and the intracavity losses are independent of the conditions of the experiment, therefore, we were able to suppress spurious photodynamic processes through preparing crystal compound, a crystal-chemical approach.

As mentioned earlier, the crystal with ytterbium can be used to generate ultrashort pulses. The obtained relation shows that for small values of the pump energy, and, consequently, the value of lasing losing more, and as a result, the leading edge of the pulse laser generation will suffer heavy losses. The trailing edge will be shortened due to saturation of the gain medium $LiY_{0,3}Lu_{0,7}F_4$:Ce^{3+}+Yb^{3+}.

5. Conclusion

As a result, we show that the $LiY_{0,3}Lu_{0,7}F_4:Ce^{3+}$ and $LiY_{0,3}Lu_{0,7}F_4:Ce^{3+}+Yb^{3+}$ crystals are promising active media of UV spectral range with low-threshold lasing (30-90 mJ/cm^2).

Due to crystal-chemical approach (additional doping by ions Yb^{3+}) we obtained the effect of suppression of spurious photodynamic processes. Intracavity losses were reduced, and as a result, significantly reduced the lasing threshold.

The active medium $LiY_{0,3}Lu_{0,7}F_4$: Ce^{3+}+ Yb^{3+} combines properties of saturable amplifier and oscillator with small saturation energy, and promising to generate pulses of ultrashort duration in the UV spectrum.

This work was supported by RFBR grant.

References

[1] Semashko V V, Dubinskii M A, Abdulsabirov R Yu, Korableva S L, Naumov A K, Nizamutdinov A S and Zhuchkov M S 2001 Photodynamic nonlinear processes in UV solid-state active media and approaches to improving material laser performance *Proc. of SPIE.* **4766** 119–126.

[2] Le T, Schowalter S J, Rellergert W, Jeet J, Lin G, Yu N A, Hudson E R 2012 Low-threshold ultraviolet solid-state laser based on a Ce 3+ :LiCaAlF 6 crystal resonator *Optics Letters* **37** **23** 4961-63

[3] Nizamutdinov A S, Semashko V V, Naumov A K, Efimov V N, Korableva S L and Marisov M A 2010 On the distribution coefficient of Ce3+ ions in LiF–LuF3–YF3 solid-solution crystals *JETP Letters* **91** 21–23.

[4] Stepanov B I 1966 *Optical quantum generators calculation methods (in russian)* (Minsk: Nauka I Tekhnika) T1 484.

[5] Akhtyamov O R, Nizamutdinov A S, Semashko V V, Naumov A K and Korableva S L 2010 Intracavity losses investigation of LiCaAlF6:Ce3+ laser *Proceedings SPIE* **7994**, 79940I-1

Journal of Physics: Conference Series **461** (2013) 012030 doi:10.1088/1742-6596/461/1/012030

Turn-on delay of QD and QW laser diodes: What is the difference?

G. S. Sokolovskii,[1] V. V. Dudelev,[1] E. D. Kolykhalova,[1,2] A. G. Deryagin,[1] M. V. Maximov,[1] A. M. Nadtochiy,[1] V. I. Kuchinskii,[1] S. S. Mikhrin,[3] D. A. Livshits,[3] E. A. Viktorov,[4] and T. Erneux[4]

[1]Ioffe Physical-Technical Institute, St. Petersburg, Russia
[2]Saint Petersburg State Electrotechnical University (LETI), St. Petersburg, Russia
[3]Innolume GmbH, Dortmund, Germany
[4]Universite´ Libre de Bruxelles, Optique Nonline´aire The´orique, Bruxelles, Belgium

gs@mail.ioffe.ru

Abstract. Turn-on delay of laser diodes with quantum-sized active media is investigated both theoretically and experimentally. In this research we show the striking difference in turn-on delay of quantum dot and quantum well laser diodes: With quantum-well lasers turn on delay tends to zero in the limit of high pumping, while with quantum dot lasers turn-on delay has the non-vanishing component which is independent of pumping.

1. Introduction

Laser diodes (LDs) with quantum-sized active media such as quantum wells (QWs) and quantum dots (QDs) are widely used and intensively studied due to their efficiency, compactness and high flexibility of properties allowing for implementation of many extremely different application-oriented constructions. In recent years, much attention is paid to the pulsed operation of LDs allowing for orders of magnitude higher pumping densities [1,2]. However, the research of dynamical properties of laser diodes, especially those based on QDs, concentrate mostly on relaxation oscillations and related phenomena such as damping of relaxation oscillations [3–5]. Except the seminal work [6], little attention is paid to study of the turn-on behavior of LDs where the pump is quickly changed from below to above threshold that allows to deeply explore the dynamical response under various lasing conditions. In this work we study the difference between the turn-on of QW and QD LDs with special attention to the nonlinear and non-instantaneous capturing of the carriers into a quantum dot which is known to strongly affect the recovery of QD material [7], but remains unresolved except for recent publication [8,9].

2. Turn-on delay of QW LD

We first consider the turn on delay in QW LD. Theory is based on the simple rate equation system:

$$\begin{cases} \dfrac{dN}{dt} = \dfrac{j}{ed} - A(N - N_{tr})S - \dfrac{N}{\tau} \\ \dfrac{dS}{dt} = \Gamma A(N - N_{tr})S - \dfrac{S}{\tau_p}, \end{cases} \tag{1}$$

where N is the carrier concentration, S is the photon density in cavity, j is the current density, e is the electron charge, d is the active layer width, A is the linear gain coefficient, N_{tr} is the transparency concentration, τ is the carrier lifetime, τ_p is the photon lifetime in cavity, Γ is the optical confinement coefficient. It is convenient to normalize the rate equations (1) before solution. It is well known that for threshold concentration N_{th} one can write:

$$\Gamma A(N - N_{tr}) = \Gamma A(N - N_{th}) + \tau_P^{-1} \tag{2}$$

Substituting (2) into the system (1) we get:

$$\begin{cases} \dfrac{dN}{dt} = \dfrac{j - j_{th}}{ed} - \dfrac{N - N_{th}}{\tau} - A(N - N_{th})S - \dfrac{S}{\Gamma\tau_p} \\ \dfrac{dS}{dt} = \Gamma A(N - N_{th})S \end{cases} \tag{3}$$

where $j_{th} = edN_{th} / \tau$ is the threshold current density. If we denote $N - N_{th} = \delta N$, then:

$$\begin{cases} \dfrac{d\delta N}{dt} = \dfrac{j - j_{th}}{ed} - \dfrac{\delta N}{\tau} - AnS - \dfrac{S}{\Gamma\tau_p} \\ \dfrac{dS}{dt} = \Gamma AS\delta N \end{cases} \tag{4}$$

Now introduce new normalized variables $D = 1 + \Gamma A\tau_p\delta N$, $J = 1 + \tau\tau_p A\Gamma\dfrac{j - j_{th}}{ed}$, $\eta = \dfrac{\tau_p}{\tau} << 1$, $I = \tau AS$, $t_1 = \dfrac{t}{\tau_p}$ and get new normalized rate equations for intensity I and carrier quantity D [10]:

$$\begin{cases} \dfrac{dI}{dt_1} = I(-1 + D), \\ \dfrac{dD}{dt_1} = \eta(J_+ - D(1 + I)), \end{cases} \tag{5}$$

with boundary conditions

$$\begin{cases} D(0^+) = J_- \\ I(0^+) = I_0 << 1 \end{cases}. \tag{6}$$

With these normalized variables the threshold current $J_{th}=1$ and threshold carrier density $D_{th}=1$. On the front edge of the pump pulse current values are: $J(0^-)=J_-<1$ and $J(0^+)=J_+>1$. In what follows we will use t instead of t_1 in all formulae for simplicity. Steady state solution of (5) at fixed current J is:

$$\begin{cases} I = 0 \\ D = J \end{cases} \quad if \;\; J < 1, \quad D = I = J - 1 \;\; \text{If}\; J \geq 1 \tag{7}$$

Eq. (7) shows that LD turns on if the carrier density reaches threshold value. The turn on delay may be found from the second equation of the system (5) if we assume that intensity below threshold is zero $I=0$:

$$\frac{dD}{dt} = \eta(J_+ - D), \quad D(0) = J_- \tag{8}$$

Solution of eq. (8) is:

$$D(t) = J_+ - (J_+ - J_-)\exp(-\eta t), \tag{9}$$

and from condition $D(\Delta t) = D_{th} = 1$ we find turn-on delay:

$$\Delta t = \frac{1}{\eta}\ln\left(\frac{J_+ - J_-}{J_+ - 1}\right). \tag{10}$$

Which for non-normalized rate equations transforms to the well known formula:

$$\Delta t = \tau \ln\left(\frac{j_+ - j_-}{j_+ - j_{th}}\right). \tag{11}$$

At strong pump $j >> j_{th}$ we get a hyperbolic function for turn-on delay:

$$\Delta t \approx \tau \frac{j_{th}}{j_+}. \tag{12}$$

Formulae (10-12) were obtained in assumption that LD turns on when the carrier concentration D reaches the threshold value D_{th}. Another type of boundary condition is that laser turns on when the intensity reaches a reference value I_{ref}. This approach is more convenient for experimental research. Solving the system (5) with new boundary conditions for turn-on by substituting I=0 into the first equation we can transform the second to:

$$I = I_0 \exp\left(\frac{1}{\eta}F(s)\right), \tag{13}$$

where $s=\eta t$ is the normalized time and:

$$F(s) \equiv (J_+ - 1)s - (J_- - J_+)(\exp(-s) - 1). \tag{14}$$

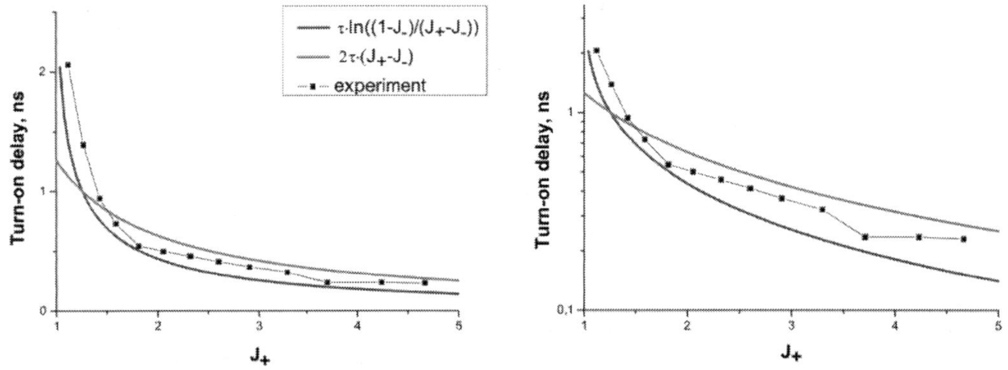

Figure 1. Turn-on delay of QW LD vs normalized current in linear (a) and log-scale (b): experiment (black rectangles), fit with eq. (10) (blue line), fit with eq. (19) (red line). Fitting values are: $J_-= 0$ and $\tau = 0.63$ ns.

We can find s_{on} from the condition $I(s_{on})=I_{ref}$:

$$F(s_{on}) = \eta \ln(I_{ref} / I_0) \tag{15}$$

with $\eta << 1$ and assuming logarithm is a slow function we can approximate:

$$F(s_{on}) \approx 0. \tag{16}$$

and in condition of strong pump $s << 1$ and $J_+ >> 1$, we simplify (14):

$$F(s) = -(1 - J_-)s + \frac{J_+}{2}s^2, \tag{17}$$

so that:

$$s_{on} \approx 2(1 - J_-)J_+^{-1}. \tag{18}$$

And now we can write expression for the turn-on delay:

$$\Delta t \approx 2\tau \frac{1-J_-}{J_+} \tag{19}$$

From here, it is easy to see that $\Delta t \to 0$ with $J_+ \to \infty$.

Experimental measurements of the turn-on delay were made with edge-emitting ridge stripe QW LD with generation wavelength 1060 nm. In our experiments we use pulses of 1 ns rise-time (measured at 10%–90% level) obtained from a high power (up to 1.5A current) pulse source. The laser output was detected using a high-speed pin detector with a cut-off frequency of 30 GHz and a 50 GHz digital oscilloscope. The turn-on delay was derived as the time difference between the rise-up of the pump current pulse measured at the laser diode and that of the signal from the photodetector. The optical and electrical path lengths were carefully estimated, and the difference between the two lengths was taken into account. Further details of the experimental technique can be found in Ref. [11]. Fig.1 shows experimental dependence of the turn-on delay time Δt on the normalized current J_+. It is clearly seen that experiment has a good agreement with (10) for low current, while for of high currents experimental dependence tends to curve corresponding to (19). In other words, the first approach for characterization of the turn-on delay is appropriate near the threshold while the second one is applicable at high pumping.

3. Turn-on delay of QD LD

In this part of the paper we will consider the turn-on delay of QD LD. The main difference between QW and QD LDs in this consideration is the nonlinear and non-instantaneous capturing of the carriers into a dot. It is known to strongly affect the recovery of QD material [7] and thus can have an impact on the laser turn-on delay [9]. In QD LDs, carriers are injected at first in to the wetting layer, and only after that they can be captured in QDs. Rate equations for this two-stage process are represented by three: the first is for the photon density in the cavity S, the second is for the QD filling probability ρ and the third is for the carrier density in the wetting layer n scaled to the two-dimensional QD density per layer [5,12-14]:

$$\begin{cases} \dfrac{dS}{dt} = -\dfrac{S}{\tau_p} + v_g g_0(2\rho - 1)S \\ \dfrac{d\rho}{dt} = \dfrac{\rho}{\tau_p} + \dfrac{2n(1-\rho)}{\tau_{cap}} - v_g g_0(2\rho-1)S\,, \\ \dfrac{dn}{dt} = \dfrac{n}{\tau} + \dfrac{j}{e} - \dfrac{2n(1-\rho)}{\tau_{cap}} \end{cases} \tag{20}$$

where τ_d is the carrier lifetime in dot, v_g is the group velocity, σ is the interaction cross-section of carriers with photons in a dot, g_0 is the differential gain, τ_{cap} is the carrier capture time in dot. Rate for the carrier capture in a dot is described by $2n(1-\rho)/\tau_{cap}$. It is proportional to the carrier density in the wetting layer n and to a probability for carrier to find a vacant dot. Factor 2 accounts for the spin degeneracy in quantum dot, while factor $1-\rho$ satisfies Pauli principle and describes nonlinear interaction between quantum dots and the wetting layer and constitutes the most important difference between QD and QW rate equations.

We normalize these rate equations similarly to the earlier QW case. New normalized variables are: $t_1 = t/\tau_p$, $I = v_g \sigma S \tau_d$, $\eta_d = \tau_p/\tau_d$, $\eta = \tau_p/\tau$ $g = \tau_p v_g g_0$, $B = \tau_d/\tau_{cap}$, $J = j\tau/q$ (pump current per dot). Substituting these variables in (20) we obtain the normalized rate equations:

$$\begin{cases} \dfrac{dI}{dt_1} = [-1 + g(2\rho - 1)]I \\ \dfrac{d\rho}{dt_1} = \eta_d[-\rho + Bn(1-\rho) - (2\rho - 1)I] \\ \dfrac{dn}{dt_1} = \eta[-n + J - 2Bn(1-\rho)] \end{cases} \tag{21}$$

In what follows we will use t instead of t_1 in all formulae for simplicity and assume that $\tau_d \approx \tau$ (hence $\eta_d \approx \eta$). The QD LD gain $g(2\rho$-1) is defined by the dot density and the normalized differential gain g. For typical value τ=1 ns and τ_{cap}=10 ps we take factor $B\approx100$. The normalized threshold current of QD LD is J_{th}=2 (for QW LD J_{th}=1). This follows from the spin degeneracy in QD energy levels. With J_-<2, laser operates in stable OFF steady state, so that:

$$\begin{cases} I = 0 \\ \rho = \rho_0 = J_- / 2 + O(B^{-1}) \\ n = n_0 = B^{-1}J / (2 - J_-) + O(B^{-2}) \end{cases} \tag{22}$$

The boundary conditions are:

$$\begin{cases} \rho(0^-) = \rho_0 \\ n(0^-) = n_0 \\ I(0^-) = I_0 << 1 \end{cases} \tag{23}$$

Turn-on delay Δt in our consideration is the lag between the momentum of time t=0$^+$, when current simultaneously becomes J_+>2, to the momentum of time when I=I_{ref}. Assuming the intensity before laser turn-on is very small we substitute J=J_+ and I=0 in the second and the third equations (21) and obtain differential equation of the second order for ρ:

$$\frac{d\rho}{ds} = B\,[(z_0 - J_+)\exp(-s) + J_+ - 2\rho](1-\rho) - \rho, \tag{24}$$

where $s = \eta t$ and $z_0 \equiv 2\rho_0 + n_0 = J_- + O(B^{-1})$. Solution of the first equation in (21) takes the form:

$$I = I_0 \exp\left(\frac{1}{\eta}G(s)\right), \tag{25}$$

where $G(s)$ is defined by:

$$G(s) \equiv -(1+g)s + 2g\int_0^s \rho(u)du. \tag{26}$$

Now we can determine numerically Δt by substituting solution for eq. (24) in eq. (26) and then substituting result in eq. (25) and finding s for I=I_{ref}. Analytically we will investigate turn-on delay in some limit case similar to QW lasers. Namely, we will consider the limit of large $B\to\infty$, and the limit of large $J_+\to\infty$. In the limit of $B\to\infty$ and 1-ρ>>B^{-1} (24) simplifies to Bernoulli equation for 1-ρ:

$$\frac{d\rho}{ds} = B[(J_- - J_+)\exp(-s) + J_+ - 2\rho](1-\rho). \tag{27}$$

which has analytical solution for ρ:

$$\rho = 1 - \frac{\exp(-BF(s))}{(1-\rho_0)^{-1} + 2B\int_0^s \exp(-BF(u))du}, \tag{28}$$

with $F(s)$ defined by:

$$F(s) \equiv (J_+ - J_-)(\exp(-s) - 1) + (J_+ - 2)s. \tag{29}$$

In the limit of $J_+\to\infty$ the approximate expression for $F(s)$ will be very useful. This approximation can be clearly seen from simplification of (29) for large J_+:

$$F(s) \approx J_+[(\exp(-s)-1)+s] \tag{30}$$

Consideration of (28) shows that it has initial level and saturates at level $\rho=\rho_s\equiv 1+O(B^{-1})$. For $\rho\approx 1$ expression (26) simplifies to:

$$G=(g-1)s \tag{31}$$

Substituting (31) in (25) we obtain:

$$I \approx I_0 \exp\left(\frac{1}{\eta}(g-1)s\right). \tag{32}$$

Then the minimum turn-on delay for QD LD is:

$$\Delta t = \frac{\tau_p}{\tau_p v_g g_0 - 1}\ln\left(\frac{I_{ref}}{I_0}\right). \tag{33}$$

This formula shows the striking difference in turn-on delay of QD and QW LDs. With quantum-well lasers turn-on delay is zero in the limit of $J_+\to\infty$. With QD lasers the situation is very different as the turn-on delay has the independent of pumping non-vanishing component given by (33).

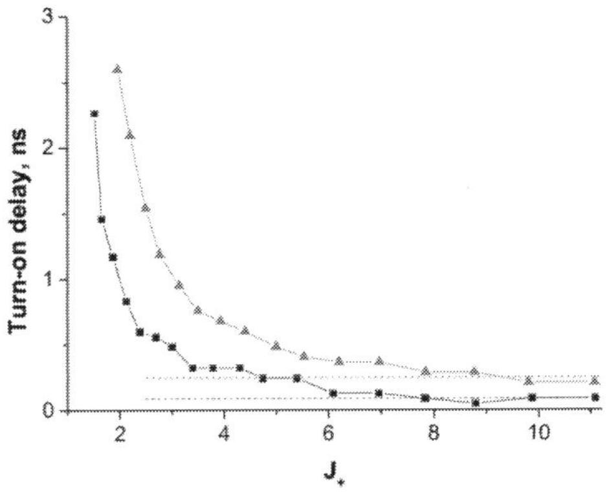

Figure 2. Experimental results for turn-on of QD LDs with 1 QD layer (red triangles) and 5 QD layers (black rectangles).

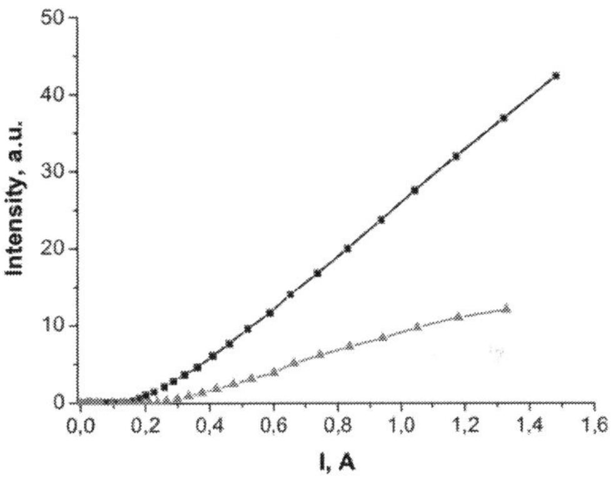

Figure 3. Light-current curves of lasers with 1 QD layer (red triangles) and 5 QD layers (black rectangles).

In our experiments we investigate lasers with 1 and 5 QD layers. Growth and basic properties of similar QD laser structures have been described elsewhere [15]. Length of all lasers was 4.5 mm and the generation line was at 1290 nm (corresponding to the ground state lasing). Experimental technique was the same as with QW LDs and results of measurements are shown in Fig. 2. *L-I* curves for these lasers are shown in Fig. 3. It is seen from fig. 2 that despite of some scattering, measurements with QD LDs clearly indicate the quasi-stable pump-independent value for higher currents. The difference of the values of the minimum turn-on delay for QD LDs with 1 and 5 QD layers is clear if the differential gain of these lasers is taken into account (see expression (33)). Difference of gain constants g_0 for these lasers is seen from almost two-fold difference of their thresholds in Fig. 3. Taking into account the input of the transparency current to the laser threshold, the three-fold difference of the minimum turn-on delay values is in a good agreement with the theoretical expression (33) for the minimum turn-on delay of QD LD.

4. Summary

Theoretical and experimental investigation of the turn-on delay of quantum well and quantum dot laser diodes is presented. We show that in contrast to the quantum wells, the nonlinear and non-instantaneous capturing of the carriers into a dot leads to the minimum turn-on delay of QD LDs being non-zero at any pumping current. The value of this non-vanishing turn-on delay depends strongly on the gain factor.

Acknowledgments

This research was supported by the Russian Ministry of Education and Science (Project 8521). E.Viktorov and T.Erneux are grateful to the Fond National de la Recherche Scientifique (Belgium) for support. The research by T. Erneux was also supported by the Air Force Office of Scientific Research (AFOSR) Grant FA8655-09-1-3068.

References

[1] G.S.Sokolovskii, D.A.Vinokurov, A.G.Deryagin, V.V.Dudelev, V.I.Kuchinskii, S.N.Losev, A.V.Lyutetskiy, N.A.Pikhtin, S.O.Slipchenko, Z.N.Sokolova, I.S.Tarasov, 2008 Tech. Phys. Lett. vol 34 p 708
[2] G. S. Sokolovskii, V. V. Dudelev, A. G. Deryagin, V. I. Kuchinskii 2012 Tech. Phys. Lett. vol 38 p 613
[3] K. Ludge, E. Scholl, E. A. Viktorov, and T. Erneux, J. 2011 Appl. Phys. vol 109 p 103112
[4] K. Ludge, M. J. P. Bormann, E. Malicr, Ph. Hovel, M. Kuntz, D. Bimberg, A. Knorr, and E. Scholl 2008 Phys. Rev. B vol. 78 p 035316
[5] T. Erneux, E. A. Viktorov, and P. Mandel 2007 Phys. Rev. A vol 76 p 023819
[6] M. Grundmann 2000 Appl. Phys. Lett. vol 77 p 4265
[7] T. Erneux, E. A. Viktorov, P. Mandel, T. Piwonski, G. Huyet, and J. Houlihan 2009 Appl. Phys. Lett. vol 94 p 113501
[8] K. Ludge and E. Scholl 2009 IEEE J. Quantum Electron vol 45 p 1396
[9] G. S. Sokolovskii, V. V. Dudelev, E. D. Kolykhalova, A. G. Deryagin, M. V. Maximov, A. M. Nadtochiy, V. I. Kuchinskii, S. S. Mikhrin, D. A. Livshits, E. A. Viktorov and T. Erneux 2012 Applied Phys. Lett. Vol 100 p 081109
[10] T. Erneux and P. Glorieux, Laser Dynamics 2010 Cambridge University Press
[11] G. S. Sokolovskii, A. G. Deryagin, and V. I. Kuchinskii 1996 Tech. Phys. Lett. vol 22 p 286
[12] M. Sugawara, K. Mukai, and H. Shoji 1997 Appl. Phys. Lett. vol 71 p 2791
[13] A. V. Uskov, Y. Boucher, J. Le Bihan, and J. McInerney 1998 Appl.Phys. Lett. vol 73 p 1499
[14] M. Ishida, N. Hatori, T. Akiyama, K. Otsubo, Y. Nakata, H.Ebe, M. Sugawara, and Y. Arakawa 2004 Appl. Phys. Lett. vol 85 p 4145
[15] M. V. Maximov, L. V. Asryan, Yu. M. Shernyakov, A. F. Tsatsul'nikov, I.N. Kaiander, V.V. Nikolaev, A. R. Kovsh, S. S. Mikhrin, V. M. Ustinov, A. E. Zhukov, Zh. I. Alferov, N.N. Ledentsov, and D. Bimberg 2001 IEEE J. Quantum Electron vol 37 p 676

Journal of Physics: Conference Series **461** (2013) 012031 doi:10.1088/1742-6596/461/1/012031

EPR method for investigation and diagnosis materials promising for quantum electronics ($PbGa_2S_4$, ZnSe)

D D Kramushchenko, H R Asatryan, V A Khramtsov, P G Baranov, I V Ilyin, V A Soltamov
Ioffe Physical-Technical Institute of the Russian Academy of Sciences, Politekhnicheskaya ul.26, St. Peterburg, 194021 Russia

E-mail: dasha.kramushchenko@mail.ioffe.ru

Abstract. Electron paramagnetic resonance (EPR) spectra of rare-earth Dy^{3+} and Ce^{3+} ions in lead thiogallate $PbGa_2S_4$ single crystals were discovered and investigated. EPR spectra were studied in X-band in the temperature range 4-30 K. Parameters of spin Hamiltonian of EPR spectra of Dy^{3+} and Ce^{3+} ions in $PbGa_2S_4$ single crystals were calculated. It was set that dysprosium and cerium ions replace lead ions in crystal lattice. Also we carried out a diagnosis of ZnSe crystals doped with different rare-earth and transition metal ions. It was shown that due to diffusion of transition elements of Cr^{3+}, Co^{2+} or Fe^{3+} along with the working admixture in the diffusion layer the EPR spectra of related admixtures are detected. The EPR diagnosis of these admixtures can be used in the development of modes, allowing minimizing the concentration of related admixtures, worsening the performance of laser materials. It was found that diffusion of Cr, Co and Fe transition metal ions leads to admixture of Mn ions. The EPR signals of Mn and Co are proposed as the ideal markers to control the crystal structure of the diffusion active layer.

1. Introduction

Dysprosium- and cerium-doped lead thiogallate single crystals are known to be promising active media for the lasers of mid-infrared range (3-5 μm). This range corresponds to so-called transparency window of the Earth's atmosphere, that leads to small losses of radiation under its propagation in the atmosphere and is very important for scientific and modern technological applications. Investigation of the optical and laser properties of $PbGa_2S_4$ single crystals activated by rare-earth ions is a subject of a number of papers [1, 2]. Nevertheless, there are no papers on the electron paramagnetic resonance (EPR) studies of $PbGa_2S_4$ crystals.

Materials based on ZnSe: Cr^{2+} are known to be the very efficient for producing compact wide continuously tunable lasers in the wavelength range of 2 - 3.4 μm at room temperature. ZnSe: Cr are used as passive shutter of holmium laser (wavelength of about 2 μm) also. ZnSe:Fe crystals are 3-5 μm tunable solid-state sources of coherent light. Using direct optical pumping at 2.698 μm ($Y_3Al_5O_{12}$:Er^{3+}) these crystals can be used as passive shutters. ZnSe:Co crystals are attractive candidates for a good saturable absorber for 1.5-μm eyesafe lasers. Thus, zinc selenide crystals doped with different ions are attractive for such applications as remote sensing, environment monitoring, mid-IR free space communications, as well as optical coherence tomography, ophthalmology and dermatology in medicine, etc.

One of the most direct techniques of the investigation of the spectroscopic properties of materials for quantum electronics is the EPR method. The EPR allows one to determine the spin and charge state of an impurity, to find the local symmetry of the impurity center and the composition of its nearest environment, the structure of the energy levels, the specificity of interaction with the crystal lattice, the presence of structural defects in the vicinity of the impurity, etc.

2. Experiment and discussion

2.1. $PbGa_2S_4$:Dy

EPR spectra of Dy^{3+} and Ce^{3+} ions in lead thiogallate ($PbGa_2S_4$) single crystals were studied in the X-band in the temperature range 4 -30K on JEOL standard radio spectrometer. The crystals were synthesized by Bridgeman-Stokbarger technique.

The observation of the EPR spectra of $PbGa_2S_4:Dy^{3+}$ in an arbitrary orientation of the external magnetic field is troublesome due to a number of specific features (anisotropy, low intensity). The EPR spectrum of $PbGa_2S_4:Dy^{3+}$ observed in the low magnetic field H || [100] at T = 4K is shown in Fig.1.

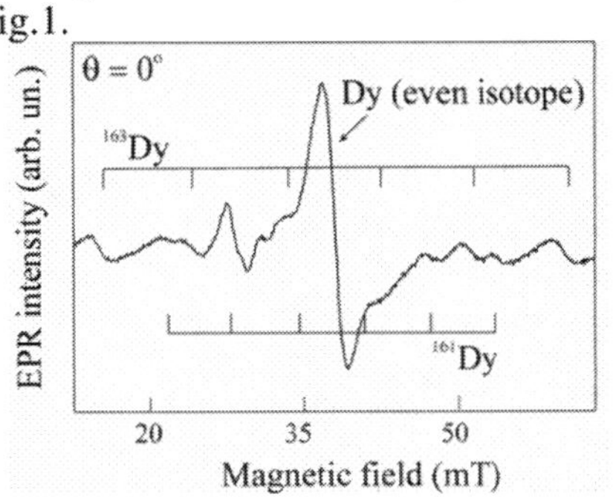

Figure 1. EPR spectrum of Dy^{3+} ions in $PbGa_2S_4$ single crystals in H || [100] orientation and T = 4K.

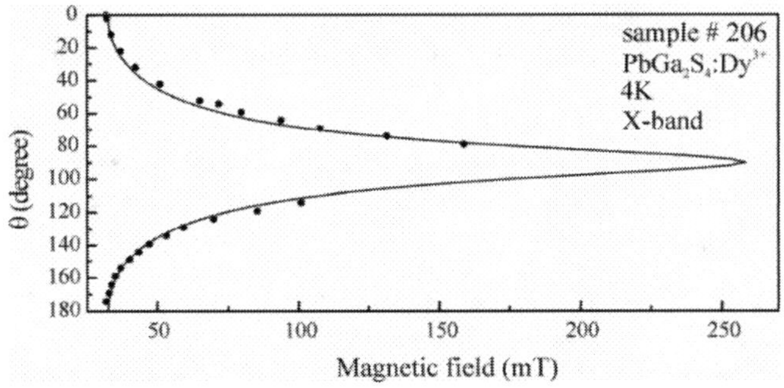

Figure 2. Calculated (curve) and experimental (circles) angular dependences of the resonance magnetic field of Dy^{3+} ions in lead thiogallate in the (010) plane.

On the basis of a detailed study of the hyperfine structure of observed EPR lines in the lead thiogallate crystals activated by dysprosium we can unambiguously state that the spectra belong to Dy^{3+} ions. Spectra consist of the intense line and two groups of low-intensity lines, each having six equidistant lines. It is worth noting, natural dysprosium has even isotope with zero nuclear spin and two odd isotopes, each having nuclear spin of five halves. The natural abundance of these isotopes (^{161}Dy and ^{163}Dy) is equal to 19.0 and 24.9 %, respectively, and the ratio of the nuclear magnetic moments ^{161}Dy/^{163}Dy is equal to 1.41. Experimental spectrum looks exactly as one would expect for dysprosium. The two groups of six lines each are the components of the hyperfine structure for the odd isotopes and the intense central line belongs to the even isotope. The ratio of the distances between the lines of the hyperfine structure in each group corresponds to the ratio of the nuclear magnetic moments of the odd isotopes, and the ratios of intensities of the lines in the spectrum correspond to the natural abundances of the isotopes.

As it could be seen from Fig. 1, a low-intensity signal is observed at the low-field side of the EPR spectrum also. This signal is split into two components at certain orientation of the magnetic field. According to the analysis, this line does not belong to dysprosium ion.

Paramagnetic Dy^{3+} centers have axial symmetry in the (001) crystal plane, i.e. the positions of the EPR lines remain unchanged under a change in the angle θ between the external magnetic field and the directions [100] and [010] within this plane. Fig. 2 shows the experimental (circles) and calculated (curve) values of the resonance magnetic field corresponding to the intense line position of the even dysprosium isotope in the (010) plane at a temperature of 4K. These angular dependences are described by an axially symmetric spin Hamiltonian

$$H = \beta \mathbf{HgS} + \mathbf{SAI}, \quad (1)$$

where the effective spin is S =1/2, **g** and **A** are the g-tensor and hyperfine interaction tensor, respectively.

The following values of the g-tensor were found from the experimental angular dependences: $g_{\parallel} = 15.06$ and $g_{\perp} = 2.47$. The local magnetic axis direction of the Dy^{3+} ion in lead thiogallate with respect to the crystallographic axes of the lattice coincides with the [001] direction. It should be mentioned that the above values of the g-factors are typical for dysprosium ions in various crystals previously studied by EPR [3].

The hyperfine structure of Dy^{3+} ions is also anisotropic. The hyperfine components are shifted toward higher magnetic fields and the distances between them are changed with a change in the orientation of the magnetic field with respect to the crystal.

The calculations yield the following values of the hyperfine constants $A_{\parallel}$ and $A_{\perp}$ for two odd dysprosium isotopes in lead thiogallate: $A_{\parallel} = 675 \times 10^{-4}$ cm^{-1} and $A_{\perp} = 111 \times 10^{-4}$ cm^{-1} for ^{163}Dy and $A_{\parallel} = 472 \times 10^{-4}$ cm^{-1} and $A_{\perp} = 77 \times 10^{-4}$ cm^{-1} for ^{161}Dy, respectively.

We come to the conclusion that the EPR spectra of dysprosium-doped lead thiogallate single crystals belong to Dy^{3+} ions localized in the Pb^{2+} sites. Only one type of the paramagnetic Dy^{3+} center was observed in the experiments, despite that there are three structurally non-equivalent positions (Pb1, Pb2, Pb3) of the lead ions in the $PbGa_2S_4$ lattice. These positions differ in anionic polyhedra of the nearest environment (in symmetry and distances to the nearest S^{2-} ions) [1, 4]. Apparently, only the Pb3 type of the lead ions corresponding to the nearest environment with the C_2 local symmetry is substituted at a low dysprosium concentration (0.05 at. %). The Pb1 and Pb2 centers have the local D_2 symmetry. Thus, the axial symmetry of the EPR spectra should not be observed in the case of substitution of these sites by paramagnetic ions. Heterovalent substitution of the ions Pb^{2+} by Dy^{3+} ion in the lead thiogallate crystal lattice requires a charge compensation, that is non-local at dysprosium concentration of 0.05 at. % in our sample.

2.2. *PbGa₂S₄:Ce*

At low temperatures, Ce^{3+}-doped $PbGa_2S_4$ single crystals exhibit an intense multi-component (more than 40 lines) EPR spectrum. The spectrum is assigned to cerium ions (g-factors are typical for cerium ions in various crystals). Angular dependences of the group of lines under the rotation in the (100) plane have been obtained. At the small angle θ between the external magnetic field and the crystallographic direction [010], four intense lines and a number of low-intensity EPR lines grouped nearby are observed. The spectrum becomes considerably more complicated with an increase in the angle ($\theta > 20^0$). The number of spectral components is increased, their intensity ratio is changed, that complicates unambiguous analysis of each separate line.

The intense and weak lines exhibit similar temperature dependences but nevertheless have different spin-lattice relaxation times. The weak lines broaden faster and their intensity decreases more abruptly with an increase of the temperature; three main centers can be distinguished. The experimental (symbols) and the theoretical (curves) resonant magnetic field angular dependences of these lines in the (100) plane at T = 26K are shown in Fig. 3. The same angular dependences of the most intense Ce^{3+} centers in lead thiogallate in the (001) plane at a temperature of 4K are shown in Fig. 4. It should be mentioned that one curve is split into two curves at the angle θ closed to 90^0 in the (001) plane. The angular dependences of all lines can be described by spin Hamiltonian

$$H = \beta \mathbf{HgS}, \quad (2)$$

where the effective spin is S =1/2.

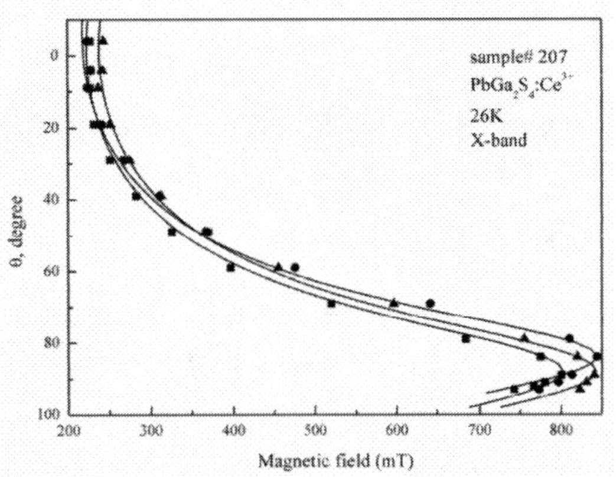

Figure 3. Experimental (symbols) and theoretical (curves) angular dependences of the intense Ce^{3+} centers in lead thiogallate at T = 26K in the (100) plane.

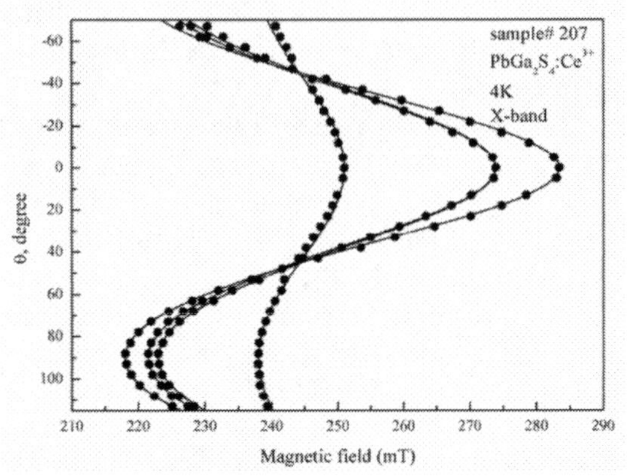

Figure 4. Experimental (dots) and theoretical (curves) angular dependences of the most intense Ce^{3+} centers in lead thiogallate at T = 4K in the (001) plane.

From the experimental angular dependences we obtained principal values of the g-tensor for the above four intense lines of cerium also (table 1). The local principal magnetic axes directions of two centers of Ce^{3+} ions in lead thiogallate with respect to the lattice crystallographic axes in the (100) plane coincide with the [001] axis. The same axis of third Ce^{3+} center deviates from the [001] axis by 4^0.

Table 1. The g-factors experimental values of four intense EPR lines of Ce^{3+} ions in $PbGa_2S_4$.

Ce center	g_x	g_y	g_z
Ce1	2.43	2.98	0.83
Ce2	2.65	2.79	0.79
Ce3	2.34	3.05	0.79
Ce4	2.43	3.00	0.83

Thus, the investigation of the angular and temperature dependences of the cerium-doped lead thiogallate single crystals EPR spectra lead to the conclusion that we observe the spectra of Ce^{3+} ions localized in the sites of Pb^{2+} in the crystal lattice.

In the case of Dy^{3+} we observed one type of paramagnetic centers despite the presence of three structurally non-equivalent positions Pb1, Pb2, Pb3 of lead ions in $PbGa_2S_4$ lattice. However, in the case of cerium, we observe much more various paramagnetic centers. Natural cerium does not have odd isotopes; therefore, the EPR spectra of these ions do not exhibit a magnetic hyperfine structure. Consequently, all observed lines are due to magnetically non-equivalent positions of Ce^{3+} in the lead thiogallate lattice.

In our opinion, a lot of lines are due to the distortion of the nearest environment of the paramagnetic center by a defect localized in the vicinity of Ce^{3+}. Heterovalent substitution of Ce^{3+} ions for Pb^{2+} ions requires compensation of the excessive positive charge. Local compensation defects can be formed by lead or gallium vacancies in the cationic environment of Ce^{3+}. The nearest environment of Pb^{2+} ions in lead thiogallate consist of 8 sulfur ions located at different distances from the center so that they form three structurally non-equivalent tetragonal antiprisms. The local symmetry of these polyhedra in the Pb3 position is C_2, and for Pb1 and Pb2 are D_2. In addition, the Pb3 site has two magnetically non-equivalent positions, whereas Pb1 and Pb2 have one per each site. The four detected intense EPR lines agree qualitatively with this picture. The next-nearest cationic shell of a Pb^{2+} ion consist of 8 lead ions and the next (gallium) shell of the lead ion is composed of 12 ions. In the case of the formation of vacancies in a lead or gallium site with all possible configurations of their positions in the polyhedra, the multiplicity of non-equivalent paramagnetic centers greatly increases. More than 40 non-equivalent positions can occur under formation of a vacancy at the sites of Pb^{2+} or Ga^{3+}. Thus, the local compensation by lead or gallium vacancies is seemingly the reason of the observed a lot of the EPR lines of Ce^{3+}. Cerium doping of lead thiogallate results in the substitution of Ce^{3+} for lead sites. In this case, some Ce^{3+} ions exhibit non-local charge compensation (intense lines), whereas the others exhibit local compensation by a lead or gallium vacancy in the nearest environment of the paramagnetic center. The observation of a lot of paramagnetic Ce^{3+} centers in $PbGa_2S_4$ also indicates the substitution for exactly the lead sites. If Ce^{3+} ions were to be localized at gallium sites, no defects were required to compensate the excessive positive charge and we observed only four EPR lines due to magnetically non-equivalent positions. In addition, the probability of the substitution of Ce^{3+} ions for Ga^{3+} ions is very small owing to a considerable difference in their ionic radii.

2.3. ZnSe

We have carried out a diagnosis of ZnSe crystals doped with different rare-earth and transition metal ions, namely ZnSe: Cr, Co, Fe, Er. The crystals are doped with transition metal ions using diffusion method or during growth process. In the case of diffusion, the impurity centers are concentrated in the near-surface region that is involved in the diffusion process. The impurity concentration in this case can be an order of magnitude greater than that obtained using the doping in the course of the ZnSe crystal growth.

At first, we consider chromium doped zinc selenide single crystals. Fig. 5 shows angular dependence of ZnSe:Cr EPR spectra detected at 9.3 GHz and T = 4K under the magnetic field rotation in the plane (110). Thick green lines are EPR spectra belonging to Cr^{2+} centers of practical interest. Fig. 6 shows the experimental (green circles) and calculated (green curves) values of the resonance magnetic field corresponding to the chromium ion. Other symbols designate the most intense EPR lines other than Cr^{2+} ions.

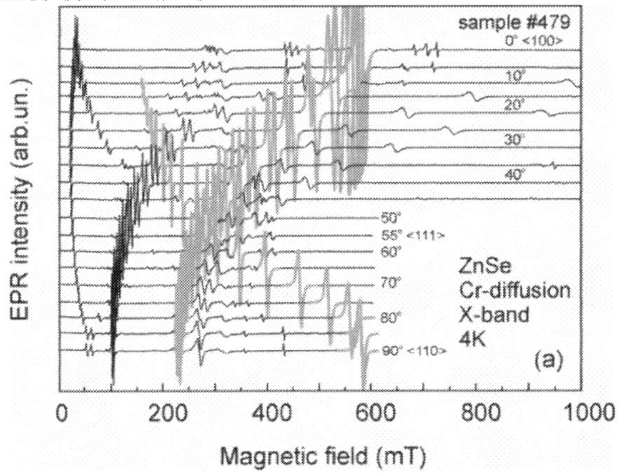

Figure 5. Angular dependence of ZnSe:Cr EPR spectra in the (110) plane at T = 4K.

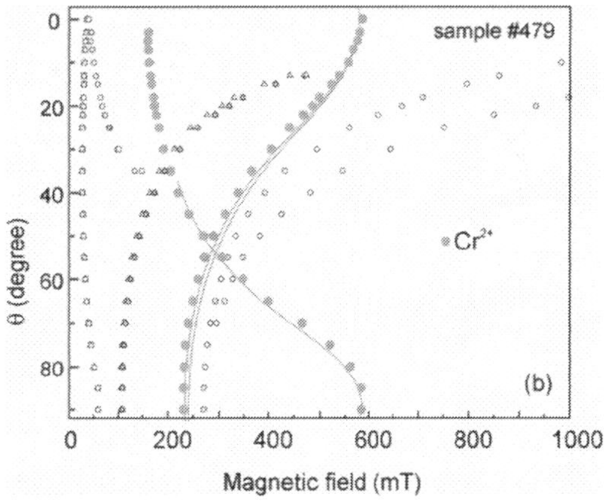

Figure 6. Experimental (green circles) and theoretical (green curves) angular dependences of the Cr^{2+} centers.

Along with chromium in the ZnSe:Cr crystal, anisotropic EPR spectra of iron, which, apparently, partly associated with other impurities, are detected. The signals except for chromium are undesirable, and the aim of further research is to minimize the related admixture content in laser materials. EPR diagnosis can be useful for solving this problem.

The study of cobalt doped zinc selenide crystals revealed the Co^{2+} and Fe^{3+} ions in a cubic environment. Co^{2+} fine structure is observed at a rotation in the (110) plane. It is worth noting, the presence of these lines in the EPR spectrum reflects the high quality of the crystal. The hyperfine interaction with the four nearest Se nuclei in the first coordination sphere has also been observed.

Journal of Physics: Conference Series **461** (2013) 012031 doi:10.1088/1742-6596/461/1/012031

In contrast to the crystals in which cobalt has been introduced during the growth process, the EPR spectra of Co^{2+} centers in the diffusion layer are by an order of magnitude more intense that indicates a high concentration of cobalt introduced by diffusion method. However, with the introduction of cobalt by diffusion along with cobalt ions of practical interest, defects worsening the performance of laser materials are generated.

In the case of ZnSe:Fe, EPR spectra of Co^{2+} and Mn^{2+} ions in a cubic environment are detected. The spectra of Fe^{2+} ions are not observed. Nevertheless, since Fe^{2+} ions are optically detected, it is important to obtain an information about the quality of the diffusion layer on the surface of the crystal. Since Co^{2+} and Mn^{2+} ions are embedded into the lattice during diffusion process as uncontrolled impurities simultaneously with iron, the EPR spectra of these ions are proposed as ideal markers to control the diffusion layer. The presence of relatively narrow lines of the manganese fine structure indicates the high quality of the crystal in the diffusion layer.

Erbium doping of the ZnSe crystals is of particular interest because Er^{3+} ions have infrared laser luminescence band. Based on detailed study of the observed EPR hyperfine structure lines we can unambiguously state that spectra belong to Er^{3+} ions. Erbium spectrum has a characteristic hyperfine structure consisting of 8 lines caused the hyperfine interaction with the nuclear magnetic moment of ^{167}Er isotope, having nuclear spin of seven halves. The natural abundance of ^{167}Er is equal to 22.9 %. Mn^{2+} ions are embedded into the lattice during diffusion also. As the manganese ions are in approximately the same diffusion layer that erbium ions, the EPR spectra can be used as markers for the diagnosis of the crystal quality. The presence of narrow and intense fine structure components in the EPR spectra of Mn^{2+} indicates good crystalline properties of the diffusion layer.

3. Conclusion

EPR spectra of Dy^{3+} and Ce^{3+} ions in $PbGa_2S_4$ single crystals were discovered and investigated. It was state that dysprosium and cerium ions replace lead ions in crystal lattice.

There is only one type of Dy^{3+} paramagnetic center (0.05 at. %.) localized at the Pb^{2+} position with the local symmetry C_2. In the case of $PbGa_2S_4$:Ce^{3+} a lot of paramagnetic centers were observed. Some Ce^{3+} ions exhibit non-local charge compensation, whereas the others exhibit local compensation by a lead or gallium vacancy in the nearest environment of the paramagnetic center. Parameters of spin Hamiltonian of EPR spectra of Dy^{3+} and Ce^{3+} ions in $PbGa_2S_4$ single crystals were calculated.

In ZnSe crystals investigated along with the working admixture in the diffusion layer EPR spectra of related admixtures worsening the performance of laser materials were detected. Introducing into the lattice during diffusion of transition metal ions of associated admixture of Mn was found. These spectra are proposed as the ideal markers for the control crystal structure of the diffusion active layer.

Acknowledgments

This work has been supported by the Programs of the Russian Academy of Sciences "Spin Phenomena in Solid State Nanostructures and Spintronics", "Fundamentals of Nanostructure and Nanomaterial Technologies" and by the Russian Foundation for Basic Research under Grant # 12-02-01011.

References

[1] Badikov V, Badikov D, Doroshenko M, Panyutin V, Chizhikov V, and Shevirdyaeva G 2008, *Opt. Mater.* **31** 184

[2] Badikov D V 2008, *Extended Abstract of Candidate's Dissertation* (Kuban State University, Krasnodar, Russia)

[3] Vlasova M V, Kakazei N G, Kalinichenko A M, and Litovchenko A S 1987, *Radiospectroscopic Properties of Inorganic Materials* (Naukova Dumka, Kiev)

[4] Kamenshchikov V N, Stefanovich V A, Gad'mashi Z P, Sidei V I, and Suslikov L M 2007, *Phys. Solid State* **49** 351

Journal of Physics: Conference Series **461** (2013) 012032 doi:10.1088/1742-6596/461/1/012032

ODMR study of ZnO single crystals containing iron impurity ions

A S Gurin[1,3], **N G Romanov**[1], **D O Tolmachev**[1], **P G Baranov**[1], **C Morhain**[2]

[1] A. F. Ioffe Physical-Technical Institute, RAS, 194021 St. Petersburg
[2] CRHEA-CNRS, Rue Bernard Grégory, F-06560 Valbonne, France

[3] E-mail: sasha.gurin@mail.ioffe.ru

Abstract. Optically detected magnetic resonance (ODMR) was studied in ZnO single crystals containing iron impurity ions. ODMR spectra of the ground state of isolated Fe^{3+} on the Zn^{2+} lattice sites were obtained by monitoring the intensity of the intra-centre Fe^{3+} luminescence together with the ODMR signals of shallow donors. The latter observation suggests that the Fe^{3+} luminescence may actually be excited due to energy transfer from the recombining donor-acceptor pairs to the impurity ions. This conclusion is supported by the presence of Fe^{3+} emission in the afterglow spectra recorded after switching off band-to band excitation light.

1. Experimental

ZnO is a very promising material for semiconductor device applications [1]. It has a direct and wide band gap in the near-UV spectral region and a large free-exciton binding energy so that exciton emission processes can persist at or even above room temperature. The availability of large single crystals is a big advantage of ZnO over GaN and allows using ZnO single crystals as substrates for MBE grown layers and nanostructures. The epitaxy of ZnO films on native substrates can result in ZnO layers with reduced concentration of extended defects and, consequently, better performance in electronic and photonic devices.

The inevitable presence of transition metals even in high-quality II-VI semiconductors is well known. A quite-common accidental contamination forming deep traps is Fe. On the other hand ZnO:Fe is considered as the interesting candidate for quantum computing applications.

The most successful experimental technique for identifying and studying the defects has often proven to be electron paramagnetic resonance (EPR) [2]. Optically detected magnetic resonance (ODMR) [3, 4] provides a huge increase in sensitivity (up to a single quantum object [5]) and can be applied for a study of spin-dependent recombination and energy transfer processes. EPR and luminescence of transition metal ions in ZnO were widely studied [6, 7] but, to our knowledge, no ODMR results were reported so far with the exception of ref. [8], where ODMR of V^{2+} was observed.

In the present paper, we report preliminary results of a study using optical detection of electron paramagnetic resonance in the photoluminescence of ZnO single crystals containing iron impurity ions.

[3] To whom any correspondence should be addressed.

2. Experimental

The samples were cut from M-plane ZnO substrates grown by hydrothermal method. Such substrates are used for MBE growth of non-polar ZnO layers and nanostructures. Photoluminescence (PL) was excited with a deuterium arc lamp with appropriate glass filters or a 405 nm semiconductor laser and detected with a grating monochromator and a PM tube. EPR was studied with a commercial X-band spectrometer. The 35 GHz ODMR spectra were recorded at 1.8 K via PL intensity either using on-off modulation of microwaves and lock-in amplification or without modulation. The samples were placed in the centre of a cylindrical H_{011} microwave cavity which had holes for excitation and emission light. The sample could be rotated about vertical axis, which allowed studying angular dependences of the spectra.

3. Results and discussion

The PL spectra of ZnO crystals under 405 nm excitation are shown in figure 1(a). The upper spectrum was recorded with low spectral resolution, which was used in most of the ODMR experiments. The high resolution luminescence spectrum is shown below. One can see that in addition to the broad band with a maximum around 600 nm, a richly structured luminescence in the spectral range of 680-800 nm is observed. In Ref. [9] this structured spectrum was unambiguously assigned to the $^4T_1(G)$- $^6A_1(S)$ transition of isolated Fe^{3+} ions on Zn^{2+} lattice sites by means of emission, excitation, and magneto-optical spectroscopy. The EPR characterization of the samples confirmed the presence of Fe^{3+} impurity ions in the sample along with traces of Mn^{2+} and Co^{2+} in lower concentration.

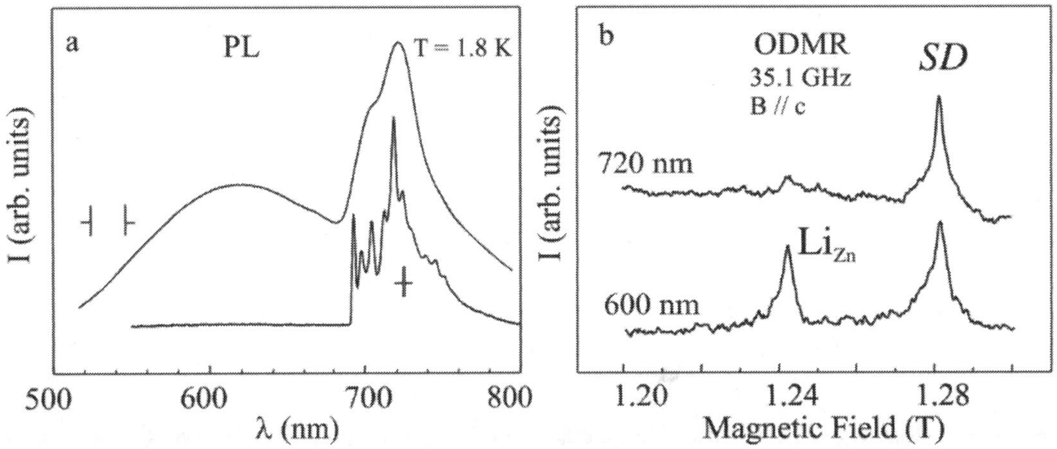

Figure 1. (a) Photoluminescence spectra recorded at 1.8 K in ZnO crystal with high spectral resolution (lower curve) and low spectral resolution used to study ODMR (upper curve). (b) Q-band ODMR spectra recorded via PL intensity without modulation of microwaves at 600 nm (lower curve) and 720 nm (upper curve). B // c.

In the ODMR spectra recorded via the intensity of the broad PL band around 620 nm the well known ODMR spectrum of shallow donors (*SD*) with *g*-factor about 1.96 [10] and lithium acceptors Li_{Zn} [11] were observed. The spectral dependence of both donor and acceptor ODMR signals obtained with microwave modulation at audio frequency and lock-in amplification is close to the broad PL band, which proves that in our samples PL in this range is dominated by the recombination *SD*-Li_{Zn}.

Figure 1(b) presents the ODMR spectra recorded on the PL intensity at 600 nm and 720 nm with cw microwaves applied. It is important that the shallow donor ODMR signal was detected not only within the D-A recombination band but within the iron intra-centre emission too. In contrast to the recombination band no ODMR signals could be seen via the Fe^{3+} emission with modulation of microwaves. The spectral dependence of the shallow donor ODMR in the spectral range of Fe^{3+}

Journal of Physics: Conference Series **461** (2013) 012032 doi:10.1088/1742-6596/461/1/012032

emission is shown in figure 2. One can see that the ODMR amplitude follows very precisely the PL spectrum corresponding to $^4T_1(G)$ - $^6A_1(S)$ transitions of Fe^{3+}.

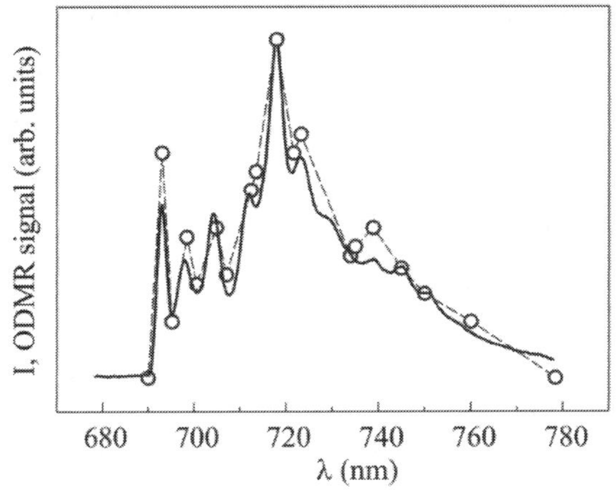

Figure 2. Spectral dependence of the shallow donor ODMR amplitude (open circles) and the PL spectrum of Fe^{3+} under 405 nm excitation (solid line) recorded with the same spectral resolution as was used for ODMR. Dashed line is a guide for eye.

Observation of the shallow donor ODMR on the iron intra-centre PL implies that the Fe^{3+} luminescence may actually be excited due to energy transfer from the recombining donor-acceptor pairs to the impurity ions.

Strongly anisotropic ODMR lines were found in the ODMR spectra recorded via the Fe^{3+} emission intensity in addition to the shallow donor (*SD*) line. Figure 3 (a) presents the ODMR spectrum for $\theta = 55°$, where θ is the angle between the magnetic field and the *c* axis. The positions of the anisotropic ODMR signals for different θ are displayed in Fig. 3 (b) (open symbols) together with the calculated angular dependence for isolated Fe^{3+} ions in ZnO. Calculations have been made with a computational package EasySpin [12]. We used a conventional axial spin Hamiltonian for axial Fe^{3+} [2] with the parameters of Fe^{3+} ions on Zn^{2+} lattice sites in ZnO taken from Ref. [13]. It is to be noted

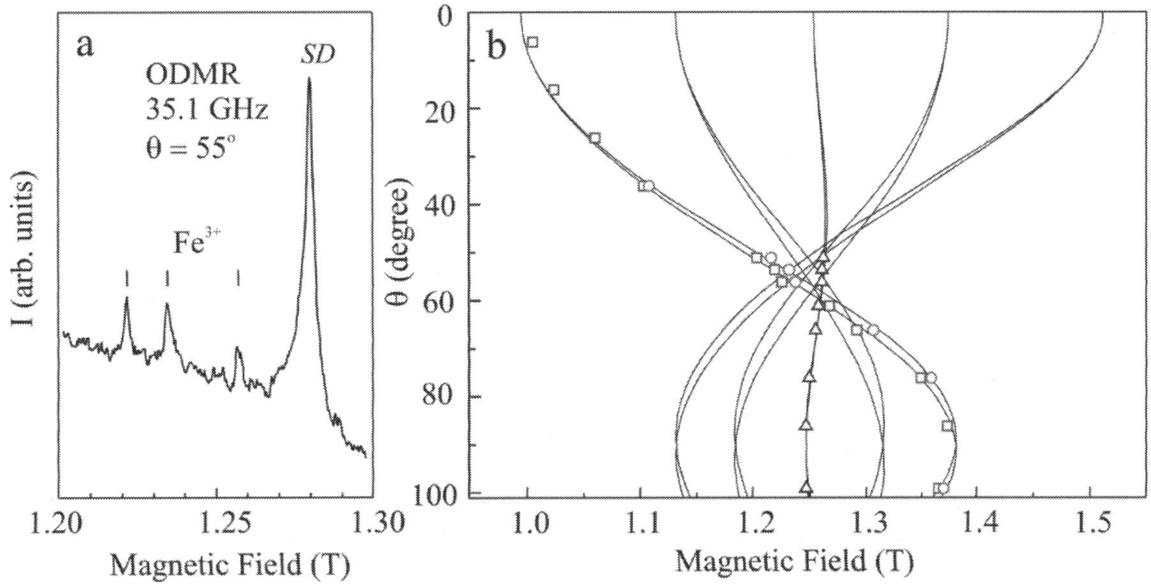

Figure 3. (a) ODMR spectrum recorded in ZnO single crystal via the 720 nm luminescence intensity at 35.1 GHz, T = 1.8 K, $\theta = 55°$ without modulation of microwaves. (b) Angular dependence of the Fe^{3+} ODMR spectrum. The open symbols are the measured ODMR positions. The lines give the calculated angular dependence using previously established parameters of EPR spectra of the axial Fe^{3+} centres.

that the fine structure splitting parameter is negative: $D = -593.7 \cdot 10^{-4}$ cm^{-1}. The calculated angular dependence is typical for axial S=5/2 paramagnetic centres. Small additional doubling of fine structure lines is due to the fact that there are two magnetically distinguishable sites in ZnO lattice for Fe^{3+} on the Zn position with slightly different interactions with the crystal field for certain orientations [13]. A very good agreement between the experimental data and the results of calculations proves that we really detect EPR in the ground state of isolated Fe^{3+} ions via the intensity the intra-centre luminescence of iron.

It should be pointed out that not all fine structure transitions of Fe^{3+} could be observed. For $\theta < 40^{o}$ we have detected only one of five possible fine structure lines that corresponds to the transition between the lowest magnetic sublevels $M_S = -5/2 \leftrightarrow M_S = -3/2$. This seems to be due to the high Boltzmann factor corresponding to EPR conditions, which leads to preferential population of the lowest levels and different intensities of different fine structure EPR lines. Similar behaviour of the fine structure components was reported for high frequency EPR of Mn^{2+} in ZnO crystals and ODMR of Mn^{2+} in CdMnSe quantum dots [14].

A major excitation mechanism of the Fe^{3+} luminescence is generally believed to be the capture of free holes by Fe^{2+} centres [9, 15]. Observation of the shallow donor ODMR via the Fe^{3+} emission in our experiments suggests an energy transfer from the recombining D-A pairs to the impurity ions.

In the samples under study a rather intense afterglow emission could be detected for up to an hour after switching off the UV light excitation. The afterglow spectrum is a wide orange band similar to that reported in [16]. ODMR of lithium acceptors and shallow donors was recorded by monitoring the afterglow intensity, which has unambiguously proved that the afterglow results from the spin-dependent recombination of these centres. In the samples with higher concentration of iron the Fe^{3+} emission was found in the afterglow spectra, which directly confirms the conclusion about energy transfer from the recombining centres to iron impurity ions.

4. Conclusions

ODMR was studied in ZnO single crystals grown by hydrothermal method and containing iron impurity ions. The broad orange band originating from the recombination of shallow donors and Li_{Zn} acceptors was observed in the luminescence spectrum together with a well structured intra-centre luminescence of Fe^{3+} corresponding to $^4T_1(G)$ - $^6A_1(S)$ transitions. Anisotropic ODMR signals were found by monitoring the intensity of the Fe^{3+} luminescence. They were unambiguously ascribed to the ground state of isolated Fe^{3+} on Zn^{2+} lattice sites. In addition, ODMR of shallow donors was observed via the intensity of Fe^{3+} intra-centre emission. This suggests that the Fe^{3+} luminescence is excited due to energy transfer from the recombining donor-acceptor pairs to the impurity ions. Observation of the Fe^{3+} emission in the afterglow spectra, which were detected at liquid helium temperatures after switching off excitation light, supports this conclusion.

Acknowledgments

This work has been supported by Ministry of Education and Science, Russia under contracts №8017 and №8568 ; the Programs of the Russian Academy of Sciences "Spin Phenomena in Solid State Nanostructures and Spintronics", "Fundamentals of Nanostructure and Nanomaterial Technologies" and by the Russian Foundation for Basic Research under Grant # 12-02-01011.

References

[1] Özgür Ü, Alivov Ya I, Liu C, Teke A, Reshchikov M A, Doğan S, Avrutin V, Cho S-J, and Morkoç H 2005 *Appl. Phys. Reviews* **98** 041301

[2] Abragam A and Bleaney B 1970 *Electron Paramagnetic Resonance of Transition Ions* (Oxford University Press, Oxford)

[3] Cavenett B C 1981 *Adv. Phys.* **30** 475
[4] Baranov P G , Romanov N G 2001 *Appl. Magn. Resonance.* **21** 165
[5] Gruber A, Dräbenstedt A, Tietz C, Fleury L, Wrachtrup J, von Borczyskowski C 1997 *Science* **276** 2012
[6] Mc Cluskey M D, Jokela S J 2009 *J. Appl. Phys.* **106** 071101
[7] Azamat D V, Debus J, Yakovlev D R, Ivanov V Yu, Godlewski M, Fanciulli M, Bayer M 2010 *Phys. Status Solidi* B **247** 1517
[8] Vlasenko L S, Watkins G D, Helbig R 2005 *Phys. Rev.* B **71**, 115205
[9] Heitz R, Hoffmann A and Broser I 1992 *Phys. Rev.* B **45** 8977
[10] Cox R T, Block D, He A, Picard R, Santier C 1978 *Solid State Communs* **25** 77
[11] Schirmer O F 1968 *J. Phys. Chem. Solids* **29** 1407
[12] Stoll S and Schweiger A 2006 *J. Magn. Reson.* **178**, 42
[13] Walsch W M Jr and Rupp L W Jr 1962 *Phys. Rev.* **126** 952
[14] Baranov P G, Romanov N G, Tolmachev D O, Babunts R A, Namozov B R, Kusrayev Yu G, Sedova I V, Sorokin S V, Ivanov S V 2008 *JETP Lett.* **88** 631
[15] Malguth E, Hoffmann A, Gehlhoff W, Gelhausen O, Phillips M R, Xu X 2006 *Phys. Rev.* B **74** 165202
[16] Zwingel D 1972 *J. Lumin.* **5** 385

Journal of Physics: Conference Series **461** (2013) 012033 doi:10.1088/1742-6596/461/1/012033

Modeling of gas flow in the cylindrical channels of high-voltage plasma torches with rod electrodes

A M Borovskoy, S D Popov and A V Surov
Institute for Electrophysics and Electric Power RAS, 18 Dvortsovaya embankment, Saint-Petersburg, 191186, Russia

E-mail: borovskoi.alex@mail.ru, sergey_popov1973@mail.ru, alex_surov@mail.ru

Abstract. The article is devoted to the calculation of gas dynamic parameters of gas flow in various areas of low-temperature plasma generator, therefore, target area's grid was built for the simulation of plasma gas flow in channels of studied high-voltage AC plasma torches and calculations of three-dimensional gas flow was made using GAMBIT and FLUENT soft-ware and Spalart-Allmares turbulence model, air flow was simulated in the tangential feed's areas, in the cylindrical channel, in the tapering nozzle chamber and in the mixing chamber of plasma torches and outside (in the environment); thus, 3D-modelling of the cold plasma-forming gas flow was performed in cylindrical channels of studied high-voltage AC plasma torches with rod electrodes for the first time.

The physical characteristics of plasma torches (thermal plasma generators) depend on the arrangement of plasma gas supply and on the interaction nature of the gas flow with electric arcs. Plasma torches with power up to 50 kW with rod electrodes, installed in cylindrical channels [1-3], were developed and tested in IEE RAS; and researches were conducted. As result external characteristics of plasma generators and physical parameters of their torches were determined [4-5]. However, the properties of electric arcs, burning in the cylindrical channels, is not fully understood; therefore an attempt was made to simulate cold gas flow at the first stage of modeling of physical processes in channels of plasma torches. The single-phase two-channel (figure 1 (a)) and three-phase three-channel (figure 1 (b)) high-voltage AC plasma torches appear to be under investigations.

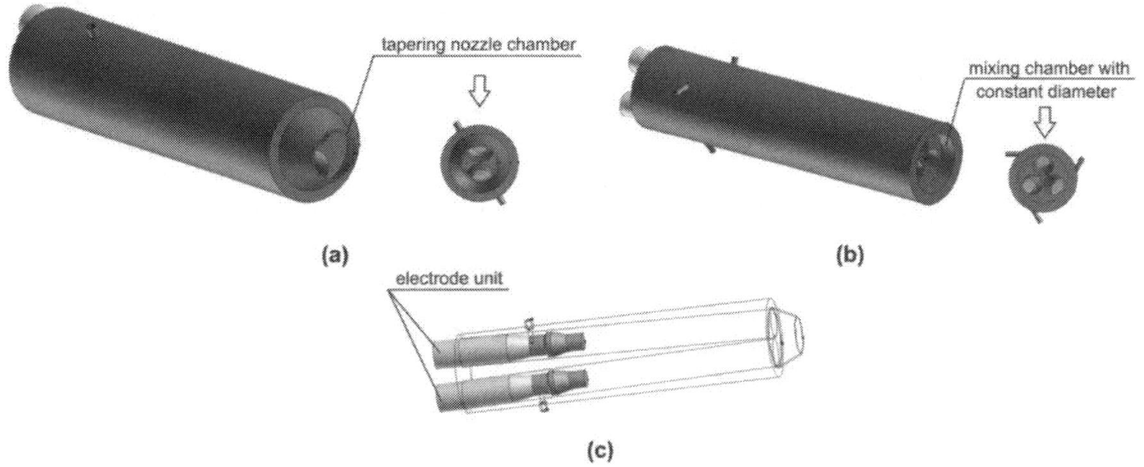

Figure 1. High-voltage AC plasma torches and their electrode units

A two-channel plasma torch has a tapering nozzle chamber (figure 1 (a)) whereas a three-channel plasma torch has a mixing chamber with constant diameter (figure 1 (b)). The investigated plasma torches are identical in design of the electrode sites, but they differ in the geometrical dimensions of channels (figure 1 (c)). The rod electrode, a part of electrode unit of these plasma torches (figure 1 (c)), made in the form of the rotating body, has a common axis with the channel, in which it is installed, and consists of conducting foundation, which is inside the insulator, and of the tip.

The calculation area's mesh was built for the simulation of plasma gas flow using GAMBIT software, the calculations of the gas flow were made using FLUENT software with Spalart-Allmares turbulence model; the gas density was determined by equation of ideal gas, and its viscosity was calculated by the Sutherland's equation with three ratios; the specific heat capacity, the thermal conductivity and the molar mass of the gas were constant. The mass flow of the plasma working gas was asked in tangential feed's inlet sections, united in one surface; the return flows, the temperature of 293,15 K and normal atmospheric pressure (101325 Pa) were asked on the surfaces of the environment, which were taken into account by the external mesh of GAMBIT software. The air flow was also simulated in the tangential feed's areas, in the cylindrical channel, in the tapering nozzle chamber and in the mixing chamber of plasma torches and outside, that is in the environment.

The contours of gas velocities, such as velocity magnitude, tangential, radial and axial velocity components, were built in the various sections on the longitudinal coordinate (figure 2) at optimum work mode (the gas mass flow of 6 and 10 g/s per all channels for two-channel and for three-channel plasma torches respectively); mass-average velocities were calculated and their dependences on the longitudinal coordinate (profiles of mass-average velocities) were built (figure 3); the gas mass flow was from 3 to 15 g/s and from 5 to 18 g/s for two-channel and for three-channel plasma torches respectively. It was revealed that the velocity takes the maximum value in the section of the channel, nearest to the tangential feed, and the velocity is from 40 to 170 m/s for two-channel plasma torch (figure 3 (c)) and from 40 to 135 m/s for three-channel plasma torch (figure 3 (d)). In the case of maximum velocity the major contribution is introduced by mass-average tangential component equal from 15 to 65 m/s and from 30 to 90 m/s for two-channel and three-channel plasma torches respectively. It is also evident that the minimum value of velocity of the working gas flow is observed in the output section of the channel and is from 1,5 to 8,4 m/s for two-channel plasma torch (figure 3 (c)) and from 1,6 to 5,5 m/s for three-channel plasma torch (figure 3 (d)). In the case of minimum velocity the major contribution is introduced by mass-average axial component equal from 1,2 to 6,2 m/s and from 1,4 to 4,9 m/s for two-channel and three-channel plasma torches respectively.

The contours of velocity magnitude and of axial velocity of plasma gas in the output section of the tapering nozzle chamber of two-channel plasma torch (z = 0,39 m) are presented in figure 2 (a) and in figure 2 (b) respectively; and also the contours of velocity magnitude and of tangential velocity were obtained in the section, nearest to the tangential feed. We have the contours of velocity magnitude and of axial velocity of plasma gas in the output section of the mixing chamber of three-channel plasma torch (z = 0,47 m) in figure 2 (c) and in figure 2 (d) respectively; and also the contours of velocity magnitude and of tangential velocity were obtained in the section, nearest to the tangential feed.

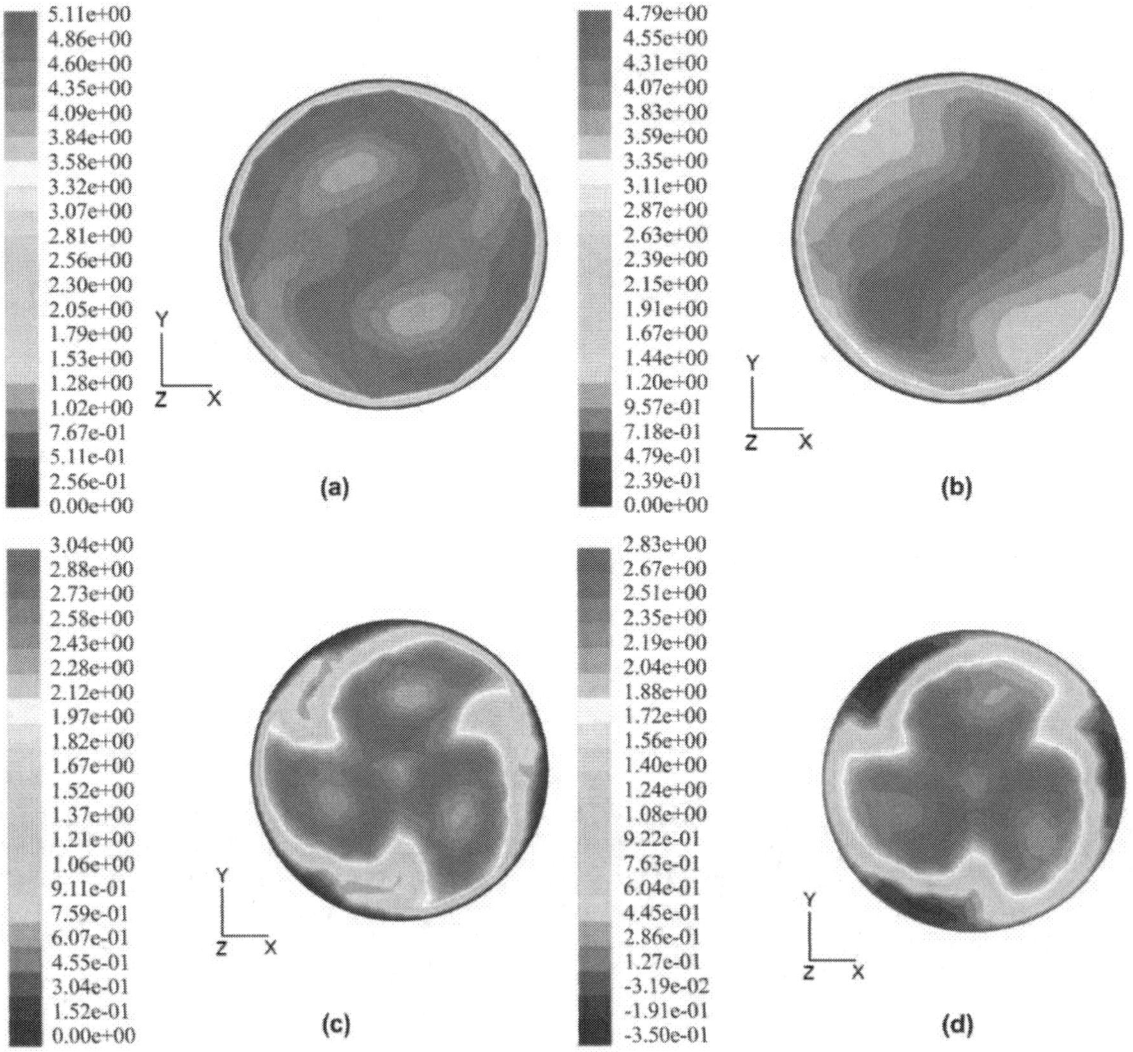

Figure 2. The contours of gas velocities in the sections of the tapering nozzle chamber and in the mixing chamber of plasma torches at optimum work mode

The profiles of mass-average velocity magnitude and of all components of plasma gas at optimum work mode in the channels of two-channel and of three-channel plasma torches are presented in figure 3 (a) and in figure 3 (b) respectively. We have the profiles of mass-average velocity magnitude of plasma gas at various work modes in the channels of two-channel and of three-channel plasma torches in figure 3 (c) and in figure 3 (d) respectively.

Journal of Physics: Conference Series **461** (2013) 012033 doi:10.1088/1742-6596/461/1/012033

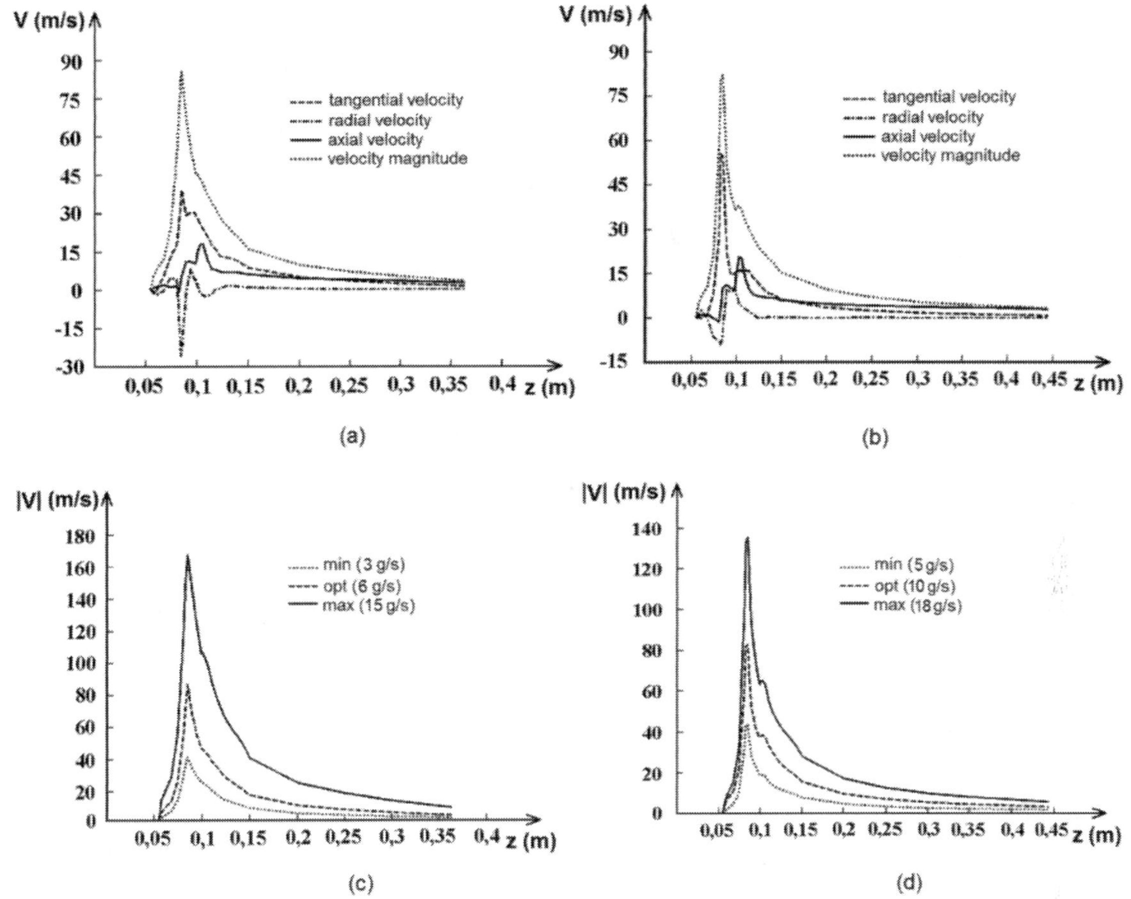

Figure 3. The profiles of mass-average velocities of working gas in the channels of plasma torches

The results of the research show the significant decrease of the mass-average velocity of gas flow in the channels of the investigated plasma torches during movement to exit and outside (in the environment), i.e. from the tapering nozzle chamber of two-channel plasma torch (figure 4 (a) for the gas mass flow of 6 g/s per all channels) and from the mixing chamber of three-channel plasma torch (figure 4 (b) for the gas mass flow of 10 g/s per all channels). The velocity takes the maximum value in the initial section of the tapering nozzle chamber and of the mixing chamber and data vary from 1,5 to 8,2 m/s for two-channel plasma torch (figure 4 (c)) and from 1,6 to 5,5 m/s for three-channel plasma torch (figure 4 (d)). In the case of maximum velocity the major contribution is introduced by mass-average axial component equal from 1,2 to 6,3 m/s and from 1,4 to 4,9 m/s for two-channel and three-channel plasma torches respectively.

The profiles of mass-average velocity magnitude and of all components of plasma gas at optimum work mode in the tapering nozzle chamber of two-channel plasma torch, in the mixing chamber of three-channel plasma torch and in the environment are presented in figure 4 (a) and in figure 4 (b) respectively. We have the profiles of mass-average velocity magnitude of plasma gas at various work modes in the tapering nozzle chamber of two-channel plasma torch, in the mixing chamber of three-channel plasma torch and in the environment in figure 4 (c) and in figure 4 (d) respectively.

Journal of Physics: Conference Series **461** (2013) 012033 doi:10.1088/1742-6596/461/1/012033

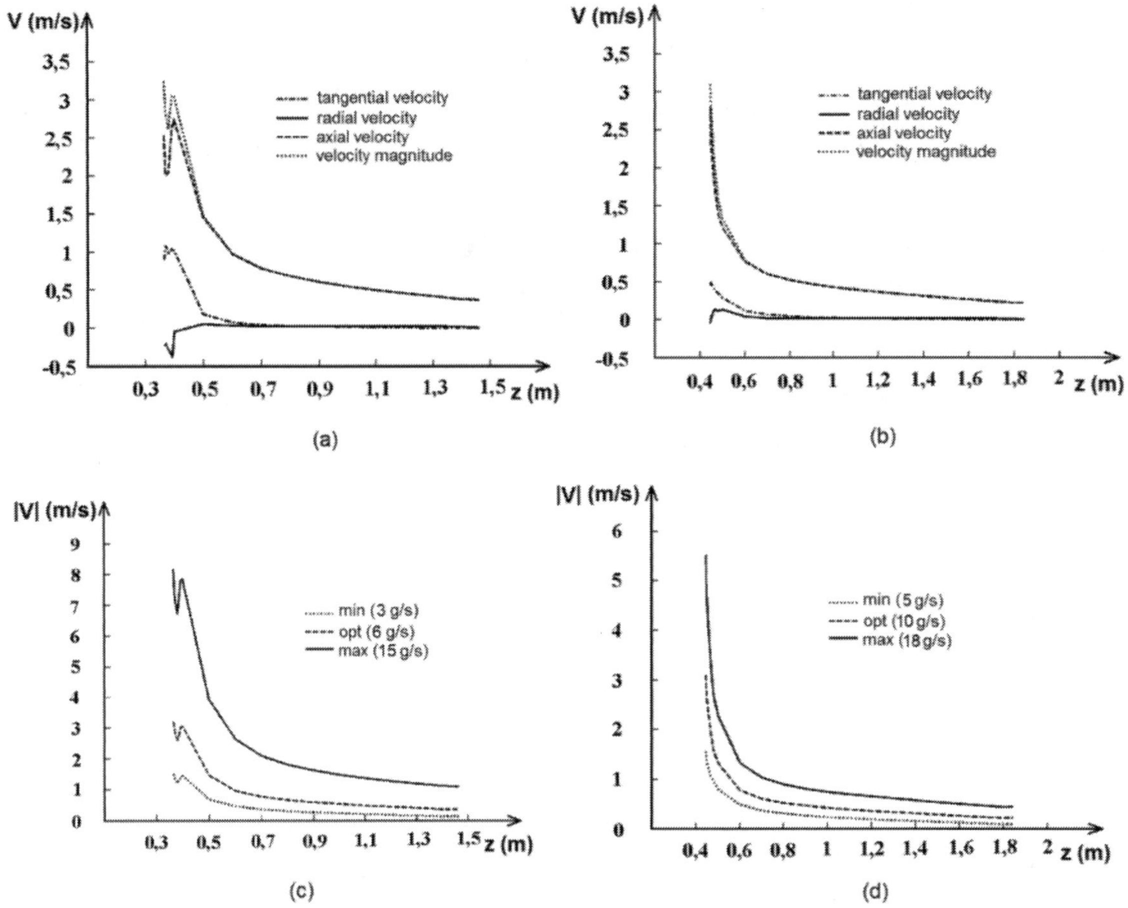

Figure 4. The profiles of mass-average velocities of working gas in the tapering nozzle chamber of two-channel plasma torch, in the mixing chamber of three-channel plasma torch and outside

Conclusion

Thus the contours of parameters of the cold gas flow were calculated in the channels and in the nozzle or in the mixing chamber of high-voltage AC plasma torches with the rod electrodes, as well as velocity profiles of the gas stream were built. It leads to further researches and improvements of the model: the burning of an arc in the channel must be taken into account and the parameters of working high-voltage AC plasma torches will be calculated; also the experimental data, obtained from a testing of this plasma torches with varying mass flow of plasma gas and power, will be used for agreement of the calculating models.

Journal of Physics: Conference Series **461** (2013) 012033 doi:10.1088/1742-6596/461/1/012033

References

[1] Rutberg Ph G, Safronov A A, Popov S D, Surov A V and Nakonechny Gh V 2005 Multiphase stationary plasma generators working on oxidizing media *Plasma Physics and controlled fusion* **47** 1681–1696

[2] Rutberg Ph G, Safronov A A, Nakonechny Gh V, Popov S D, Serba E O, Spodobin V A and Surov A V 2007 AC plasma torches with a rod electrode with the capacity from 5 to 50 kW for plasma chemical applications *Izv Physics Application* **№ 9** 77–79 (in Russian)

[3] Nikonov A V, Ovchinnikov R V, Pavlov A V, Popov S D, Safronov A A and Surov A V 2008 Single-phase AC plasma torch as a source of low-temperature air plasma at atmospheric pressure *Izvestia of higher educational institutions PHYSICS* **51 №11/3** 25–30 (in Russian)

[4] Rutberg Ph G, Safronov A A, Surov A V, Pavlov A V, Popov S D, Spodobin V A and Rutberg A Ph 2009 Spectral investigations of electric arc alternating current plasma generators with power to 600 kW *Journal of High Temperature Material Processes* **13** 195–203

[5] Rutberg Ph G, Pavlov A V, Popov S D, Sakov A I, Spodobin V A and Surov A V 2009 Spectral measurements of the gas and electron temperatures in the flame of a single-phase AC plasma generator *Thermophysics of high temperatures* **47 №2** 175–180

Journal of Physics: Conference Series **461** (2013) 012034 doi:10.1088/1742-6596/461/1/012034

Water flow in micro- and nanochannels. Molecular dynamics simulations.

N I Podolska[1,2,3] **and A I Zhmakin**[1,2]

SPb Branch of Joint Supercomputer Centre of the Russian Academy of Sciences Polytekhnicheskaya str. 26, 194021, Saint-Petersburg, Russia

Ioffe Physical-Technical Institute of the Russian Academy of Sciences, Polytekhnicheskaya str. 26, 194021 Saint-Petersburg, Russia

Saint-Petersburg Academic University of the Russian Academy of Sciences Khlopina str. 8/3, 194021, Saint-Petersburg, Russia

E-mail: n.i.podolska@gmail.com

Abstract. The flow in carbon nanotube is analysed using mathematical model that accounts for a depletion layer with reduced viscosity near the wall. The model by the author of the paper [Myers T G 2011 *Microfluid Nanofluid* **10** 1141] is corrected. Moreover, the structure of the water flow inside 1 nm diameter and 1.5 nm length nanotubes is examined using molecular dynamics simulations. Molecular dynamics simulations of water flow velocity and flow rate though carbon nanotubes are reported too.

1. Introduction

Progress in micro- and nanotechnology in recent decades has led to the allocation and progress in new research areas in which combined existing science fields. The study of liquid behaviour at micro- and nanoscale (microfluidics and nanofluidics, respectively) is of great importance for various applications - biology, including cryobiology (transport of water, ions and macromolecules through pores, ice crystallization, etc), materials science (wettable surfaces, materials for artificial biological tissues, etc), microanalytical systems (micro Total Analysis Systems - TAS, MicroFluidicSystems – μ MFS, Lab-on-a-Chip – LOC, etc). The key issues here are effects of space restrictions, wettability and surface morphology on liquid properties and behaviour.

Methods used to study behaviour and properties of liquids in bulk (macro scale) are not applicable to micro- and nanoscales, but molecular dynamic method (simulations) is applicable, and is used for theoretical and numerical studies [1-11]. Density profile shows high oscillations in liquid layers adjacent to micro- and nanochannel walls, and as a consequence, there is an inhomogeneity of the liquid flow, which is not included in the classical hydrodynamic approach. In this case, predicted by Navier-Stokes velocity profiles and viscosities are violated at critical channel widths and diameters. In the case of water, these critical width or diameter of the channel corresponds to approximately 1.4 nm for the velocity profile and 5.1 nm for the viscosity [12].

Recent experiments [13-15] and molecular dynamics simulations [16-20] have shown that water is transported through carbon nanotubes (CNTs) at unexpectedly high flow rate. The flow rates of pressure-driven water through membranes of 1.6 and 7 nm diameter carbon nanotubes [21, 22] are two to five orders-of-magnitude greater than those predicted by the continuum-based no-slip Hagen-Poiseuille relation. These findings have a great effect on the molecular sieving, chemical detection, and drug delivery fields, where such high flow rates would significantly increase device efficiency, accuracy, and throughput [23]. The contained fluid structure has also been shown to depend upon the CNT diameter: single-file molecule chains at the smallest diameters and high-density structures at larger diameters. The flow of water inside CNTs of diameters below 1.66 nm is noncontinuum: the problem cannot be accurately described using conventional continuum fluid mechanics with its associated linear constitutive relations and no-slip boundary conditions [11, 20].

2. Models of flow in nanoscale channels

An important step towards understanding liquid flow in nanoscale systems is to predict the transition from continuum to subcontinuum transport as the flow area decreases. In a continuum system, the behaviour of a liquid can be described in terms of infinitesimal volume elements that are small compared to the flow domain but have well defined thermophysical properties. Applying Newton's second law to a system of volumetric elements gives rise to the Cauchy and Navier-Stokes equations, which can be used to derive the Poiseuille and other continuum-level flow relations [24]. In a system where the size of a liquid molecule is comparable to the size of the flow domain, however, the notions of a representative volumetric element and continuum-based relations are invalid. Within such ''subcontinuum'' systems, the movement of individual molecules must be considered when predicting mass and momentum transport [25].

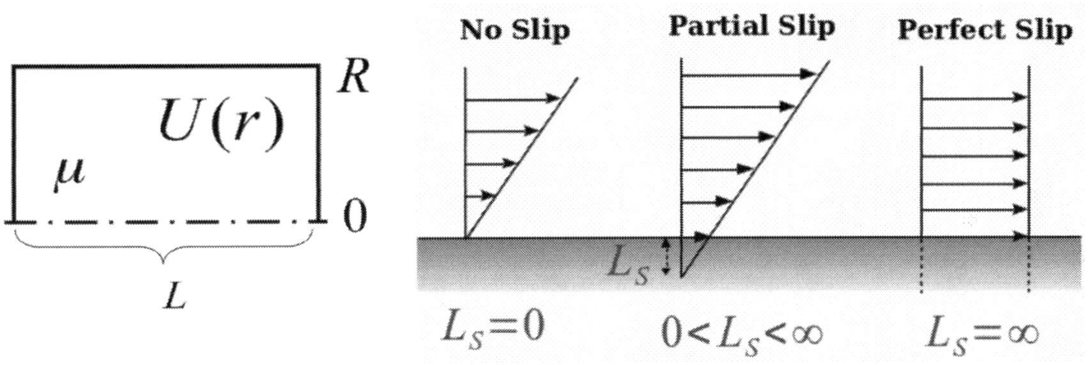

(a) (b)

Figure 1. The Hagen–Poiseuille equation, describes the classical model (a) and slip-length model (b) for flow in a circular cylindrical pipe.

2.1. *Boundary Conditions: No Slip-Length Model*

The Hagen–Poiseuille equation, describes the classical model for flow in a circular cylindrical pipe (see Figure 1a), is

$$u_{HP} = -\frac{R^2}{4\mu}\frac{\Delta P}{\Delta z}\left(1-\frac{r^2}{R^2}\right), \tag{1}$$

where u_{HP} is the velocity in the z direction, $\frac{\Delta P}{\Delta z}$ is the pressure gradient along the pipe, R is the radius and μ is the fluid viscosity. The corresponding flux is given by

$$Q_{HP} = 2\pi\int_0^R u_{HP}(r)rdr = -\frac{\pi}{8\mu}\frac{\Delta P}{\Delta z}R^4 , \tag{2}$$

and boundary conditions are

Journal of Physics: Conference Series **461** (2013) 012034 doi:10.1088/1742-6596/461/1/012034

$$\left.\frac{\partial u}{\partial r}\right|_{r=0}=0 \quad \text{and} \quad u(r)\big|_{r=R}=0\ . \tag{3}$$

2.2. Boundary Conditions: Slip-Length and Depletion Layer Models
In carbon nanotubes, the flow is enhanced and the real value of the flux is significantly higher than this classical value. A popular approach to explain this enhancement is to introduce a slip-length (see Figure 1b) into the mathematical model, that is, the no-slip boundary conditions are replaced by

$$\left.\frac{\partial u}{\partial r}\right|_{r=0}=0 \quad \text{and} \quad u(r)\big|_{r=R}=-L_S\left.\frac{\partial u}{\partial r}\right|_{r=R}, \tag{4}$$

where L_S is the slip length.

The high values of slip-length in CNT studies have led some authors to question the validity of the slip modified Hagen–Poiseuille model [25, 26]. The authors of Ref.[27] state that the slip-length should have a single value independent of the tube radius and much less than those quoted in the literature. They attribute some of the high experimental values to contamination by hydrophobic particles. An alternative explanation to the slip-length is based on the fact that CNTs are hydrophobic [28-30] and that some experiments have shown the existence of depletion layers between water and hydrophobic surfaces [18, 27, 31, 32].

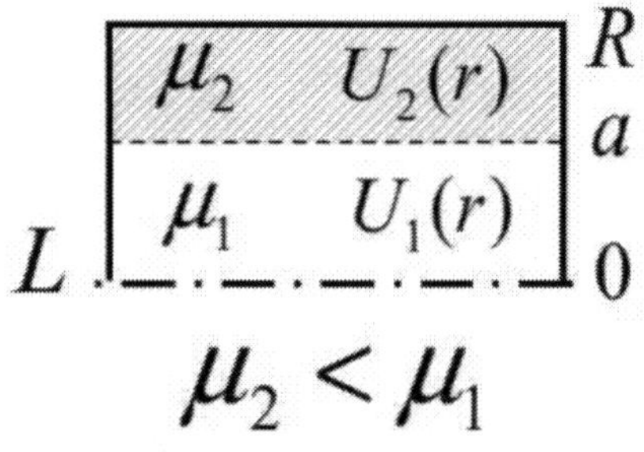

Figure 2. Flow in a circular cylindrical pipe including a region of low viscosity (depletion layer) near the tube wall.

A mathematical model for flow including a region of low viscosity (depletion layer) near the tube wall was investigated in the paper [33].

Consider a pipe of cross-section R occupied by two fluids (Figure 2). In the bulk flow region, defined by $0 \le r \le a$, the author of Ref. [33] imposes a viscosity μ_1. In the annular region near the wall, defined by $a \le r \le R$, we impose a viscosity μ_2. The assumption of two regions with different viscosities leads to what is commonly termed a bi-viscosity model in the non-Newtonian flow literature. In this case, the boundary conditions are

$$\left.\frac{\partial u_1}{\partial r}\right|_{r=0}=0 \quad \text{and} \quad u_2(r)\big|_{r=R}=0 \tag{6}$$

$$u_1(r)=u_2(r)\big|_{r=a} \quad \text{and} \quad \mu_1\frac{\partial u_1}{\partial r}=\mu_2\left.\frac{\partial u_2}{\partial r}\right|_{r=a}$$

and pressure driven flow equations in cylindrical system are

$$\frac{\partial p_1}{\partial z}=\frac{\mu_1}{r}\frac{\partial}{\partial r}r\frac{\partial u_1}{\partial r} \quad \text{and} \quad \frac{\partial p_2}{\partial z}=\frac{\mu_2}{r}\frac{\partial}{\partial r}r\frac{\partial u_2}{\partial r} \tag{7}$$

So, the velocities are

$$u_1(r)=\frac{1}{4\mu_1}\frac{\Delta p}{\Delta z}\left(r^2-a^2\right)+\frac{1}{4\mu_2}\frac{\Delta p}{\Delta z}\left(a^2-R^2\right) \tag{8}$$

$$u_2(r)=\frac{1}{4\mu_2}\frac{\Delta p}{\Delta z}\left(r^2-R^2\right)\ .$$

In the paper [33] there are mathematical errors in equations for fluxes Q_1, Q_2 and $Q = Q_1 + Q_2$ leading to the wrong expressions for the flow rate enhancement and ratio of viscosities. The correct expression for the total flux Q is:

$$Q(r) = \frac{\pi}{8}\frac{\Delta p}{\Delta z} a^4 \left(\frac{1}{\mu_2} - \frac{1}{\mu_1} \right) - \frac{\pi}{8}\frac{\Delta p}{\Delta z} R^4 \frac{1}{\mu_2} \tag{8}$$

or taking into account the expression (2):

$$Q(r) = Q_{HP}(R,\mu_2) + Q_{HP}(a,\mu_1) - Q_{HP}(a,\mu_2) , \tag{9}$$

where $Q_{HP}(a,\mu_1) = -\frac{\pi}{8\mu_1}\frac{\Delta p}{\Delta z} a^4$, $Q_{HP}(R,\mu_2) = -\frac{\pi}{8\mu_2}\frac{\Delta p}{\Delta z} R^4$

and $Q_{HP}(a,\mu_2) = -\frac{\pi}{8\mu_2}\frac{\Delta p}{\Delta z} a^4$.

So, the flow rate enhancement and ratio of viscosities are

$$\varepsilon = \frac{Q(r)}{Q_{HP}(R,\mu_2)} = 1 - \frac{a^4}{R^4}\left(\frac{\mu_2}{\mu_1} - 1 \right) \quad \text{and} \quad \frac{\mu_2}{\mu_1} = 1 + \frac{R^4}{a^4}(1-\varepsilon) . \tag{10}$$

Since $\frac{\mu_2}{\mu_1} > 0$ and $\varepsilon > 0$, then $0 < \varepsilon < 1 + \frac{a^4}{R^4}$. In this case, since $R \geq a$ (see Figure 2), the range of flow rate enhancement to be $\varepsilon \in (0;2)$. For the slip model (4), the corresponding enhancement is

$$\varepsilon_{slip} = 1 + \frac{4L_S}{R} . \tag{11}$$

The wrong flow rate enhancement and ratio of viscosities in Ref. [33] are

$$\varepsilon = \frac{a^4}{R^4}\left(1 - \frac{\mu_1}{\mu_2} \right) + \frac{\mu_1}{\mu_2} \quad \text{and} \quad \frac{\mu_2}{\mu_1} = \frac{1 - \frac{a^4}{R^4}}{\varepsilon - \frac{a^4}{R^4}} . \tag{12}$$

2.3. Depletion Layer Model Verification

To verify this model (10) the author of Ref.[33] considered an experiment described in Ref.[15]. Their flow enhancement indicates a slip length of 30–40 nm for pipes of radius R=20 nm. Setting L_S = 35 nm and R = 20 nm determines their enhancement factor (11) as $\varepsilon_{slip} = 8$. To obtain $\mu_2/\mu_1 = 0.018$, the author of [33] set $\varepsilon = 8$ and also taken $a = 19.3$ nm in (12). Using the same data and the expression (10), we obtain $\mu_2/\mu_1 = -4.8$ (using $\varepsilon_{slip} = 8$, R=20 nm and a=19.3 nm) and $\varepsilon = 1.85$ (using $\mu_2/\mu_1 = 0.018$, R=20 nm and a=19.3 nm), but μ_2/μ_1 must be >0. At the same time, $\mu_2/\mu_1 = 0.02$ from (10) at $\varepsilon = 1.85$, R=20 nm and a=19.3 nm and this value supports the depletion layer model (the viscosity of oxygen in water is also approximately 0.02 [33]).

3. Enhancement. Molecular Dynamics Simulations

We use molecular dynamics (MD) simulation to investigate pressure-driven (28 pN per each water molecule) water flow through single wall CNTs with diameter 1 nm and length 1.5 nm. The configuration of used pressure-driven flow simulation domains are shown in Figure 3. The rigid TIP3P water model and Lennard-Jones CNT model without electrostatic are used. Water density is

1000kg/m^3. Our MD simulations are performed using NAMD [34] and VMD [35], a parallelized non-equilibrium molecular dynamics solver, that is open-source and available to download from Theoretical and Computational Biophysics Group website (http://www.ks.uiuc.edu/).

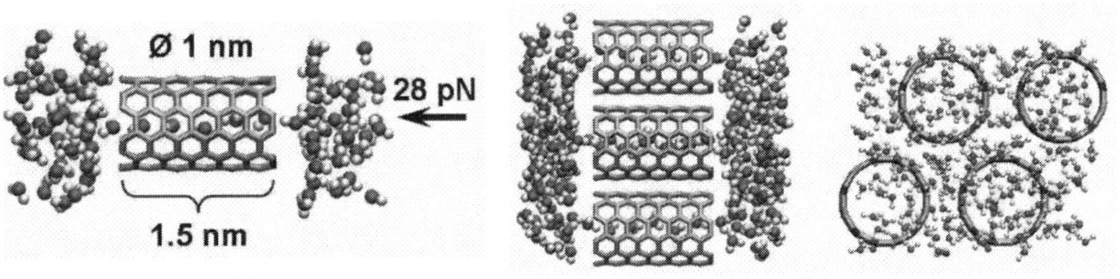

(a) (b)

Figure 3. Modelling CNTs: (a) single CNT and 101 water molecules (b) four CNTs and 248 water molecules.

Some simulation results are presented in Figure 4: (a) flow rate through the four CNTs as a function of simulation time; (b) enhancement factor of the water single chain as a function of simulation time too. One can see that the simulation time of about 30 ns is enough to obtain the convergence of flow rate and, consequently, enhancement factor. So, the flow rate is about 130 molecules per ns, the enhancement factor is 4.3 and the water molecule velocity is about 50 m/s.

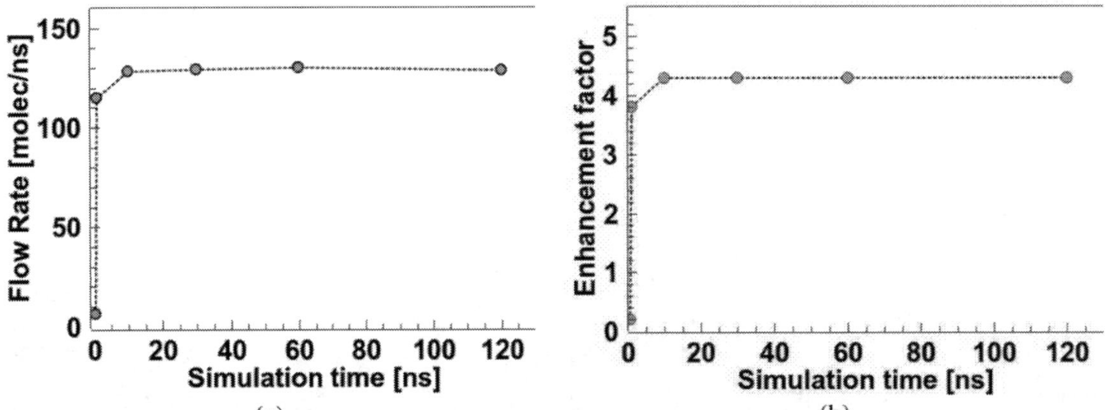

(a) (b)

Figure 4. Simulation results for the flow through four CNTs: flow rate [molecules/ns] (a) and enhancement factor (b).

4. Conclusions

In conclusion, the depletion layer model is promising and should be further developed, e.g., by incorporating an approximation of the viscosity profile by second-degree polynomial. Molecular dynamics simulations, providing reliable information for concrete flow cases, also could produce benchmark solutions for the assessment of approximate theories.

References

[1] Squires T M and Quake S R 2005 *Rev. Mod. Phys* **77** 977
[2] Eijkel J C T and van den Berg A 2005 *Microfluid Nanofluid* **1** 249
[3] Whitesides G M 2006 *Nature* **442** 368
[4] Phillips J C, Braun R, Wang W, Gumbart J, Tajkhorshid E, Villa E, Chipot C, Skeel R D, Kale

L and Schulten K 2005 *J. Comput. Chem.* **26** 1781
[5] Bizzarri A R and Cannistraro S 2002 *J. Phys. Chem. B* **106** 6617
[6] Aman K, Lindahl E, Edholm O, Hakansson P and Westlund P-O 2003 *Biophys. J.* **84** 102
[7] Senapati S and Berkowitz M L 2003 *J. Chem. Phys.* **118** 1937
[8] Bostick D L 2004 *Molecular dynamics simulation of biomembranes in aqueous solution* (PhD thesis, University of North Carolina, Chapel Hill)
[9] Heslot F, Fraysse N and Cazabat M 1989 *Nature* **338** 640
[10] Varshney A, Brooks F P and Wright W V 1994 *IEEE Computer Graphics and Applications* **14** 19
[11] Thomas J A and McGaughey A J H 2009 *Phys. Rev. L* **102** 184502
[12] Sofos F, Karakasidis T E and Liakopoulos A 2009 *Contemporary Engineering Sciences* **2** 283
[13] Majumder M, Chopra N, Andrew R and Hinds B J 2005 *Nature* **438** 44
[14] Holt J K, Park H G, Wang Y, Stadermann M, Artyukhin A B, Grigoropoulos C P, Noy A and Bakajin O *Science* , **312(5776)** 1034
[15] Whitby M, Cagnon L, Thanou M and Quirke N 2008 *Nano Letters* **8(9)** 2632
[16] Thomas J A, McGaughey A J H and Kuter-Arnebeck O 2010 *International Journal of Thermal Sciences* **49(2)** 281
[17] Corry B 2008 *Journal of Physical Chemistry B* **112(5)** 1427
[18] Joseph S and Aluru N R 2008 *Nano Letters* **8(2)** 452
[19] Suk M E, Raghunathan A V and Aluru N R 2008 *Applied Physics Letters* **92(13)** 133120
[20] Nicholls W D, Borg M K and Reese J M 2011 *14th NSTL Nanotech* Conference (Boston, Massachusetts, 2011.06.13 - 2011.06.16)
[21] Holt J K, Park H G, Wang Y, Stadermann M, Artyukhin A B, Grigoropoulos C P, Noy A and Bakajin O 2006 *Science* **312** 1034
[22] Majumder M, Chopra N, Andrews R and Hinds B J 2005 *Nature* **438** 44
[23] Noy A, Park H G, Fornasiero F, Holt J K, Grigoropoulos C P and Bakajin O 2001 *Nanotoday* **2** 22
[24] Landau L D and Lifshits E M 2003 *Theoretical Physics* (*Hydrodynamics* vol VI) (Moscow: Fizmatlit) pp 736
[25] Verweij H, Schillo M C and Li J 2007 *Small* **3(12)** 1996
[26] Thomas J A and McGaughey A J H 2008 *Nano Lett* **8(9)** 2788
[27] Cottin-Bizonne C, Cross B, Steinberger A and Charlaix E 2005 *Phys Rev Lett* **94** 056102
[28] Werder T, Walther J H, Richard L J, Halicioglu T, Noca F and Koumoutsakos P 2001 *Nano Lett* **1** 697
[29] Hummer G, Rasaiah J C and Noworyta J P 2001 *Nature* **414(8)** 188
[30] Noy A., Park H G, Fornasiero F, Holt J K, Grigoropoulos C P and Bakajin O 2007 *Nano Today* **2** 22
[31] Neto C, Evans D R, Bonaccurso E, Butt H-J and Craig V S J 2005 *Rep Prog Phys* **68** 2859
[32] Alexeyev A A and Vinogradova O I 1996 *Colloids Surf A: Physicochem* **108** 173
[33] Myers T G *Microfluid Nanofluid* 2011 **10** 1141
[34] Phillips J C, Braun R, Wang W, Gumbart J, Tajkhorshid E, Villa E, Chipot C, Skeel R D, Kale L and Schulten K J 2005 *Comput. Chem.* **26** 1781
[35] Hamphrey W, Dalke A and Schulten K J 1996 *Molec. Graphics* **14** 33

15th Russian Youth Conference on Physics and Astronomy (PhysicA.SPb) IOP Publishing
Journal of Physics: Conference Series **461** (2013) 012035 doi:10.1088/1742-6596/461/1/012035

The use of supercritical endothermic fuel

Mikhaylov AM[1,2,3]

[1] PHD student, Saint-Petersburg State Polytechnical University, Saint-Petersburg, Russia
[2] Chief engineer, Hypersonic Research Institute,
212 Moskovsky Prospekt, Saint-Petersburg, Russia
[3] Main expert, FSUE "RSC "Applied Chemistry", Saint-Petersburg, Russia

E-mail: A.M.Mikhaylov@gmail.com

Abstract. Endothermic reactions as a method of cooling are considered by many researchers due to the necessity of removal and recovery of large heat fluxes. The main applications of endothermic fuels technology are considered hypersonic aircraft. The increased interest in hydrogen energy, already implemented in specific technical developments, is pushing for consideration of existing developments not only in the spectrum of highly specialized technology of thermal protection, but also in as the hydrogen energy technology. It is proposed the use of endothermic catalytic steam reforming reaction of hydrocarbon fuel in a supercritical state. Summarizes the results achieved in detail both theoretically and experimentally investigated thermal and chemical aspects of the implementation of technology in terms of low pressure in the gaseous or liquid reagents in the light of the supercritical state of the reactants.

1. Introduction

The use as reagents of different type of hydrocarbons allows, except removal of heat by the endothermic effect of the reaction to obtain a hydrogen-containing fuel having better energy characteristics. In stand-alone hydrogen energy plants traditionally to produce hydrogen from hydrocarbon fuels regarded exothermic reaction of partial oxidation of hydrocarbons in the water vapor presence [1;2].

We have previously detailed theoretically and experimentally investigated the thermophysical and chemical aspects of the implementation of the technology in terms of low pressure in the gaseous or liquid state reagents [3;4].

2. Chemical absorption of heat

Removing heat main factors are heat capacity and the heat of phase transitions of refrigerant. A feature of the first factor is the uniform distribution of the heat capacity for heating, change in heat flow leads to a proportional increase in temperature. The second factor is different in that the heat is phase transition absorbed at a certain temperature, but for the technology used in refrigerants, these temperatures are low.

If you need to absorb large heat flows while maintaining high temperatures of cooling facilities, the best factors would be cool with a maximum heat capacity at the required temperatures usually defined by the constructional limit of the material. In addition to phase transitions of (evaporation, the transition to the supercritical state) as such may be endothermic reactions. The chemical reaction rate increases by 2-4 times temperature increases up to 10 degrees, but as the temperature change is a function of the difference between the incoming and chemically absorbed heat flux:

$$dT = f(E_{incoming} - E_{chemical\ absorbed}),$$

and absorbed heat flux is proportional to the reaction rate and, therefore, temperature::

$$E_{chemical\ absorbed} \sim V_{reaction} = f(T)\ ,$$

it balances the heat absorption heat flux increases with its growth.

3. Steam reforming, as the basis of chemical heat recovery

Developed technology of chemical heat recovery based on the steam conversion reaction of hydrocarbons [5]. Comparison of the thermal characteristics of the reaction by the example of

Journal of Physics: Conference Series **461** (2013) 012035 doi:10.1088/1742-6596/461/1/012035

kerosene T1 cooling power of water, hydrogen, and thermal decomposition reaction of kerosene for different temperatures are shown in Table 1. In addition, the table shows the relative concentrations of hydrogen and carbon condensate products. Initial temperatures of all components of 393.15 K, except cryogenic hydrogen, for it is the initial temperature T <T boil = 22K.

Table 1. Heat consumption for the cooling liquid heating and expansion, composition of the products when heated to different temperatures.

	Cooling Method	*T*=700K			*T*=1000K			*T*=1373,15K		
		X H2	X carbon	ΔH kJ / kg	X H2	X carbon	Δ*H* kJ / kg	X H2	X carbon	Δ*H* kJ / kg
1	H_2O heat capacity boiling	0	0	3251	0	0	3910	4.24e-5	0	4818
2	T1 thermal cracking heat capacity boiling	0.09	0.52	472	0.40	0.52	3236	0.48	0,51	5415
3	43.58% T1 56.41% H_2O steam conversion heat capacity boiling	0.11	0.16	2246	0.51	0.08	5940	0.65	0	9258
4	H2 heat capacity boiling	1	0	9860	1	0	14570	1	0	20255

When heating kerosene T1, consisting of a mixture of different hydrocarbons, at a temperature of 700K in the decomposition products contained up to 9.3 mol% of hydrogen, and when to 1373.15K - up to 48 mol% H2. Maximum heat consumption for heating a mixture of kerosene and water take place at the maximum formation of hydrogen in the product mixture. For example, at T = 1373.15 K composition of 43.58% T1 and 56.41% H20 gives the greatest amount of hydrogen in the mixture of degradation products, providing the highest heat input 9258 kJ / kg, with no soot.

Loss of heat to heat the hydrogen is much more than evaporation, heating and all other conversion of the substances. Characteristics of hydrogen are shown for comparison. The specifics of the use of hydrogen and hydrocarbon fuel are significantly different from each other, which do not allow considering them as alternative refrigerants. Thermodynamic equilibrium calculations were performed using the program, which is adaptation of the method described in [6;7], using the thermodynamic functions of the bank IVTANTHERMO [8] . Calculations were carried out for the pressure P = 1 bar.

The data presented in Table 1 do not reflect all the features of hydrocarbon conversion reactions as a factor in cooling. To illustrate the uneven distribution of heat absorption during the growth were calculated equilibrium compositions for a larger number of temperatures that were converted to mass fractions of the components.

When mixtures of known composition on the basis of data on the specific heat of components, taken from [8], [9] and [10], to calculate the "net" isobaric heat capacity at different temperatures. Results are presented in the form of graphs in Figure 1.

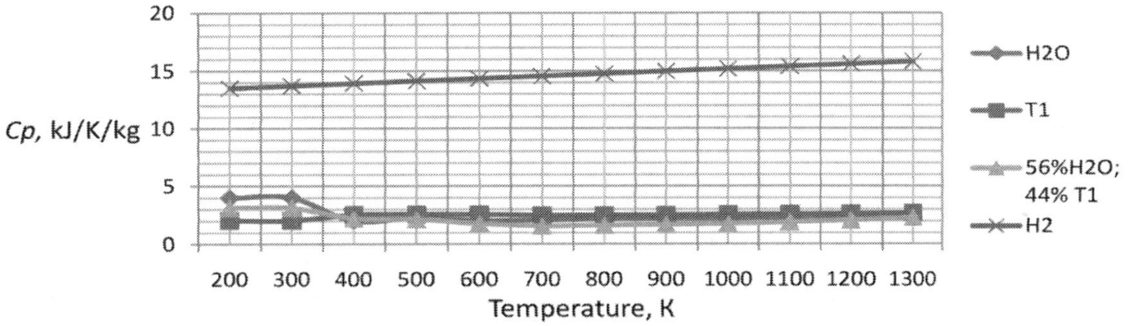

Figure 1. Refrigerant heat capacity at different temperatures.

As seen refrigerant heat much lower than that of hydrogen, even the presence in the products of the products of kerosene and steam reforming of large molar fraction of hydrogen does not improve the situation, because due to the low density of the mass fraction of hydrogen in the products is low.

To estimate the contribution to the heat-absorbing capacity of the phase transitions and chemical reactions was calculated modified indicator thermal capacity *Cp'*:

$$Cp'(T) = \frac{\Delta H}{\Delta T} = Cp(T) + Qreaction * \frac{\Delta\alpha(T)}{\Delta T};$$

Qreaction – heat of chemical reaction per kg reacted; $\alpha(T)$ – conversion degree at temperature.

Since the latent heat of the phase transition is absorbed at a certain temperature, the right to display it in the same dimension that was used in Figure 1 is not possible. In order not to lose the clear presentation of the graph in dimensions kJ/K/kg of K, the latent heat of the phase transition has been allocated to the temperature range.

To estimate the contribution of the heat of the chemical reaction has been calculated changes in the equilibrium composition of refrigerant temperature increases and these changes are calculated thermal effects. The result of these operations is shown in Figure 2.

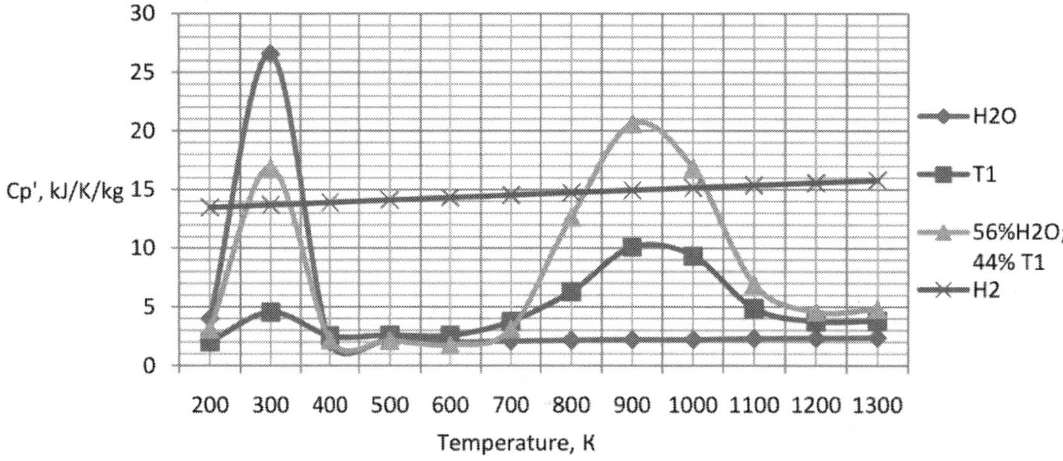

Figure 2. Modified thermal capacity indicator of refrigerants.

Thus, the use of steam reforming reaction in solving the problems has a number of features. Cooling surface temperature is maintained as close as possible to the temperature limit of heat-resistant structural materials, which has a positive impact on the flow processes stability in a refrigerated facility. High heat-absorption capacity in excess of heat-absorption capacity of hydrogen in the working area (850 K-1050 K), provides stabilization of the temperature cooled surface at this level.

4. Experimental research

This section presents the experimental results on the methane conversion at moderate heat fluxes q ≤ 100 kW/m^2. Schematic diagram of the setup is shown in Fig.3. The object of the pilot study is the structure of 592 mm length. A high-temperature jet of nitrogen, heating heat-transfer walls of the two slot channels was fed into the central rectangular channel of the reactor from plasmotron via adapter. Preheated mixture of hydrocarbons and water vapor was fed to the channels which actually were the reactive zone with 571 mm length, 60 mm width, and 4 mm height. It is the cooling jacket to the inner channel. The walls of TCR are made out of stainless steel used for the manufacture of flame tubes of the combustion chambers. The lengths of the reactor are 4 paired thermocouple and gas samplers. They are placed in 153 mm, 332 mm, 421 mm and 501 mm from the inlet to the reactor. 65 mm from the inlet to the reactor the single thermocouple is.

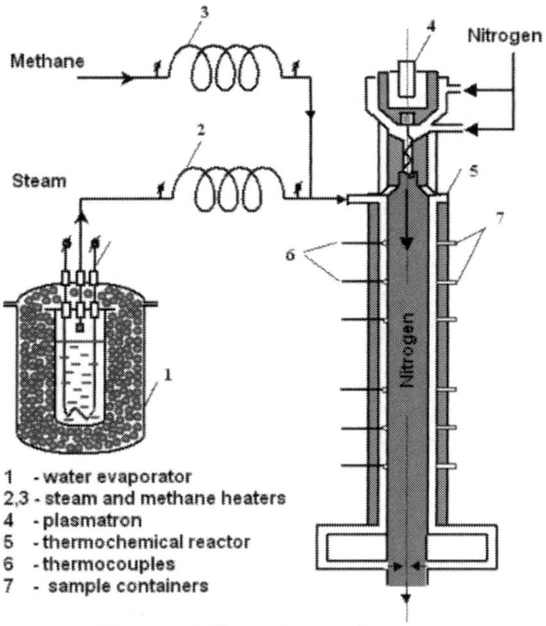

Figure 3. Experimental set-up

First, experiments were performed in the reactor with smooth walls, which have a natural catalytic activity. At flow rate of steam-methane mixture G = 0.8 g/s and a wall temperature Tw = 1100 ° C methane conversion degree did not exceed 6 % vol. Reduction of flow rate to 0.2 g/s resulted in the increase of conversion degree up to 37 % vol. due to the increase in the residence time of the mixture in the reactor. However, a qualitative change in the situation occurred when a volumetric nickel catalyst with a developed surface was placed into the reactor. The degree of conversion at that increased up to 92 %, as shown in Fig.4. Hydrogen yield at G = 0.8 g/s and Tw = 1100 ° C made up 45% vol. and at Tw =1300 °C increased up to 65 % by volume.

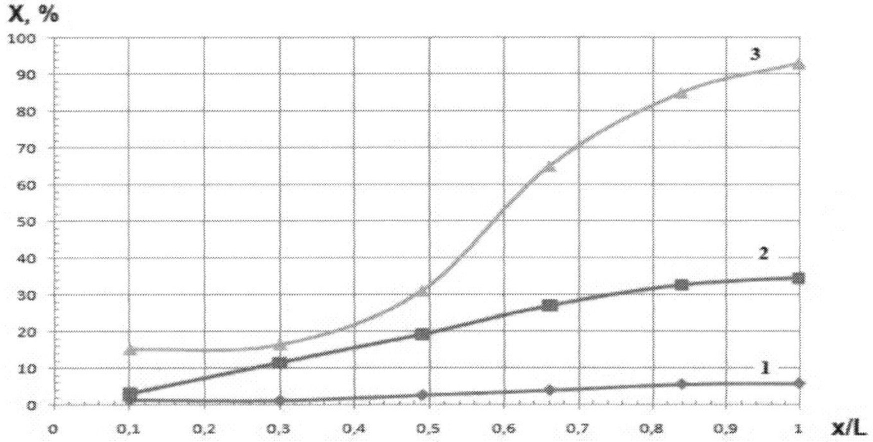

Figure 4. Steam conversion change along the reactor's length.
1 - G = 0.8 g/s, smooth wall 2 - G = 0.2 g/s, smooth wall 3 - G = 0.8 g/s, volumetric catalyst

Our experiments qualitatively confirmed the results of calculations of TCR modeling in slender channel approximation. Flow rate of chemically reacting mixture and the temperature of the wall where the reaction takes place have the greatest effect on the characteristics of the process under study. Some patterns can be seen in the detailed consideration of the results of the third stage. Figure 5 shows the composition of the samples, as in Figure 6 the temperature of gases. Sampler 1 corresponds to thermocouple 2, sampler 2 - thermocouple 3, and so on.

Journal of Physics: Conference Series **461** (2013) 012035 doi:10.1088/1742-6596/461/1/012035

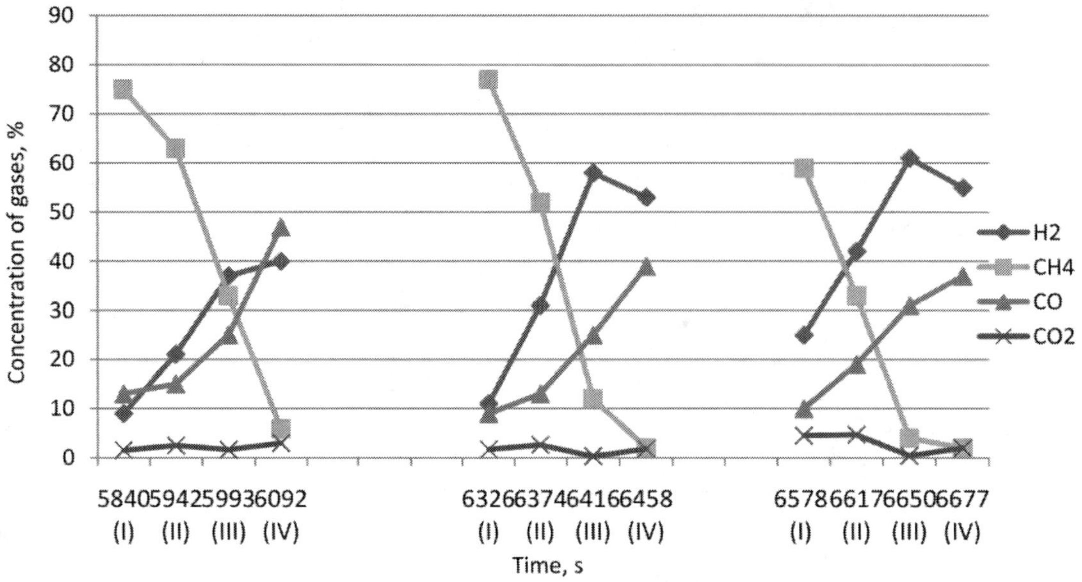

Figure 5. Gas concentration in samples.

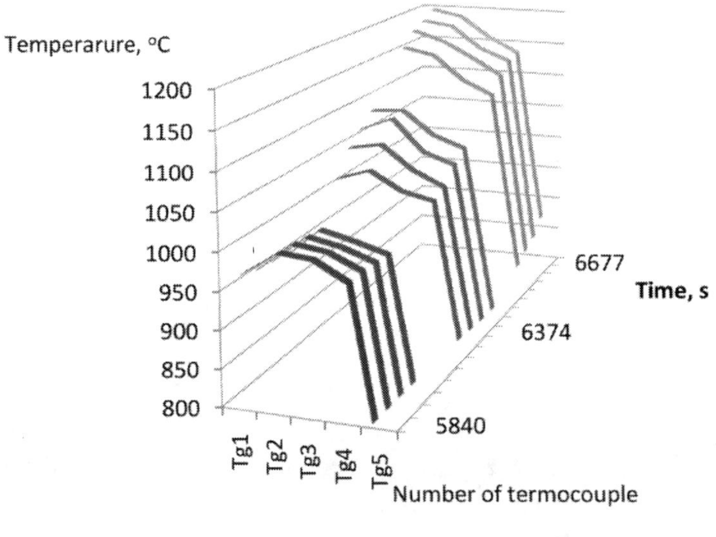

Figure 6. Gas temperature in points of sampling

The graph in Figure 5 shows that in the first mode (5840 s – 6092 s) is reached and the maximum degree of conversion of the hydrogen concentration reaches 40%. In the second (6326 s – 6458 s) and third (6578 s – 6677 s) mode, the concentration was achieved in 58% and 61% respectively. However, the drop is observed in hydrogen concentration to 53% and 55% respectively. In this case, the graph shows that the temperature at all points monotonically increasing. Accordingly, the drop is the result of the conversion course of internal processes.

Next was calculated removable heat flux and it was about 3 kW or 44 average specific kW/m3. The main reasons for the low heat flux removes include enough high heat transfer and reaction rate. Solvable problem has a dual purpose - the removal of a large heat and high conversion. For a high degree of conversion is necessary to ensure the long duration of stay in the reaction zone, since the limiting factor is the speed of the reaction, for removal of the heat flux to provide a large share of consumption and, again, the high speed of the reaction.

5. Conclusion

Obvious solution to this situation is to use a more active catalyst having a large specific surface area. But this approach several drawbacks. First, the active catalysts are expensive and mechanically weak. Secondly these catalysts are very sensitive to poisons. Thirdly, in our case is difficult to use the excess weight of the catalyst, in the case of chemical industrial plants provides long trouble free and stable operation.

As an alternative to using a larger amount of the catalyst is proposed to use an endothermic catalytic steam reforming reaction of hydrocarbon fuel in a supercritical state. Supercritical state, a separate state of aggregation, is characterized by high density and high reactivity, which distinguishes it from the gas and liquid, respectively [11]. Based on laboratory studies of hydrocarbon conversion by supercritical water [12,13], we can identify a number of advantages of this approach:

1. Separate flow conversion of hydrogen generation in the reactor heat source can use any high-temperature apparatus, in the case of the combustion chamber as a heat source; you can ensure optimal combustion in it, in contrast to the auto-thermal partial oxidation reactor, where oxidation and conversion are common.
2. Ability to raise the pressure in the reactor of hydrogen generation that the technology of the partial oxidation is difficult because of the reaction of partial oxidation sustainability and, as a consequence, the possibility of transition to a supercritical state.
3. As the density increases the residence time of reagents from the catalyst that helps intensify the reaction and reduce the size of the reactor. As a result, the limiting factor is the heat of reaction
4. Suppression of the production side products. Due to the separate reactions becomes possible to provide optimum conditions for maximum selectivity and depth conversion.
5. Formation of products of high pressure allows the use of various methods of separating the components of treatment and further storage without the cost of compression.
6. Thermal isolation is a consequence of the lack of the need for cooling, as the main target reaction is endothermic and there is no need to introduce additional refrigerant.

6. References

[1] Chentsov MS, Sokolov VS and Prokhorov NS 2006 *Alternative Energy and Ecology AEE* The concept of hydrogen reforming of diesel fuel in the composition of the atmosphere independent power plant with electrochemical generatoromi for non-nuclear submarines.**11 (43)**

[2] Wabash River Coal Gasification Repowering Project Final Technical Report. *Work performed under Cooperative Agreement DE-FC21-92MC29310.* (West Virginia: The U.S. Department of Energy / Office of Fossil Energy / National Energy Technology Laboratory / Morgantown)

[3] Kuranov AL, Korabelnikov AV and Mikhailov AM 2012 *Int. Symp. on "Thermochemical processes in plasma aerodynamics" St. Petersburg* Experimental study of steam reforming of methane in the thermochemical reactor

[4] Kuranov AL, Korabelnikov AV and Savarovsky AA 2012 *Int. Symp. on "Thermochemical processes in plasma aerodynamics" St. Petersburg* Numerical study of heat and mass transfer in a thermochemical reactor.

[5] Kuranov AL, Korabelnikov AV and Kuchinsky VV et al. 2000 *The 2nd Workshop on Magneto-Plasma - Aerodynamics in Aerospace Applications Moscow IVTRAN* Modern Condition of Development and Perspective Researches Hypersonic of Technologies Under the Concept "AJAX"

[6] Kitain MM and Katin E 1984 *Combustion and explosion* On the determination of the equilibrium composition of the combustion products **2**

[7] Molchanov O, Kitain MM, Suhovitskaya EA and Zemtseva JT 1989 *Theoretical method for calculating the energy and basic physical and chemical properties of the flow* (Leningrad: GIPKh)

[8] Automated system IVTANTHERMO 1985 (Moscow: ICT Academy of Sciences)

[9] Vargaftik NB 1972 *Handbook of thermophysical properties of gases and liquids* (Moscow: Nauka)

[10] Aleksandrov AA and Grigoriev BA 1999 *Tables of thermophysical properties of water and steam: A Handbook*. Rec. State. standard reference data service (Moscow: Publishing MEI)

[11] Galkin AA and Lunin VV 2005 *Russian Chemical Reviews* The water in the sub-and super-critical conditions - the universal medium for chemical reactions **74 (1)**

[12] Boukis N, Diem V, Habicht W and Dinjus E 2003 *Ind. Eng. Chem. Res.* Methanol Reforming in Supercritical Water **42** 728-735

[13] Pairojpiriyakul T, Kiatkittipong W, Assabumrungrat S and Croiset E 2012 *ISCRE* Hydrogen production from catalytic supercritical water reforming of glycerol with cobalt based catalysts **22**

Journal of Physics: Conference Series **461** (2013) 012036 doi:10.1088/1742-6596/461/1/012036

Increasing the emission power of LEDs (λ = 1.7-2.4 μm) by changing the light direction in the GaSb/GaInAsSb/GaAlAsSb heterostructure

A.V. Zolotukhin, V.V. Sherstnev, K.S. Savelieva, E.A. Grebenshchikova, O.Yu. Serebrennikova, N.D. Ilinskaya, S.A. Slobozhanuk, E.V. Ivanov, Y.P. Yakovlev.

Ioffe Physical-Technical Institute of the Russian Academy of Sciences, St. Petersburg 194021, Russian Federation

E-mail: azolvs@gmail.com

Abstract. Development of a curved reflective surface on the bottom side of the LED chip based on the GaSb/GaInAsSb/GaAlAsSb heterostructure increases the emission power of LEDs in 1.9-2.0 times compared with power of the tradition LED chip. This effect was investigated for the wavelengths of 2.0-2.3 μm. We believe that the improving of LED efficiency is the result of changes of light direction by the reflection from the curved surface formed by the hemispherical pits in the crystal.

1. Introduction

Semiconductor light emitting diodes (LEDs) for the mid-IR spectral range of 1.7 - 2.4 μm are promising for ecological monitoring and medical diagnostics [1]. One of the basic materials for development of LEDs for the spectral range of 1.7 - 2.4 μm is GaSb/GaInAsSb/GaAlAsSb heterostructure system. Though the internal quantum efficiency in the heterostructures can reaches 70% [2], the large refractive index of the semiconductor (n = 3.5 - 3.7) complicates the radiation output from the LED chip volume. This is explained by the fact that only a small portion of the light generated in the crystal falls on the outer surface at an angle less than the total internal reflection angle. As a result, only 8% of the radiation can go beyond of the parallelepiped shape crystal. Therefore one possibility to improve the LEDs efficiency is to find methods to increase the portion of light coming out the crystal.

The different ways of texturing the semiconductor surface are discussed in literature [3-6]. Texturing results in a change in the direction of the light flux within the crystal. As a result, the probability that the light flux will incident on the top side of the chip at a suitable angle is increased. This method was used to increase the output radiation of LEDs both for visible and near-IR spectral range.

Different methods of texturing the surface of semiconductor crystal are also used to increase the external quantum efficiency of LEDs in the mid-IR range (2-5 μm), though multipass is less effective in narrow-band materials than in wideband materials. Low efficiency is explained by the fact that the absorption of radiation by free carriers is proportional to the square of the radiation wavelength.

We investigated the GaSb-based LEDs for the spectral range of 1.8 - 2.4 μm with mesas of cubic and step-pyramid shape in ref. [7]. Theoretical and experimental data have shown that the external quantum efficiency is proportional to the S/V ratio, where S is the area of the surface through which the light is coming out and V is the total volume of the crystal. Once the crystal form is changed from cubic to pyramidal one, the coefficient of radiation coupling can be increased by a factor of 1.5. If LEDs are fabricated in a conical form with a concave surface, the radiation output can be also enhanced by a factor of 1.5, as discussed in [8].

2. Experimental samples

This investigation is a continuation of our investigations on the development of spontaneous sources for the mid-IR spectral range. The aim of this investigation is to study the possibility of increasing the optical power of LEDs based on GaInAsSb/GaAlAsSb heterostructures by changing the direction of the light beams. Such a changing of the light beam direction is achieved by multiple reflections of the photons on the curvilinear surfaces created in the crystal.

We used a GaSb (100) substrate to produce LEDs. It was doped by Tellurium to the carrier concentration of $1.2\cdot10^{18}$ cm^{-3}. The n-GaInAsSb 4 μm thick active layer was grown on the substrate by liquid phase epitaxy. The layer was doped by Tellurium to the carrier concentration $n{\sim}1.4\cdot10^{17}$ cm^{-3}. 3 μm thick wide-bandgap emitter was grown on the active layer. The Emitter was doped by Germanium to the carrier concentration $p{\sim}(2\text{-}3)*10^{18}$ cm^{-3}.

Three types of LED chips were developed on the basis of GaSb/GaInAsSb/GaAlAsSb heterostructure (series AV-54) by contact photolithography and wet chemical etching. Designs of these three chip types are shown schematically in Figure 1a. The chips had a square form with the size of 950 x 950 μm. The circular contact was formed by deposition of three layers of Cr/TeAu/Au in the center of the chip on the top side of the structure. The external contact diameter was 770 μm; the inner diameter of the contact was 740 μm.

Three types of LED chips differed by the design of the bottom side contact only. Lateral section schemes are shown in Figure 1 b, c, d.

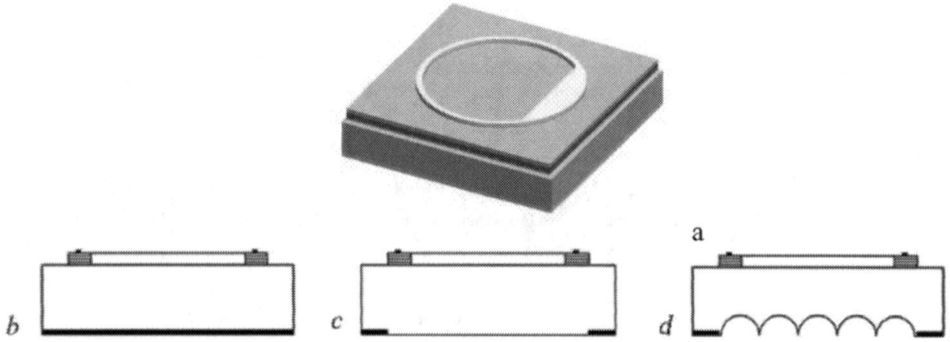

Figure1. LED chips with different geometry of bottom surfaces.
a - LED chip with a top circular contact.
(b, c, d) – lateral view of the LED chips with:
b – solid ohmic bottom contact;
c – bottom contact with a 880 μm diameter window;
d – bottom contact with 120 μm diameter hemispherical pits in the window.

Thus, one portion of the radiation from the active layer was directed outside the crystal through the near surface region (top side) of the chip, whereas the other portion of the radiation was directed towards the bottom side of the chip through the substrate 200 μm thick transparent for radiation.

Let us consider three types of the chip design.

The chip of the first type is characterized by a solid ohmic contact formed on the bottom side of the chip (Figure. 1 a). After alloying the contacts eutectic alloy was formed on the surface of semiconductor. This alloy adsorbs part of radiation directed to the bottom side of the chip.

The light flux is passed through the substrate transparent for radiation from active area to the bottom side of the chip. The large portion of the light flux is adsorbed at the interface between the substrate and ohmic contact in the eutectic alloy.

To form the chip of the second type a window with the diameter of 880 μm was formed in the ohmic contact (Figure. 1 b). Light flux impinged on the bottom side of the chip was reflected from the area free of metallization. Such a reflection allowed to increase the light flux reflected from the bottom side of the chip inside the crystal. However, light flux cannot change the angel of incidence and the reflection angle in the crystal.

In the LED chips of the third type the hemispherical pits with the radius of 60 μm were formed on the bottom side in the area free of metallization. In this case the radiation generated in the crystal is reflected repeatedly from the curved surface formed by the hemispherical pits (Figure. 1 c). As a result, the direction of light flux is changed by reflection from a curvilinear surface and much of it was extracted through surface on the top side of the chip.

LED chips were mounted on the TO-18 frame. Bottom side of the chip was adjoined to the diode frame.

3. Measuring the characteristics and discussion

Spectral characteristics were studied by an automated setup based on a monochromator DK-480 (CVI Laser Corp.) with 300g/mm diffraction lattice. An optical radiation was detected on the monochromator exit by InSb photo detector (Judson Technologies) with sensitive platform with the diameter of 4 mm. The signal from the photo detector was detected by digital selective amplifier SR-810 (Stanford Research Systems) after pre-amplification. The resulting digital signal was transmitted to a computer. This computer implemented a control of the measurement process.

An electroluminescence was excited by rectangular current pulses with a filing factor of 50 % and a 521 Hz repetition rate. Pulse amplitude was varied from 20 mA to 220 mA. Negative potential was applied to the GaSb substrate and positive potential was applied to the covering layer.

The dependence of the emission spectra and radiation power on the current density flowing though the LED chip was investigated. Emission spectra for three types of LED chips are shown in Figure 2. These spectra were measured at a current of 250 mA. All the emission spectra obtained have similar shape but different intensity of radiation. Maximum radiation intensity was observed in the LED chips with the hemispherical depressions formed on the bottom side of the chip. Minimal radiation intensity was observed in the chips containing a solid ohmic contact. Dependence of the output optical power on the current for three types of the LED chips is shown in Figure 3. The steepest dependence of the radiation power on the current was observed for the LED chips containing the hemispherical depressions on the bottom side of the chips. The most flat dependence was observed for the chip containing the solid ohmic contact. The increase in output power by a factor of 1.7 was observed for LED chips of the second type compared with the chips of the first type (at the current of 200 mA). The double increase in the output power was observed for the chips of the third type compare with the chips of the first type. The increase of the output power for LED chips with spherical etched pits on the bottom side of the chip provides a changing of the direction of the light beams by reflections on a curved surface.

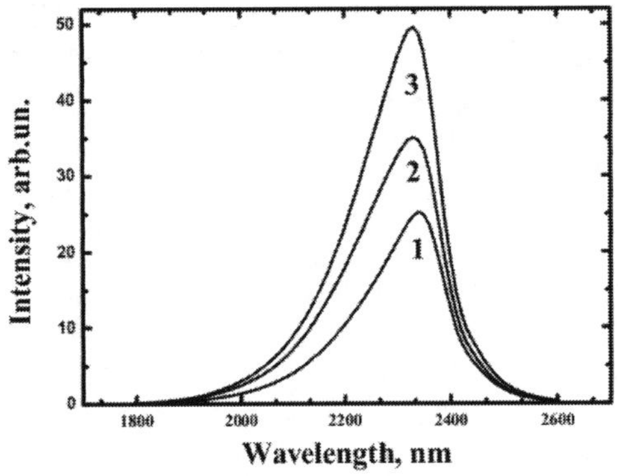

Figure 2 Emission spectra of LED chips with:
1 – solid ohmic bottom contact;
2 – bottom contact with a 880 μm diameter window;
3 – bottom contact with 120 μm diameter hemispherical pits in the window.

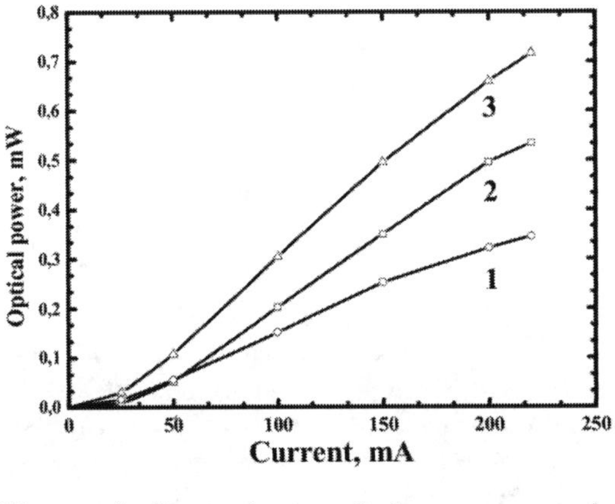

Figure 3. Dependence of the output optical power on the current for the LED chips with:
1 – solid ohmic bottom contact;
2 – bottom contact with a 880 μm diameter window;
3 – bottom contact with 120 μm diameter hemispherical pits in the window.

4. Conclusion

Thus, in this study a new way of increasing the emission power of LEDs for mid-infrared spectral range was proposed. For the LED chips of the third type with structured bottom surface an increase of the emission power by a factor of 1.9 is observed in the entire spectral range of 1.7-2.5 μm.

Implementation of the curved reflective surfaces in developed LED chip is an effective method for increasing the emission power of LEDs for the mid-IR spectral range.

This work was partially supported by the program of the President of the RAS № 24 "Fundamentals of Technology of Nanostructures and Nanomaterials", RFBR 10-02-93110-NTsNIL_a, 10-20-00548-a, and state contract № 11705.004.11.001.

5. References

[1] A.Kriev, H.H.Gao, V.V.Sherstnev, Y.P.Yakovlev. // Electron. Lett., 1999, V.35, N.19, P. 1665-1667.

[2] B.V. Tsarenkov, Vodovozova, S.A. Dokhman, A.N. Imenkov, L.M. Kogan, L.A. Titova, Y.P.Yakovlev., Pat. № 414950, Ser. 12,10, 1973.

[3] I.Schnitzer, E.Yablonovitch, C.Caneau, T.J.Gmitter, A.Scherer. // Appl. Phys. Lett., 1993, V. 63, N. 16, P. 2174-2176.

[4] R.Windishc, P.Heremans, P.Kiessel, G.H.Doller, B.Dutta, G.Dorhs. // Appl. Phys. Lett., 1999, V.74, N.16, P. 2256 - 2258.

[5] N.V. Zotova, N.D.Ilinskaya, S.A. Karandyshev, B.A. Matveev, M.A. Remennyy, N.M. Stus // Sov.Phys. Semicond. 2006. V.40. N.6. P. 717 - 723.

[6] A.A. Andaspaeva, A.N. Baranov, A.A. Guseinov, A.N. Imencov, N.M. Kolchanova, Y.P.Yakovlev, // Sov.Phys. Semicond, 2003, V.24, N.10, P. 1708-1721.

[7] E.A.Grebenshchikova, A.N. Imencov, B.E. Zhurtanov, T.N. Danilova, A.V. Chernyaev, N.V. Vlasenko, Y.P.Yakovlev, // Sov.Phys. Semicond, 2003, V.37, N.12, P. 1465-1472.

[8] E.A.Grebenshchikova, A.S. Golovin, A.N. Imencov, S.S. Kizhaev, Y.P.Yakovlev, // Pisma,Zh.Tekh.Fiz, 2010, V.36, N.3, S. 104-110.

Journal of Physics: Conference Series **461** (2013) 012037 doi:10.1088/1742-6596/461/1/012037

Exchange enhancement of quasiparticle and ESR spin-gap in symmetric and asymmetric narrow-gap quantum wells

S S Krishtopenko[1,2], K P Kalinin[1], A V Malyzhenkov[1], K V Maremyanin[1], V I Gavrilenko[1], M Goiran[2] and Yu G Sadofyev[3]

[1] Institute for Physics of Microstructures, Russian Academy of Sciences, GSP-105, 603950, Nizhny Novgorod, Russia
[2] Laboratoire National des Champs Magnétiques Intenses (LNCMI-T), CNRS UPR 3228 Université de Toulouse, 143 Avenue de Rangueil, F-31400 Toulouse, France
[3] P.N. Lebedev Physical Institute, Russian Academy of Sciences, 119991 Moscow, Russia

E-mail: gavr@imp.sci-nnov.ru

Abstract. We report a study of electron-electron (*e-e*) interaction effect on spin-gap in *n*-type narrow-gap quantum well (QW) heterostructures. By using the Hartree-Fock approximation (HFA) and generalized single-mode approximation (GSMA) based on the 8-band **k·p** Hamiltonian we demonstrate that spin-orbit interaction and the mixing between |*S*>- and |*P*>-states in the conduction and valence bands affect significantly many-body corrections to the spin-gap in symmetric and asymmetric QWs in the integer and fractional Quantum Hall regime. The spin-gap values estimated for 2D electron gas (2DEG) in InAs/AlSb QWs are compared with the experimental results obtained by magnetotransport and by electrically detected electron spin resonance (ESR).

1. Introduction

The recent focus on making use of the spin degree of freedom has also revived an interest in basic studies of the enhanced g-factor in QW heterostructures. The g-factor values in 2DEG obtained in magnetotransport and 'magnetooptic' experiments greatly differ from each others that result from the different contribution of *e-e* interaction effects into 'magnetostatic' and 'magnetooptic' phenomena. The 'magnetostatic' effects are related to the exchange renormalization of quasiparticle energy spectrum [1] determined by the 2DEG ground state. The 2DEG investigation techniques involving magnetotransport, capacitance spectroscopy of Landau levels (LLs) [2, 3], thermal capacity measurements [4, 5] and resonant tunneling in a magnetic field [6] allow one to directly determine the spin-gap at the Fermi level that characterizes quasiparticle g-factor.

In 'magnetooptic' experiments with 2D systems the resonant absorption of an electromagnetic wave gives rise to the quasiparticle-pair excitations formed by electrons that are excited onto an unfilled or partially filled LL, and quasiholes appearing simultaneously at the electron's vacated level, which cause the 2D system state to change from the ground state to the excited one. The simplest 2DEG excitation in a magnetic field is a spin exciton (spin wave) arising at the electron transition between the LLs spin-split through Zeeman interaction. The long-wave limit of the spin exciton energy determines the resonant frequency of absorption, measured in ESR. The Larmor theorem states that *e-e* interaction does not affect ESR energy in a 2D system with rotation invariance in the spin space. So, if the effects related to spin-orbit interaction (SOI) in QW heterostructures are rather weak, the g-factor determined from the ESR measurements is the 'single-electron' one, i.e. unaffected by the *e-e* interaction [7].

Since typical narrow-gap QW heterostructures based on InSb, HgTe and InAs are characterized by pronounced SOI, which leads to breaking of rotational symmetry in spin space, the Larmor theorem is violated in these structures [8]. Our earlier studies [8-11] have shown the SOI and mixing between $|S>$- and $|P>$-states in the Γ_6, Γ_7 and Γ_8 bands to largely influence the *e-e* interaction effects such as quasiparticle and ESR g-factor enhancement in 2DEG. It was shown that exchange enhancement of the quasiparticle g-factor in narrow-gap QWs arises for *any* LL filling factors [9-11] in contrast to the QWs with parabolic electronic subbands [1]. In [8] we demonstrated that the ESR g-factor and quasiparticle g-factor, measured in magnetotransport, coincide at *even-valued* LL filling factors in this heterostructures.

In this paper we extend the limits of previous works to theoretical study of exchange enhancement of spin-gap observed in ESR and magnetotransport in asymmetric narrow-gap QWs within Hartree-Fock approximation (HFA). The spin-gap values calculated for 2DEG in InAs/AlSb QW heterostructures [12-18] are compared with the experimental data obtained by magnetotransport [17, 18] and by ESR in moderate magnetic field. We also predict additional correlation-induced enhancement of ESR spin-gap for polarized Fractional Quantum Hall states (FQHS) as compared with the values obtained within the HFA. A non-linear dependence of ESR energies on LL filling factor in ultra-quantum limit is demonstrated.

2. Theory

To take into consideration SOI and the mixing of the conduction band (Γ_6) with the light- and heavy-hole bands (Γ_8) and with the split-off band (Γ_7) exactly, the 8-band **k·p** Hamiltonian as a single-particle kinetic energy operator is used [9]. It directly takes into account the influence of nonparabolicity, lattice-mismatch deformation and SOI on electron energy spectrum. As all calculations in this paper are performed for InAs/AlSb QW heterostructures, we also take into consideration the 'built-in' electric field of ionized donors in GaSb cap layer [14-16] in single-particle Hamiltonian. The Coulomb Green function was obtained by solving the electrostatic problem in three-layered system with different permittivity values in QW and in barriers [9].

2.1. Quasiparticle spin-gap versus ESR energy

Many-body corrections to quasiparticle spin-gap were obtained in the 'screened' Hartree-Fock approximation (HFA) [9]. To take into account the LL broadening under the influence of a random potential caused by defects in actual structures, we used a Gaussian profile for the DOS of each LL:

$$D(E)=\frac{1}{2\pi a_B^2}\frac{2}{\sqrt{2\pi}\Gamma}\exp\left(-2\frac{E^2}{\Gamma^2}\right), \tag{1}$$

where a_B is the magnetic length ($a_B^2 = c\hbar/eB$), Γ is the width of the DOS. By analogy with the Born approximation for the δ-correlated random potential [19] we assume that

$$\Gamma=\Gamma_0\sqrt{\frac{B}{B_0}}, \tag{2}$$

where $B_0 = 0.5\ T$ is the magnetic field in which the Shubnikov-de Haas oscillations show up [17, 18] and Γ_0 is free parameter.

Figure 1 illustrates the quasiparticle and ESR spin-gap calculations versus magnetic field in InAs/AlSb heterostructures with a 15 nm wide QW. The step-like features observed at even-valued LL filling factors appear due to the electronic subband nonparabolicity in InAs/AlSb QW [8-11]. As the magnetic field increases, the Fermi level 'jumps' from one pair of spin-split LLs to the lower-lying pair with a greater Zeeman splitting. This causes a sharp rise of the Zeeman energy at the Fermi level. It is seen that the single-particle spin-gap in a symmetric QW is determined by the Zeeman splitting of

LLs demonstrating a linear dependence on magnetic field in the region of weak magnetic fields. The Rashba spin splitting in an asymmetric QW leads to distortion of the monotonic dependence of single-particle spin-gap on magnetic field and appearance of a pronounced minimum in the low magnetic fields region. In this case the spin splitting at the Fermi level is determined by two contributors: the Rashba- and the Zeeman splitting [20]. As the magnetic field increases from zero, the single-particle spin-gap quickly drops, going smoothly over to the Zeeman splitting. Note that for high LL indices n (in weak magnetic fields) the single-particle spin-gap can be obtained analytically [21]:

$$\Delta_0 \approx \left[\left(\hbar\omega_C - g_{(1e)}\mu_B B\right)^2 + \Delta_R^2\right]^{1/2} - \hbar\omega_C, \qquad (3)$$

where $g_{(1e)}$ is a single-particle g-factor [5, 6], $\omega_C = eB/m^*c$, Δ_R and m^* are the Rashba splitting and the electron effective mass at the Fermi wavevector in a zero magnetic field, respectively. Thus, the single-particle spin-gap decreases linearly with B in weak magnetic fields.

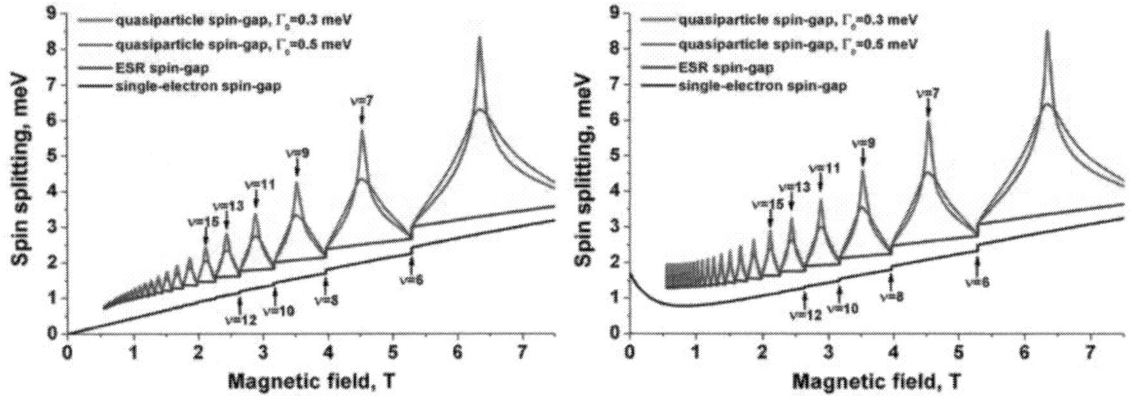

Figure 1: Spin-gap in a 15 nm wide InAs/AlSb QW with symmetric (left panel) and asymmetric (right panel) profile versus magnetic field at 2DEG concentration of $7.65\cdot10^{11}$ cm^{-2}. Black curve is the single-particle spin-gap calculated within the Hartree approximation. Red and green curves correspond to the quasiparticle spin-gap for LL broadening values $\Gamma_0 = 0.3$ meV and 0.5 meV respectively. Blue curve is ESR spin-gap. The arrows indicate the magnetic field values corresponding to the integer-valued LL filling factors.

Strong SOI and mixing between |*S*>- and |*P*>-states in the conduction and valence bands in the InAs/AlSb QWs also leads to considerable enhancement of quasiparticle and ESR spin-gaps. It is clear that the Rashba spin splitting in the asymmetric QWs resulted in a noticeable effect on the enhanced spin gap (quasiparticle and ESR) in weak magnetic fields only. It should be noted that, in asymmetric QWs in low magnetic fields, both spin-gaps enhanced by exchange interaction tend to Rashba splitting; in symmetric QWs, the exchange enhanced spin-gaps tend to zero. It can be shown that HFA corrections to the energy of spin-gaps at even-valued LL filling factors in symmetric [8-11] and asymmetric [18] narrow-gap QWs have a square-root dependence on the magnetic field resulting in the g-factor divergence in low magnetic field region [8, 10].

2.2. ESR spin-gap for FQHS

HFA completely ignores the correlation interaction between 2D electrons, and it is a fairly good approximation for spin-gap calculation at integer LL filling factors or in the case of large number occupied LLs only. To take into account the correlated nature of 2DEG ground state the generalized single-mode approximation (GSMA) instead of HFA should be used [22, 23]. In this section by using GSMA, we perform the calculations ESR spin-gap at filling factors $\nu = 1/M$ (where M is an odd integer), corresponding to the polarized Laughlin states and at $\nu = 1/2$ for polarized Composite Fermi Sea (CFS) and Pfaffian states [24]. By using the 8-band **k·p** Hamiltonian for single-particle states for narrow-gap QW heterostructures, the many-body corrections to the ESR spin-gap for the FQHS have the following form [25]:

$$\Delta_{ESR}^{(GSMA)} = \int \frac{d^2\vec{q}}{(2\pi)^2} \widetilde{h}_0^{(a)}(q)\left[\widetilde{V}_{-1,0,-1,0}^{(b,a,b,a)}(\vec{q}) - \widetilde{V}_{0,0,0,0}^{(a,a,a,a)}(\vec{q}) + \widetilde{V}_{0,-1,-1,0}^{(a,b,b,a)}(\vec{q})\right], \tag{4}$$

where $\widetilde{V}_{-1,0,-1,0}^{(b,a,b,a)}(\vec{q})$, $\widetilde{V}_{0,0,0,0}^{(a,a,a,a)}(\vec{q})$, $\widetilde{V}_{0,-1,-1,0}^{(a,b,b,a)}(\vec{q})$ are the matrix elements of *e-e* interaction explicitly given in [8] and $\widetilde{h}_n^{(i)}(q)$ is defined by the Fourier component of the pair distribution function $g_n^{(i)}(r)$ for the electrons in LL (n, i):

$$\widetilde{h}_{n'}^{(i')}(q) = \nu_{n'}^{(i')} \int d^2\vec{r} \exp(-i\vec{q}\vec{r})\left[g_{n'}^{(i')}(r) - 1\right], \tag{5}$$

where $\nu_n^{(i)}$ is filling factor for LL (n, i) [8-10]. To calculate the ESR spin-gap values for the FQHS, we use the parameterization of $g_{n'}^{(i')}(r)$ proposed by Girvin for the Laughlin states [22], and the parameterization for $\nu_{n'}^{(i')} = 1/2$ obtained in [26] by fitting the pair distribution functions for polarized CFS and Pfaffian states [24]. By employing the particle-hole symmetry that takes place in homogeneous 2D systems, we find $\Delta_{ESR}^{(GSMA)}$ for the conjugate states $\nu_{n'}^{(i')} = 1 - 1/M$ as well.

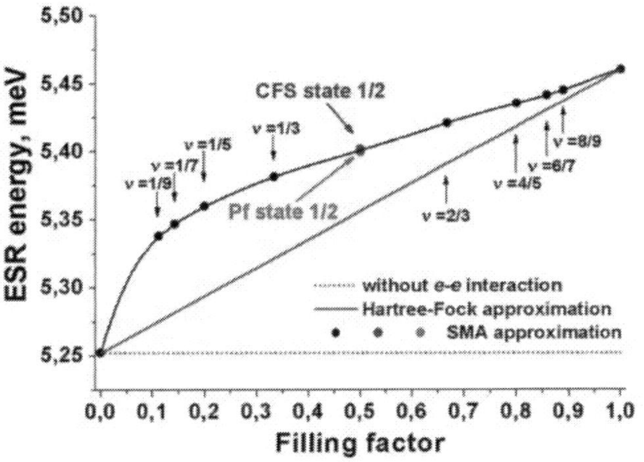

Figure 2: ESR spin-gap in a 15 nm wide InAs/AlSb QW with symmetric profile, calculated at a fixed magnetic field of 10 T. Brown dotted curve corresponds to Hartree approximation. Green curve is the result of calculations taking into account the *e-e* interaction within HFA. Black symbols correspond to the ESR spin-gap obtained in the GSMA for Laughlin states *1/M* and *1-1/M* (*M=1,3,5,7,9*). The corresponding energies for polarized CFS and Pfaffian states are marked in blue and red, respectively.

One can also see from figure 2 that, by taking into account the correlation interaction in the GSMA calculations, we obtain not only a greater, as compared to the 'mean-field' HFA values, enhancement of the ESR spin-gap for FQHS, but also a nonlinear dependence of the ESR energy on an LL filling factor. Note that the correlation corrections are the dominating contributor in the ESR spin-gap for FQHE at LL filling factor close to zero. For the filling factors approaching unity both approximations, HFA and GSMA, yield the same results.

3. Experiment

The InAs/AlSb QW heterostructures to be studied were grown by molecular beam epitaxy (MBE) on a semi-insulating GaAs (001) substrate with a metamorphic GaSb buffer layer. The order and conditions of growth of the layers are described in detail elsewhere [13, 17]. Magnetotransport measurements were performed at the temperature 250 mK, with sweeping the magnetic field. In electron transport measurements, a low-frequency (9 Hz) alternating current of about 1 μA was passed through the B824 sample. The parameters of the B824 structure at liquid-helium temperature are given in [17]. The sample in the form of a Hall bar was placed at the center of a superconducting solenoid in a dilution cryostat. To measure the longitudinal resistance (R_{xx}), we used the standard locked-in detection scheme. Figure 3 is the experimental and calculation results of quasiparticle spin-gap studies performed in B824 sample. It can be seen that the experimental data for the B824 sample are in good agreement with the results of theoretical calculations performed for the case of the most asymmetrical QW profile. Agreement between the theory and experiment supports the conclusion [14, 15] that the main 'suppliers' of 2D electrons to the InAs QW are surface donors in the GaSb coating layer and suggests that, in InAs/AlSb heterostructures, the 'built-in' electric field is highly asymmetric.

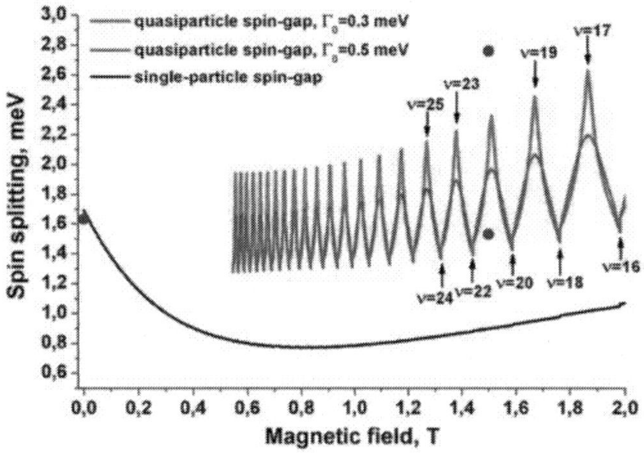

Figure 3: Quasiparticle spin-gap in the B824 sample (d_{QW} = 15 nm, n_S = $7.67{\cdot}10^{11}$ cm^{-2}). Black curve is the single-particle spin-gap calculated within the Hartree approximation. Red and green curves correspond to the quasiparticle spin-gap for LL broadening values Γ_0 = 0.3 meV and 0.5 meV respectively. The arrows indicate the magnetic field values corresponding to the integer-valued LL filling factors. The blue symbols are experimental values obtained from magnetotransport data [18].

ESR measurements were carried out at 4.2 K using electrical detection technique [7]. Magnetic field was oriented perpendicular to the sample plane. As a radiation source the backward-wave tube which covers the range of 250-500 GHz was used. The measurements were performed on two samples T340 and T338 with a QW width of 15 nm and 18 nm, respectively. The 2DEG concentration and mobility in the samples at T = 4.2 K were $6.70{\cdot}10^{11}$ cm^{-2} and $4.5{\cdot}10^{5}$ cm^2/V·s for T340 sample and $7.65{\cdot}10^{11}$ cm^{-2} and $4.6{\cdot}10^{5}$ cm^2/V·s for T338 sample. The experimental values are presented in figure 4 by symbols. One can see a good agreement between theoretical predictions and experimental data if the asymmetric 'built-in' electric field of the surface donors in GaSb cap layer [14, 15] resulting in the Rashba splitting [16] is taken into account. A small discrepancy between calculated curve and experimental data can be attributed, in our opinion, to the influence of dynamic polarization (Overhauser shift [27]) on the ESR line shape and position.

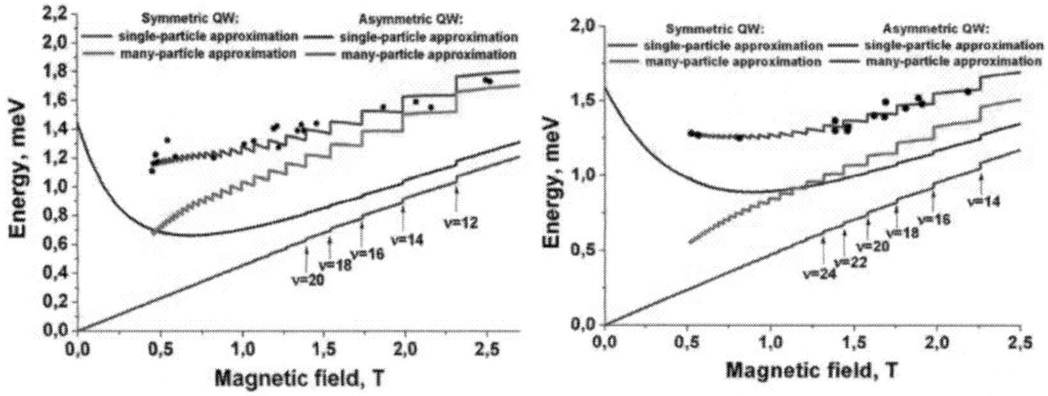

Figure 4: ESR spin-gap in T340 (left panel) and T338 (right panel) samples. The symbols are experimental data in the samples. Magnetic fields corresponding to even-valued LL filling factors are indicated by arrows. Black and brown curves correspond to the Hartree approximation in a asymmetric and symmetric QWs. Red and blue curves are the calculation within HFA in InAs/AlSb QW with a symmetric and asymmetric profile, respectively.

This work demonstrates that subband structure related to the single-particle wave functions and SOI affects significantly the exchange enhancement of quasiparticle and ESR spin-gap in symmetric and asymmetric narrow-gap QWs both in the case of integer and fractional Quantum Hall regime. We show that ESR can serve as a tool to probe the *e-e* interaction effects in narrow-gap QWs. We hope that our results will stimulate further experimental 'magnetostatic' and 'magnetooptic' studies of narrow-gap QW heterostructures and will provide an in-depth understanding of the physical mechanisms responsible for the exchange enhancement of spin-gap in 2DEG.

4. Acknowledgements

This work is partially supported by the non-profit Dynasty Foundation, by the RFBR (Grants 11-02-93111, 11-02-00952, 12-02-00940), by the Russian Academy of Sciences, and by Russian Ministry of Education and Science (Grant НШ-4756.2012.2, Contract P805). The authors are grateful to A. V. Ikonnikov for his helpful discussions on this work.

References

[1] T Ando and Y Uemura, *J. Phys. Soc. Jpn* **37**, 1044 (1974).
[2] T P Smith, B B Goldberg, P J Stiles and M Heiblum, *Phys. Rev. B* **32**, 2696 (1985).
[3] E E Mendez, L Esaki and W I Wang, *Phys. Rev. B* **33**, 2893 (1986).
[4] E E Mendez, H Ohno, L Esaki and W I Wang, *Phys. Rev. B* 43, 5196 (1991).
[5] E E Mendez, J Nocera and W I Wang, *Phys. Rev. B* **47**, 13937 (1993).
[6] O E Dial, R C Ashoori, L N Pfeiffer and K W West, *Nature* **448**, 176 (2007).
[7] M Dobers, K von Klitzing and G Weimann, *Phys. Rev. B* **38**, 5453 (1988).
[8] S S Krishtopenko, V I Gavrilenko and M Goiran, *J. Phys.: Condens. Matter* **24**, 252201 (2012).
[9] S S Krishtopenko, V I Gavrilenko and M Goiran, *J. Phys.: Condens. Matter* **23**, 385601 (2011).
[10] S S Krishtopenko, V I Gavrilenko and M Goiran, *J. Phys.: Condens. Matter* **24**, 135601 (2012).
[11] S S Krishtopenko, V I Gavrilenko and M Goiran, *Solid State Phenomena* **190**, 554 (2012).
[12] A V Ikonnikov, S S Krishtopenko, V I Gavrilenko, Yu G Sadofyev, Yu B Vasilyev, M Orlita and W Knap, *J. Low Temp. Phys.* **159**, 197 (2010).
[13] S S Krishtopenko, A V Ikonnikov, K V Maremyanin, K E Spirin, M Sadowsky, V I Gavrilenko, M Goiran, Yu G Sadofyev and Yu B Vasilyev, *J. Appl. Phys.* **111**, 093711 (2012).
[14] V I Gavrilenko, A V Ikonnikov, S S Krishtopenko, A A Lastovkin, K V Maremyanin, Yu G Sadofyev and K E Spirin, *Semiconductors* **44**, 616 (2010).
[15] K E Spirin, K P Kalinin, S S Krishtopenko, K V Maremyanin, V I Gavrilenko and Yu G Sadofyev, *Semiconductors* **46**, 1396-1401 (2012).
[16] V I Gavrilenko, S S Krishtopenko and M Goiran, *Semiconductors* **45**, 110 (2011).
[17] V Ya Aleshkin, V I Gavrilenko, A V Ikonnikov, S S Krishtopenko, Yu G Sadofyev and K E Spirin, *Semiconductors* **42**, 828 (2008).
[18] S S Krishtopenko, K P Kalinin, V I Gavrilenko, Yu G Sadofyev and M Goiran *Semiconductors*, **46**, 1163 (2012).
[19] T Ando, A B Fowler and F Stern, *Rev. Mod. Phys.* **54**, 437 (1982).
[20] P Pfeffer and W Zawadzki, *Semicond. Sci. Technol.* **19**, 1 (2004).
[21] Yu A Bychkov and E I Rashba, *J. Phys. C* **17**, 6039 (1984).
[22] S M Girvin, A H MacDonald and P M Platzman, *Phys. Rev. B* **33**, 2481 (1986).
[23] J P Longo and C Kallin *Phys. Rev. B* **47**, 4429 (1993).
[24] K Park, V Melik-Alaverdian, N E Bonesteel and J K Jain, *Phys. Rev. B* **58**, 10167 (1998).
[25] S S Krishtopenko (to be published).
[26] Stefano Chesi and Daniel Loss, *Phys. Rev. Lett.* **101**, 146803 (2008).
[27] A Berg, M Dobers, P R Gerhardts and K von Klitzing, *Phys. Rev. Lett.* **64**, 2563 (1990).

Journal of Physics: Conference Series **461** (2013) 012038 doi:10.1088/1742-6596/461/1/012038

Cyclotron resonance in HgCdTe-based heterostructures in strong magnetic fields.

M S Zholudev[1,2], F Teppe[1], M Orlita[3], V Ya Aleshkin[2], A V Ikonnkov[2], V I Gavrilenko[2], W Knap[1], N N Mikhailov[4], S A Dvoretskii[4]

[1] Laboratoire Charles Coulomb (L2C), UMR CNRS 5221, GIS-TERALAB, Universite Montpellier II, 34095 Montpellier, France
[2] Institute for Physics of Microstructures, Russian Academy of Sciences, Nizhny Novgorod, Russia
[3] Laboratoire National des Champs Magnetiques Intenses, CNRS-UJF-UPS-INSA, 25, avenue des Martyrs, FR-38042 Grenoble, France
[4] A.V. Rzhanov Institute of Semiconductor Physics, Siberian Branch, Russian Academy of Sciences, Novosibirsk, 630090 Russia

E-mail: zholudev@ipmras.ru

Abstract. Cyclotron resonance (CR) study of HgCdTe-based quantum wells with both inverted and normal band structures in quantizing magnetic fields was performed. In normal-band sample in addition to electron CR transitions a strong absorption line presumably related to impurity transitions was discovered. Anticrossing of electron and hole Landau levels was observed in inverted-band sample. In semimetallic HgTe quantum wells with inverted band structure, a hole CR line was observed for the first The obtained results were interpreted within the Kane 8×8 model, the valence band offset of CdTe and HgTe, and the Kane parameter EP being adjusted.

1. Introduction

HgTe/CdTe-based quantum wells (QWs) exhibit a number of remarkable properties. HgTe has inverted band structure and is said to have negative gap about 0.3 eV while CdTe has normal band structure and gap about 1.6 eV. At the critical HgTe QW thickness (6.3 to 7 nm depending on Cd content in the barrier), the forbidden gap is absent and both electrons and holes are characterized by the linear energy-momentum law of massless Dirac fermions [1, 2]. When HgTe QW width exceeds this critical value, the energy band structure is inverted (the conduction band states are formed by p-type wavefunctions while s-type wavefunctions form the valence band states; see, e.g., [1, 3] and references therein).

In the inverted band structure regime, HgTe QWs are shown to be two-dimensional (2D) topological insulators that have attracted a great fundamental interest [1, 2, 4, 5]. It was demonstrated [4] that a quantum spin Hall insulator state exists in such systems that can be destroyed by magnetic field due to crossing of Landau levels of different bands [6]. Actually, these two levels have recently shown to display the effect of the avoided crossing [7, 8].

In wide HgTe QWs (width above 12 nm) the energy of side valence band maxima exceed the energy of conduction band bottom. Such structures are said to be in semimetallic state where

Journal of Physics: Conference Series **461** (2013) 012038 doi:10.1088/1742-6596/461/1/012038

2D electrone and hole gases exist simultaneously [9]. In inverted QWs hole-like symmetry of conduction-band Bloch functions enhances spin-dependent effects like the Rashba splitting that has been shown to achieve 30 meV [3, 10]. All these properties allows considering HgTe/CdTe-based QWs as perspective base for spintronic devices. On the other hand, narrow HgTe QWs have been proposed as a material for detectors of THz radiation since they possess certain advantages over bulk HgCdTe solid solutions that are widely used for mid-infrared (IR) photodetectors.

In this work we present the results of cyclotron resonance measurements theoretical calculations for three HgCdTe-based samples of different types (normal, inverted and semimetallic) that display some of the properties described above.

2. Experiment

We present the results obtained with three HgTe/$Cd_yHg_{1-y}Te$ samples (see table 1). The structures were grown by molecular beam epitaxy on semi-insulating GaAs(013) substrates [11]. The ZnTe and thick relaxed CdTe buffer layers, a 100-nm $Cd_yHg_{1-y}Te$ lower barrier layer, a HgTe QW, and a 100-nm $Cd_yHg_{1-y}Te$ upper barrier layer were grown one by one, followed by a 50-nm CdTe cap layer.

Table 1: Parameters of HgTe/$Cd_yHg_{1-y}Te$ quantum wells

Sample number	Growth number	QW width (nm)	y_{Bar}	Band structure	Doping
1	110623	5.6	0.62	normal	ND
2	091222-1	8	0.7	inverted	In, $1.3 \cdot 10^{11}$ cm^{-2}
3	110615	20.2	0.65	semimetallic	ND

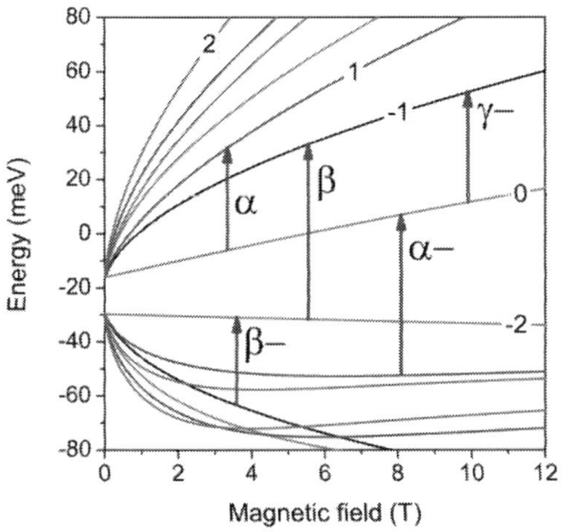

Figure 1: Landau levels for sample 1.

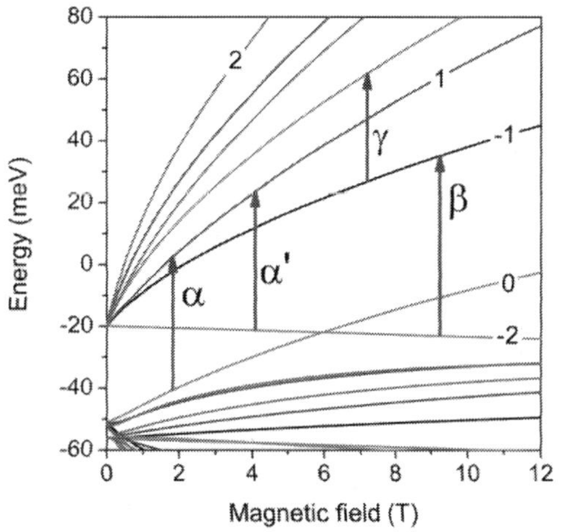

Figure 2: Landau levels for sample 2.

There was no intentional doping in samples 1 and 3. In the sample 2 barriers were symmetrically doped with donors (indium). Sample 1 has normal band structure. Samples 2 and 3 have inverted band structure and sample 3 was in semimetallic state.

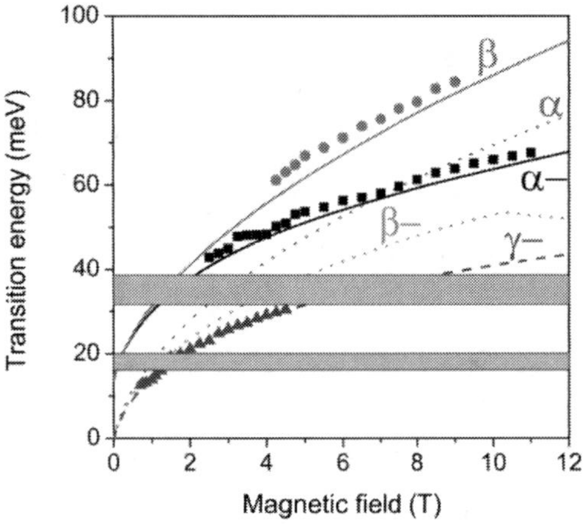

Figure 3: Theoretical transition energies (lines) and experimental absorption line positions (markers) for sample 1. Dot lines represent the transitions not observed. Dash line shows the transition presumably involving impurity states. Grey bars show the restrahlen bands.

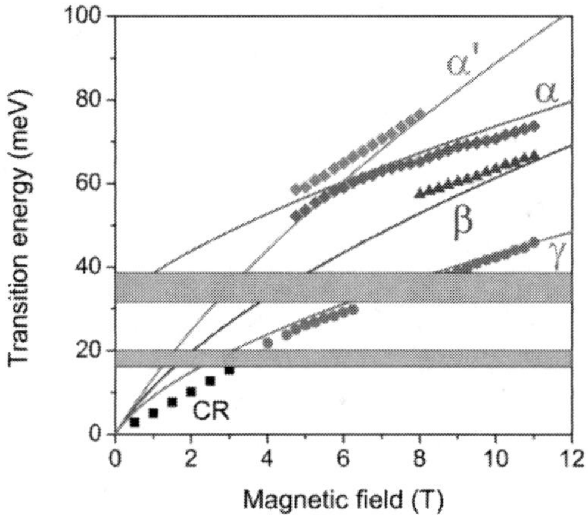

Figure 4: Theoretical transition energies (lines) and experimental absorption line positions (markers) for sample 2. Grey bars show the restrahlen bands.

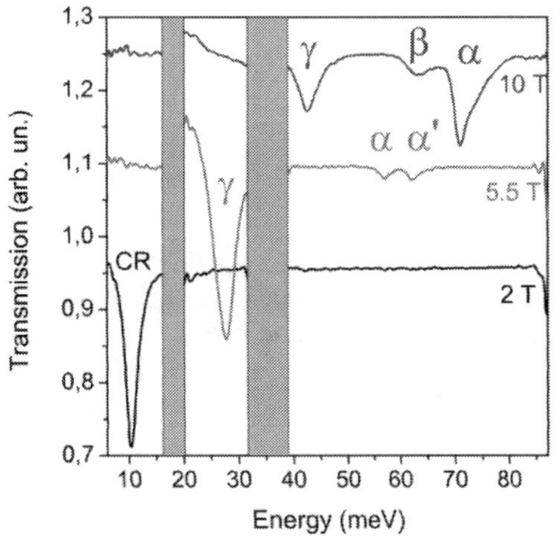

Figure 5: Selected transmission spectra for sample 2. Grey bars show the restrahlen bands.

Table 2: Transitions between LLs. The first number is the LL number n. The letter "c" or "v" means conduction and valence band respectively. The number next to the letter is the number of the level of certain n in the band (counted from the band gap).

Transition	Normal	Inverted
α	0,c1 $\rightarrow$ 1,c1	0,v1 $\rightarrow$ 1,c1
α'	—	-2,c1 $\rightarrow$ 1,c1
$\alpha-$	1,v1 $\rightarrow$ 0,c1	1,v1 $\rightarrow$ 0,v1
β	-2,v1 $\rightarrow$ -1,c1	-2,c1 $\rightarrow$ -1,c1
$\beta-$	-1,v1 $\rightarrow$ -2,v1	-1,v1 $\rightarrow$ -2,c1
γ	-1,c1 $\rightarrow$ 0,c2	-1,c1 $\rightarrow$ 0,c1
$\gamma-$	0,c1 $\rightarrow$ -1,c1	0,v1 $\rightarrow$ -1,c1

In order to measure the infrared transmittance, the sample was exposed to the radiation of a Globar, which was analyzed by a Fourier transform spectrometer and delivered to the sample via light-pipe optics. The transmitted light was detected by a composite bolometer, which operated at 4.2 K and was placed directly below the sample. The magneto-optical spectra were measured

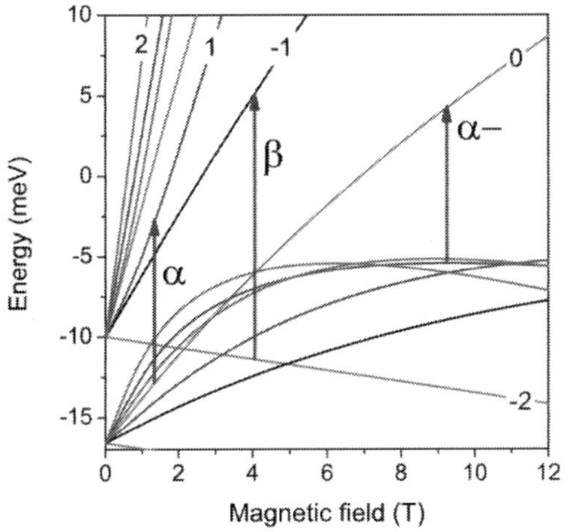

Figure 6: Landau levels for sample 3.

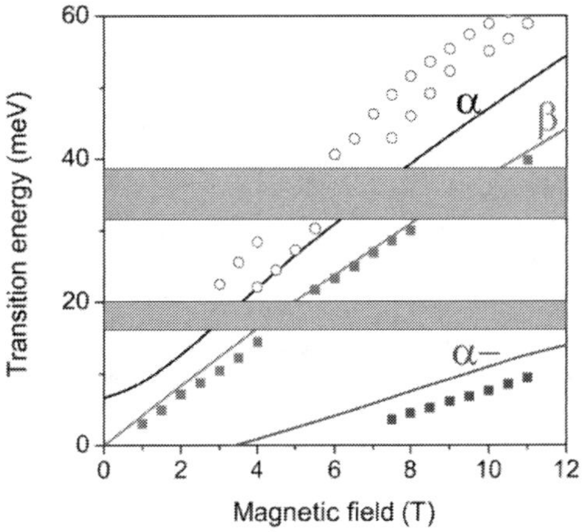

Figure 7: Theoretical transition energies (solid and doted lines) and experimental absorption line positions (markers) for sample 3. Empty markers represent lines of lower intensity. Grey bars show the restrahlen bands.

in Faraday configuration using a superconducting coil (up to 11 T). All the spectra presented here were normalized by the sample transmission at zero field and then divided by the rate of reference signals (signal without sample) at non-zero and zero magnetic field. The latter have been done to eliminate slope caused by the influence of the field on the bolometer.

3. Theory

The band structure in the absence of the magnetic field and the Landau levels (LLs) in the QWs under study were calculated in the axial approximation in the same way as described in [8, 12, 13] in the four-band model. The calculation is based on the envelope function method proposed by Burt [14]. The envelope functions were found as the solutions of the time-independent Schrodinger equation with the 8×8 Hamiltonian taking into account built-in strain. To calculate the envelope functions and the corresponding values of the electron energy, the structure was approximated by a superlattice of weakly interacting QWs. The lattice period was chosen such that the interaction between the wells would not significantly affect the energy spectrum. The calculation was performed by expanding the envelope functions in plane waves. The expression for the Hamiltonian of the heterostructure grown on the (013) plane was derived by the method described in [15]. Zeeman term was found to be negligible for our structures and dropped. The components of the built-in strain tensor were calculated with the use of the formulas from [8, 12]. The band parameters of the materials used in the calculation are taken from [13, 16].

In order to get better agreement of the calculations and experimental data on transition energies we have adjusted some parameters of the model [8]. The first one is the valence band offset (VBO) of CdTe and HgTe. Since there is no way to measure this parameter directly it is usually estimated by matching experimental results and calculations. According to [16] VBO is 570 ± 60 meV. We have used the value 620 meV to fit our results. The second adjusted parameter is the Kane parameter P (or $E_P = 2m_0P^2/\hbar^2$). According to the model we use, the wave function of the electron is expanded over the same basis functions in whole heterostructure [14]. Hence parameter E_P depending only on basis functions is also the same for all layers and can be freely adjusted since the reference crystal is chosen arbitrarily. The point one should take care is to keep other parameters in agreement with E_P so that bulk conduction band effective mass m_c not change. The E_P value used was 20.8 eV (instead of 18.8 eV [13]).

4. Results and discussion

Calculated magnetic field dependencies of the LLs for the samples are presented on figures 1, 2 and 6. Color of the levels corresponds to their numbers n. We follow numbering convention used in [13] i.e. quantum number n has values from -2 to infinity. Selection rules for Faraday configuration allows transition only between levels whose numbers n differs by one. The transitions that are expected to be observed are represented with arrows and marked with Greek letters. This notation differs for normal and inverted bands and is described in table 2.

The first sample has normal band structure. It was not intentionally doped and has low carrier concentration. In these conditions we can observe the transitions involving the levels closest to the energy gap as shown in figure 1. The transitions that fit the experimental points are $\alpha-$, β and $\gamma-$ 3. Even though $\gamma-$ transition is allowed by selection rules calculation of the oscillator strengths shows that this transition should not be observed in this geometry. On the other hand the line we see in experiment can not be approximated by any other transitions (see Figure 3). We suppose that in this sample transition between impurity states bound to the related Landau levels takes place.

The Landau levels for the second sample are plotted in Figure 2. It was doped with donors and has finite concentration of electrons in conduction band. This sample has inverted band structure which is notable by the inverted layout of LLs with $n = 0$ and $n = -2$ closest to band gap with respect to normal band structure (Figure 1). These levels display the effect of avoided crossing [7, 8] which can not be described by axial model we use. This effect results in admixture of $n = 0$ in the wavefunction of the level with $n = -2$. Due to this mixing the transition α' between LL with $n = -2$ and lowest LL with $n = 1$ in conduction band is allowed near the crossing point.

The transitions from the crossing levels (α and α') are observed as two close peaks (figure 5) one of which vanishes when magnetic field is far from the crossing point (figure 4). We have also observed points corresponding to β and γ transitions. The cyclotron resonance (CR) lines marked with black squares on figure 4 demonstrate linear behavior with magnetic field as expected.

Sample number 3 has inverted band structure. During the experiment was in semimetallic state since its valence band side maxima are above the bottom of conduction band. One can see on figure 6 that because of this feature all Landau levels from the first valence subband have parts lying above the starting point of conduction band LLs. Each level with $n \geq 1$ have maximum at some magnetic field that is decreasing with n.

In this sample we observe two intensive lines (see figure 7). The first one is in a good agreement with β transition between conduction band levels. The other one goes alongside $\alpha-$ transition between valence band levels. Agreement is not good here probably because axial description is less accurate for valence band than for conduction band. Interband α transition is not observed above 6 T since there should be no carriers on the corresponding LLs in these fields. Few points of low intensity can be attributed to this transition between 4 and 6 T. Weak high-frequency lines probably resulted from some interband transitions (cyclotron or impurity). At the moment, it is difficult to identify them because of the great number of allowed transitions between valence and conduction band LLs in this frequency range.

Thus fields about 8 T we observe electron and hole intraband transitions simultaneously which is only possible for semimetallic structure.

5. Conclusion

We have experimentally studied far infrared magnetoabsorption spectra under quantizing magnetic fields in a series of HgCdTe-based heterostructures including normal-band, inverted-band and semimetallic samples. The results obtained are interpreted on the basis of Landau level calculations within the axial Kane 8×8 model. Impurity related absorption lines have been revealed in addition to electron CR ones. We have confirmed anticrossing of electron and hole Landau levels in inverted band structure. In semimetallic sample electron and hole CR transitions were observed simultaneously.

6. Acknowledgments

This work was supported by CNRS and GDR-I project "Semiconductor sources and detectors of THz frequencies", and by the Scientific Council of Montpellier II University via the "Nanosciences" grant. We acknowledge the Region of LanguedocRoussillon through the "Terahertz Platform" project. The work was also supported by Russian Foundation for Basic Research (grants 11-02-00958, 11-02-97061, 11-02-93111) and by Russian Academy of Sciences.

References

[1] Bernevig A, Hughes T and Zhang S C 2006 *Science* **314** 1757–1761
[2] Buttner B, Liu C X, Tkachov G, Novik E G, Brune C, Buhmann H, Hankiewicz E M, Recher P, Trauzettel B, Zhang S C and Molenkamp L W 2011 *Nat. Phys.* **7** 418–422
[3] Gui Y S, Becker C R, Dai N, Liu J, Qiu Z J, Novik E G, Schafer M, Shu X Z, Chu J H, Buhmann H and Molenkamp L W 2004 *Phys. Rev.* B **70** 115328
[4] Konig M, Wiedmann S, Brune C, Roth A, Buhmann H, Molenkamp L W, Qi X L and Zhang S C 2007 *Science* **318** 766–770
[5] Konig M, Buhmann H, Molenkamp L W, Hughes T, Liu C X, Qi X L and Zhang S C 2008 *J. Phys. Soc. Jpn.* **77** 031007
[6] Schultz M, Merkt U, Sonntag A, Rossler U, Winkler R, Colin T, Helgesen P, Skauli T and Lovold S 1998 *Phys. Rev.* B **57** 14772–14775
[7] Orlita M, Masztalerz K, Faugeras C, Potemski M, Novik E G, Brune C, Buhmann H and Molenkamp L W 2011 *Phys. Rev.* B **83** 115307
[8] Zholudev M, Teppe F, Orlita M, Consejo C, Torres J, Dyakonova N, Czapkiewicz M, Wrobel J, Grabecki G, Mikhailov N, Dvoretskii S, Ikonnikov A, Spirin K, Aleshkin V, Gavrilenko V and Knap W *Phys. Rev.* B In press
[9] Kvon Z D, Olshanetsky E B, Kozlov D A, Mikhailov N N and Dvoretskii S A 2008 *JETP Letters* **87** 502–505
[10] Spirin K E, Ikonnikov A V, Lastovkin A A, Gavrilenko V I, Dvoretskii S A and Mikhailov N N 2010 *JETP Lett.* **92** 63–66
[11] Dvoretsky S, Mikhailov N, Sidorov Y, Shvets V, Danilov S, Wittman B and Ganichev S 2010 *J. Electron. Mater.* **39** 918–923
[12] Ikonnikov A V, Zholudev M S, Spirin K E, Lastovkin A A, Maremyanin K V, Aleshkin V Y, Gavrilenko V I, Drachenko O, Helm M, Wosnitza J, Goiran M, Mikhailov N N, Dvoretskii S A, Teppe F, Diakonova N, Consejo C, Chenaud B and Knap W 2011 *Semicond. Sci. Technol.* **26** 125011
[13] Novik E G, Pfeuffer-Jeschke A, Jungwirth T, Latussek V, Becker C R, Landwehr G, Buhmann H and Molenkamp L W 2005 *Phys. Rev.* B **72** 035321
[14] Burt M G 1992 *J. Phys.: Condens. Matter* **4** 6651
[15] Los J, Fasolino A and Catellani A 1996 *Phys. Rev* B **53** 4630
[16] Becker C R, Latussek V, Pfeuffer-Jeschke A, Landwehr G and Molenkamp L W 2000 *Phys. Rev.* B **62** 10353–10363

Journal of Physics: Conference Series **461** (2013) 012039 doi:10.1088/1742-6596/461/1/012039

Photoconductivity of InGaN/GaN multiple quantum well heterostructures

M V Baranovskiy and G F Glinskii

Department of Micro- and Nanoelectronics, Saint Petersburg Electrotechnical University "LETI", 197376, St. Petersburg, Russia

maxim.bmw.spb@gmail.com

Abstract. Photocurrent and photoconductivity of InGaN/GaN multiple quantum well heterostructures as a function of applied reverse bias is investigated. Optical excitation was carried out in blue and violet regions of the spectrum, and temperature was ranging from 10 to 300 K. We observed characteristic features related to consequently moving space charge boundary through the quantum wells. For each quantum well there is a range of reverse bias with negative differential conductivity when excited by blue light. Frequency and temperature measurements revealed the presence of at least two different mechanisms that determine the photoconductivity of the structures.

1. Introduction

At the present time semiconductor heterostructures are widely used for the creation of electronic devices for various purposes. This was contributed by the significant progress made in recent years in the theory of such systems [1], and in growth technologies. The especial interest to InGaN/GaN multiple quantum well (MQW) structures is due to possibility of their use for light-emitting diode (LED) lighting. However, there are a number of obstacles to significant increase in the brightness of LEDs based on these structures. Among them there are reduction of efficiency as the injected current density increases (so-called efficiency droop), and non-uniform properties of the structure with depth, which prevents an increase in the intensity by increment of number of quantum wells in the active region. As a result at the present time much attention is paid to the problems of investigation and diagnostics of InGaN/GaN multiple quantum well structures. Today methods of admittance spectroscopy [2] and the capacitance-voltage (*C-V*) profiling [3] are widely used due to their high informational content. They can be utilized to determine the distribution of free charge carriers in the depth of heterostructure, *i.e.* obtain information about location and quality of each quantum well. A more informative method is admittance measurements, combined with optical excitation of heterostructure at a specific wavelength. If the photon energy is larger than band gap of the quantum well InGaN, but less than band gap of the barrier GaN, the absorption of light and the generation of excess carriers occur only in quantum wells. As a result, the response of each quantum well is increased. Thus, in [4], the method of capacitance-voltage profiling in combination with the optical excitation was used for determination of well-to-well non-uniformity in InGaN/GaN multiple quantum well structures.

In this work investigations of photoelectric phenomena in quantum well structures are described. The samples were commercial grade blue emitting LEDs based on InGaN/GaN MQW heterostructure. The photocurrent and differential photoconductivity as functions of reverse voltage applied to the *p-n*–

junction are investigated. The excitation was carried out by the light from LEDs emitting in blue and violet regions of the spectrum. We clearly observed characteristic features in photocurrent and photoconductivity related to consequently moving space charge boundary through the quantum wells.

2. Samples under investigation and experimental technique

The samples were commercial grade blue emitting InGaN/GaN LEDs. The wavelength at the maximum of luminescence was 465 nm. The area of *p-n*–junction, measured using an optical microscope, was 0.067 mm^2. From the capacitance-voltage measurements conducted with the modulation frequency of 1 MHz (Fig. 1a), it was determined that structures under investigation contained five quantum wells at a distance of approximately 18 nm from each other (Fig. 1b).

The measurements were performed on the automated setup [5], consisting of the helium cryostat Janis CCS-150/204N, the temperature controller LakeShore 325, the immittance meter E7-20, the power supply Agilent E3643A and a computer. Blue and violet LEDs were used as an exciting light source. The wavelength of the peak of luminescence spectrum for the blue LED was 465 nm and for the violet - 420 nm. The width of spectrum for both LEDs was about 30 nm, and the optical power was about 1 mW. The intensity of the optical excitation was constant, but was not normalized.

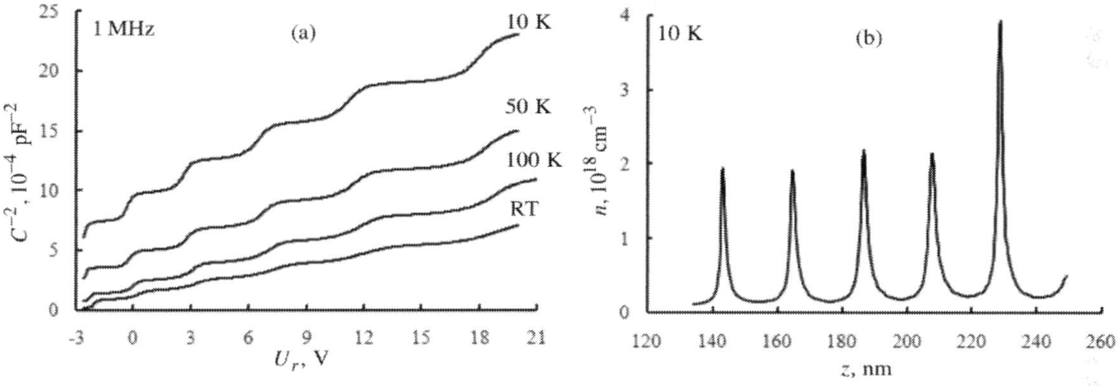

Figure 1. (a) – Capacitance of samples to the power of (-2) as a function of reverse voltage, measured at the frequency of 1 MHz at different temperatures; (b) – Charge carriers concentration profile, obtained by *C-V*-profiling at T=10 K.

Samples were measured as follows. The sample under investigation was held in a cryostat chamber and connected to the immittance meter, which allowed measuring electrical parameters of the sample such as the capacitance, the conductivity and the current. Temperature of the sample in the cryostat was ranging from 10 to 300 K. The excitation light was constantly focused on the sample through an optical window in the measuring chamber.

In measurements of photocurrent a constant voltage was applied to the samples, providing the bias of *p-n*–junction, and the through current was recorded. Measurements were carried out under illumination and in the dark, after which the difference between them was obtained.

In measurements of differential photoconductivity a complex signal was applied to samples, which is the amount of DC voltage, providing the bias of *p-n*–junction, and AC voltage with constant amplitude of 40 mV. Active and reactive components of admittance, the differential conductance and differential capacitance, were determined. Measurements of differential conductivity under illumination in the blue region of the spectrum and in the dark were conducted, after which the difference between them was obtained. Measurements were made at frequencies from 30 Hz to 10 kHz. The measurements were not made at higher frequencies, because there is a growth of dark conductivity as the frequency increases, against which the effects associated with the photoelectric phenomena become insignificant. This dark conductivity is associated with lagged recharge of deep

Journal of Physics: Conference Series **461** (2013) 012039 doi:10.1088/1742-6596/461/1/012039

traps and bound states in the quantum wells when the external voltage is rapidly changing. In this work, this phenomenon is not discussed in detail.

3. Experimental data and discussion

The experimental investigation of the dependence of photocurrent in InGaN/GaN MQW heterostructures on the reverse bias revealed the presence of characteristic features. In Fig. 2 the photocurrent I_{ph} as a function of reverse bias U_r is plotted. These curves were obtained by optical excitation in violet (curve 1) and blue (curve 2) regions of the spectrum. A comparison of these curves with the data of capacitance-voltage measurements (dashed line in Fig. 2) suggests that the photocurrent features characterize the internal structure of samples. They occur at the voltages at which the boundary of the space charge region crosses the quantum wells. Indeed, in the case where the quantum well is in the quasi-neutrality area, where the electric field of the *p-n*–junction is absent, the flows of charge carriers generated by light in all directions are the same, and the total photocurrent in this case is zero. Conversely, if the quantum well is located inside of the space charge region, the electric field of the *p-n*–junction leading to efficient spatial separation of the light-generated electrons and holes, which have overcome the potential barrier of a quantum well by thermal emission. In this case the photocurrent is not zero. Thus, with increasing reverse bias, when the boundary of the space charge region crosses a quantum well, the photocurrent should rise sharply, as is observed in the experiment. Curve 1 in Fig. 2 clearly shows the stepwise increase of photocurrent.

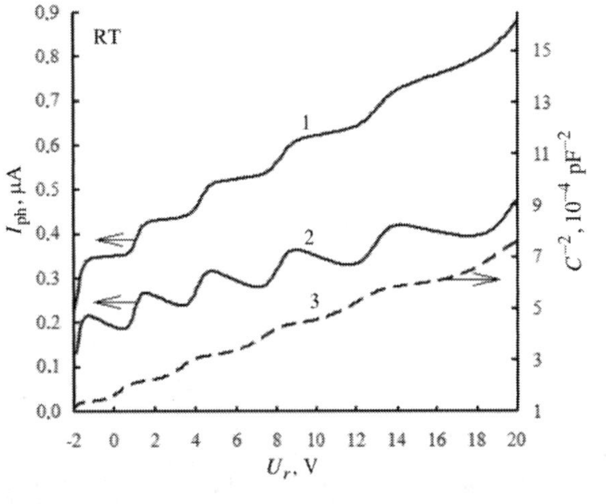

Figure 2. Photocurrent as a function of reverse voltage, measured with optical excitation in the violet (curve *1*) and blue (curve *2*) regions of the spectrum at room temperature. For comparison, on the second axis the capacitance-voltage dependence is plotted (dashed curve *3*).

On the curve 2 there are areas of photocurrent decrease with increasing reverse bias. An essential condition for the emergence of this effect was that the optical excitation was carried out in the spectral region corresponding to the edge of the optical absorption of quantum wells InGaN. As described in [6], this effect is probably due to the decrease of the absorption of light as a result of the expansion of the space charge. At small reverse bias when InGaN quantum well is in the quasi-neutrality region, there is a strong built-in piezoelectric field inside the well. Due to quantum-confined Stark effect, the energy gap between the ground states of electrons and holes in a quantum well is much less than in the quantum well of the same width and depth, but without the piezoelectric field. With increasing reverse bias quantum well gets into the space charge region of *p-n*–junction, the electric field is partially compensated piezoelectric field inside the quantum well. This results in an increase of the energy gap between the ground states of the electron and the hole, and the optical absorption edge is shifted to shorter wavelengths. As a result, the number of absorbed photons is significantly reduced.

In Fig. 3 the dependence of photocurrent on the reverse bias at different temperatures is plotted. It can be noted that the features of the photocurrent are most clearly observed at room temperature and at low temperatures they are significantly smoothed. This is due to the fact that charge carriers generated by light in quantum wells, to participate in the photocurrent have to overcome the potential barrier of the quantum well. The most likely mechanism for this process is the thermal activation, so the increase in temperature leads to an increase in the photocurrent as well as enhancement of the features in the current-voltage dependence.

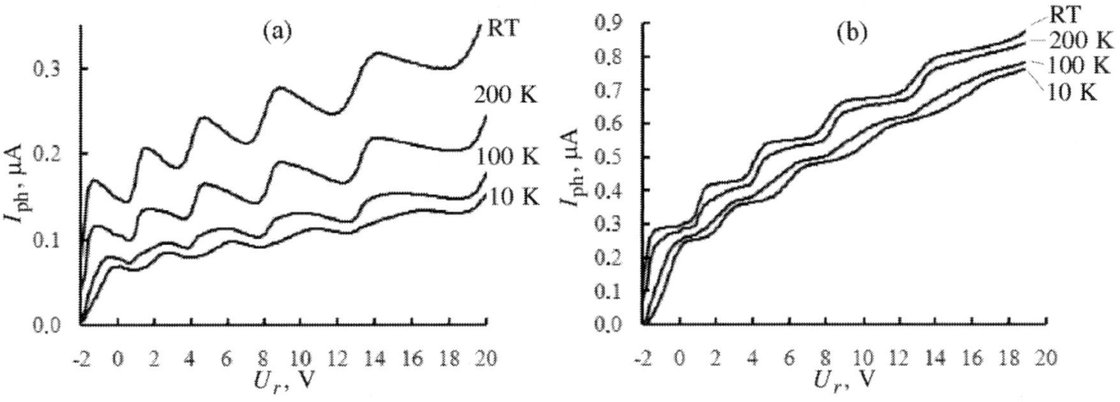

Figure 3. Photocurrent as a function of reverse voltage, measured under illumination in blue (a) and violet (b) regions of the spectrum at different temperatures.

More detailed information about the structure can give differential photoconductivity investigations at different frequencies and temperatures. Such studies allow the analysis of the rate of various mechanisms that determine the photoconductivity. In Fig. 4 a differential photoconductance vs. reverse bias is plotted. Fig. 4a shows the data obtained at a frequency of 300 Hz at different temperatures. Fig. 4b shows the dependencies obtained at different frequencies at 150 K. The sharp peaks in the conductance at reverse voltages 1.1, 4.2, 8.3, 13.2 and 20 V related to abrupt increases in the photocurrent when the boundary of the space charge crosses quantum wells. The amplitude of the peaks decreases with decreasing temperature and is almost independent of frequency.

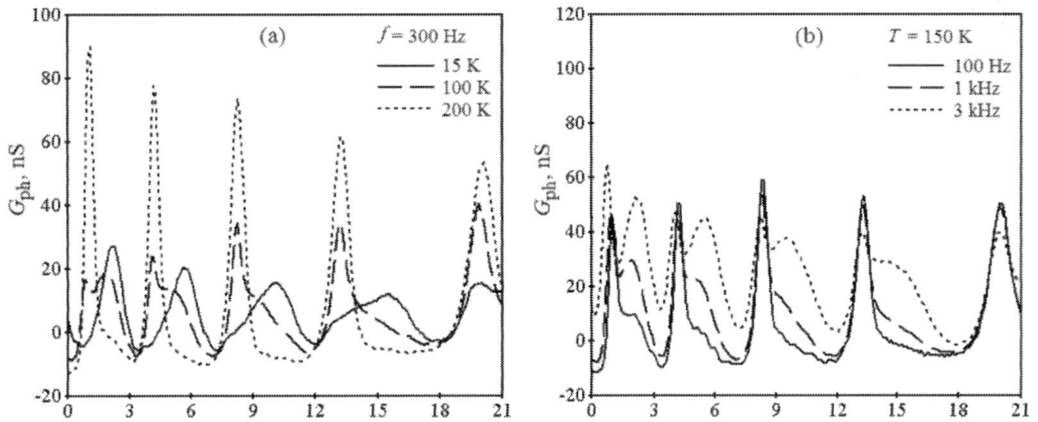

Figure 4. Differential photoconductance as a function of reverse voltage.

This behavior is indicative of the fact that the photoconductivity depends on the process of thermal emission of carriers from quantum wells into the barriers. For each quantum well, there is also an additional peak observed at higher voltages than the first peak. Additional peak exists only at low temperatures, and its amplitude increases with decreasing temperature and increasing frequency. The origin of this phenomenon is not explained. These data suggest that the formation of the photocurrent is probably attended by several competing mechanisms at different rate; their probabilities depend on temperature and frequency.

4. Conclusion

The investigation of InGaN/GaN multiple quantum well heterostructures have shown that the dependence of photocurrent on the reverse bias characterize the internal structure of the samples. At voltages when the boundary of *p-n*–junction space charge crosses quantum wells, the photocurrent sharply increases, that is related to spatial separation of the light-generated charge carriers in the electric field of the *p-n*–junction. It is experimentally shown that when the optical excitation is carried out at a wavelength close to the absorption edge in quantum wells, there are areas of negative differential conductivity. It is assumed that their presence is associated with the shift of the optical absorption edge in quantum wells to the shortwave region, caused by compensation of piezoelectric field in the quantum wells by the electric field the *p-n*–junction. In the case of excitation by light with high energy photons negative differential conductivity disappears.

It is experimentally shown that features of the photocurrent are most pronounced at room temperature and are significantly smoothed at low temperatures. It is explained by the fact that for participation in photocurrent the charge carriers must overcome the potential barrier of the quantum well. The mechanism of this process is the thermal activation, the probability of which is maximal at high temperatures. However, investigations of the differential photoconductivity at different frequencies and temperatures revealed the presence at least one another mechanism that influences the photoconductivity. It is more probable at low temperatures and high frequencies. The origin of this phenomenon is not explained.

Based on these results, we can conclude that the research of dependence of photocurrent on the reverse bias can be used as an effective non-destructive method for diagnostics of semiconductor heterostructures [7], in particular, LEDs based on InGaN/GaN MQW structure. There are several advantages of this approach in relation to commonly used admittance methods. Firstly, the implementation of this method requires relatively simple equipment: a voltage source, a current meter and a source of optical excitation, *e.g.* LED or laser. Secondly, the measurements of a sample can be made for hundredths of a second. It can be achieved because the instantaneous value of current is measured. Unlike this, for the capacitance measurements the lock-in amplifiers are used. Finally, the observed features in photocurrent associated with quantum wells are most pronounced at high temperatures, so the cryostat is not required for the measurements.

References

[1] Glinskii G F 2008 *Semiconductors and semiconductor nanostructures: symmetry and electronic states* (St. Petersburg: "Technolit")

[2] Zubkov V I 2007 *Diagnostics of semiconductor nanoheterustructures with admittance spectroscopy methods* (St. Petersburg: "Elmor")
Zubkob V I, Kapteyn C M A, Solovonov A V and Bimberg D 2005 *J. Phys.: Condens. Matter* **17** 2435-42

[3] Zubkov V I, Melnik M A, Solomonov A V, Tsvelev E O, Bugge F, Weyers M and Trankle G 2004 *Phys. Rev.* B **70** 75312
Zubkov V I, Melnik M A, Solovonov A V, Pikhtin A N and Bugge F 1999 *Semiconductors* **33** 858-61

[4] Kim T S, Ahn B J, Dong Y, Park K N, Lee J G, Moon Y, Yuh H K, Choi S C, Lee J H, Hong S K and Song J H 2012 *Appl. Phys. Lett.* **100** 71910

[5] Baranovskiy M V and Glinskii G F 2012 *Izvestiya SPbGETU "LETI"* **4/2012** 3-7

[6] Baranovskiy M V Glinskii G F and Mironova M S 2013 Photoelectric method for diagnostics of multiple quantum well InGaN/GaN heterostructures *Semiconductors* **47** (to be published)

[7] Baranovskiy M V and Glinskii G F 2012 The device for monitoring of the quality of semiconductor quantum well heterostructures *Patent* **RU 117714** issued June 27, 2012

Journal of Physics: Conference Series **461** (2013) 012040 doi:10.1088/1742-6596/461/1/012040

Inter-valley mixing in Si/SiO_2 heterostructures

G F Glinskii and M S Mironova
Department of Micro- and Nanoelectronics, St. Petersburg State Electrotechnical University, 5 Professora Popova Street, St. Petersburg 197376, Russia

E-mail: mironova.m.s@gmail.com

Abstarct. We present a theory for heterostructures based on diamond structured semiconductors, with an emphasis of calculating inter-valley coupling. The theory allows taking into account microscopic symmetry of heterointerface within the envelope-function approach. To find out corrections to the effective Hamiltonian we use methods of group theory. We calculated valley-orbit splitting and electron wave functions for SiO_2/Si/SiO_2 finite square quantum wells. The results are in good agreement with those obtained by other authors.

1. Introduction.

Due to long electron spin relaxation and spin coherence times silicon is considered to be very attractive material for spintronic devices. However, the presence of six equivalent minima (valleys) in the conduction band can lead to decoherence of the electronic states. For this reason, long-lived spin states can be obtained only if energy levels formed by these valleys are split. This situation can be realized in quantum well structures [1].

There are two main approaches to describe inter-valley mixing in heterostructures: microscopic (the tight-binding method, the empirical pseudopotential method) and the envelope-function method. Earlier attempts to take into account inter-valley mixing within the envelope-function approach were made in [2,3]. It is shown [2] that for an infinitely deep quantum well the splitting of the electronic ground state $\Delta E \propto \sin(2k_0 L)$, where k_0 is location of the conduction band minimum in the Brillouin zone, L is the quantum well width. In [4-6] the effects of inter-valley mixing in SiGe/Si/SiGe quantum wells are described by both the tight-binding method and the envelope-function approach. Within the envelope-function approach the potential of heterointerface is treated as a δ-function and taken into account within the perturbation theory. This approach is strictly phenomenological and does not consider the microscopic structure of the heterointerface.

In this paper, we develop a theory which allows taking into account the microscopic symmetry of the heterointerface in structures based on indirect-gap semiconductors. At the heart of our method lies numerical calculation of the Schrödinger equation for the wave functions defined in the crystal lattice nodes. Heterostructure on the basis of A/B materials is assumed as bulk material A, some atoms of which are substituted by atoms B. The solution of the Schrödinger equation is sought in the form of an expansion in the multi-valley Kohn-Luttinger basis. Independent non-zero matrix elements are determined using methods of group theory.

Journal of Physics: Conference Series **461** (2013) 012040 doi:10.1088/1742-6596/461/1/012040

2. Hamiltonian of heterostructure

We considered heterostructure based on two diamond structured semiconductors A and B. As an approximation, we assume that the lattice parameters of bulk materials are the same. Hamiltonian describing the motion of electrons in the bulk material A can be expressed as

$$\hat{H}_A = \frac{\hat{\mathbf{p}}^2}{2m_0} + \sum_i \sum_{\mathbf{a}} V_A(\hat{\mathbf{x}} - \mathbf{a} - \chi_i), \tag{1}$$

where $\hat{\mathbf{p}}$ is the momentum operator; m_0 is the electron rest mass; $\mathbf{a}$ are direct lattice vectors, indexing unit cells; χ_i are vectors indicating a position of an atom of i-th sublattice in a unit cell, $i = 1,2$; $V_A(\mathbf{x})$ is a local potential produced by the atom A and defined within a unit cell.

To describe the behavior of an electron in the heterostructure, we introduce the characteristic functions $f_i(\mathbf{a})$ indicating the replacement of atoms in the i-th sublattice[7]

$$f_i(\mathbf{a}) = \begin{cases} 1, & \textit{if the unit cell with number } \mathbf{a} \textit{ occupied by atom B} \\ 0, & \textit{if the unit cell with number } \mathbf{a} \textit{ occupied by atom A} \end{cases}.$$

Thus, the Hamiltonian of the heterostructure A/B can be written as follows

$$\hat{H} = \frac{\hat{\mathbf{p}}^2}{2m_0} + \sum_i \sum_{\mathbf{a}} V_A(\hat{\mathbf{x}} - \mathbf{a} - \chi_i) + \sum_i \sum_{\mathbf{a}} f_i(\mathbf{a}) \Delta V_{AB}(\hat{\mathbf{x}} - \mathbf{a} - \chi_i), \tag{2}$$

where $\Delta V_{AB}(\mathbf{x}) = V_B(\mathbf{x}) - V_A(\mathbf{x})$ is the difference between the local potentials produced by atoms A and B. The solution of the Schrödinger equation with the Hamiltonian (2) is sought in the form of an expansion in the multi-valley Kohn-Luttinger basis $e^{i\mathbf{k}\hat{\mathbf{x}}}\left|{}^{\star}\alpha, n, \mathbf{k}_j^0\right\rangle$, where n and j are indices numbering the states, which transform according to the irreducible representation $D^{({}^{\star}\alpha)}$ of wave-vector star ${}^{\star}\mathbf{k}^0$ of full space group symmetry of O_h^7, $\mathbf{k}$ is a wave vector close to extreme points $\mathbf{k}_j^0$. Bloch states $\left|{}^{\star}\alpha, n, \mathbf{k}_j^0\right\rangle$ determining the energy spectrum $E_A^{({}^{\star}\alpha)}\left(\mathbf{k}_j^0\right)$ of a bulk semiconductor A at points $\mathbf{k}_j^0$ satisfy the Schrödinger equation with the Hamiltonian (1)

$$\hat{H}_A\left|{}^{\star}\alpha, n, \mathbf{k}_j^0\right\rangle = E_A^{({}^{\star}\alpha)}\left(\mathbf{k}_j^0\right)\left|{}^{\star}\alpha, n, \mathbf{k}_j^0\right\rangle$$

Using the commutation relation $\hat{\mathbf{p}}e^{i\mathbf{k}\hat{\mathbf{x}}} = e^{i\mathbf{k}\hat{\mathbf{x}}}(\mathbf{p} + \hbar\mathbf{k})$, the Hamiltonian (2) in the chosen basis can be rewritten as

$$\left\langle {}^{\star}\alpha, n, \mathbf{k}_j^0 \right| e^{-i\mathbf{k}\hat{\mathbf{x}}} \hat{H} e^{i\mathbf{k}'\hat{\mathbf{x}}} \left| {}^{\star}\alpha', n', \mathbf{k}_{j'}^0 \right\rangle$$
$$= \left\{ \left[E_A^{({}^{\star}\alpha)}\left(\mathbf{k}_j^0\right) + \frac{\hbar^2 k^2}{2m_0} \right] \delta_{\alpha\alpha'} \delta_{nn'} + \frac{\hbar}{m_0} \mathbf{k} \mathbf{p}_{nn'}^{{}^{\star}\alpha {}^{\star}\alpha'}\left(\mathbf{k}_j^0\right) \right\} \delta_{jj'} \delta_{\mathbf{k}\mathbf{k}'} + \Delta U_{nn'}^{{}^{\star}\alpha {}^{\star}\alpha'}\left(\mathbf{k}_j^0 + \mathbf{k} - \mathbf{k}_{j'}^0 - \mathbf{k}'\right) \tag{3}$$

The first term in (3), which is diagonal in the indices j and j', $\mathbf{k}$ and $\mathbf{k}'$, is the same as ordinary *kp*-Hamiltonian of bulk semiconductor A, where

$$\mathbf{p}_{nn'}^{{}^{\star}\alpha {}^{\star}\alpha'}\left(\mathbf{k}_j^0\right) = \left\langle {}^{\star}\alpha, n, \mathbf{k}_j^0 \right| \hat{\mathbf{p}} \left| {}^{\star}\alpha', n', \mathbf{k}_j^0 \right\rangle.$$

The second term, $\Delta U_{nn'}^{{}^{\star}\alpha {}^{\star}\alpha'}\left(\mathbf{k}_j^0 + \mathbf{k} - \mathbf{k}_{j'}^0 - \mathbf{k}'\right)$, determines both intra-valley and inter-valley scattering of electrons at the heterointerface

$$\Delta U_{nn'}^{\alpha\alpha'}\left(\mathbf{k}_j^0 + \mathbf{k} - \mathbf{k}_{j'}^0 - \mathbf{k}'\right) = \sum_i \sum_{\mathbf{a}} f_i(\mathbf{a}) \left\langle {}^{\star}\alpha, n, \mathbf{k}_j^0 \right| e^{-i\mathbf{k}\hat{\mathbf{x}}} \Delta V_{AB}(\hat{\mathbf{x}} - \mathbf{a} - \chi_i) e^{i\mathbf{k}'\hat{\mathbf{x}}} \left| {}^{\star}\alpha', n', \mathbf{k}_{j'}^0 \right\rangle \tag{4}$$

Because materials A and B have the same space group symmetry O_h^7, potential $\Delta V_{AB}(\mathbf{x})$ is an invariant under point-group symmetry of a node T_d. If we restrict ourselves to linear in $(\mathbf{k} - \mathbf{k}')$

terms, the calculation of intervalley mixing is reduced to the determination of the following matrix elements

$$\begin{gathered}\left\langle {}^{\star}\alpha,n,\mathbf{k}_j^0\right|\Delta\hat{V}_i^{\Gamma_1}\left|{}^{\star}\alpha',n',\mathbf{k}_{j'}^0\right\rangle \\ \left\langle {}^{\star}\alpha,n,\mathbf{k}_j^0\right|\Delta\hat{V}_i^{\Gamma_1}\hat{\mathbf{x}}\left|{}^{\star}\alpha',n',\mathbf{k}_{j'}^0\right\rangle\cdot(\mathbf{k}-\mathbf{k}')\end{gathered} \quad (5)$$

3. Inter-valley mixing in SiO_2/Si/SiO_2 quantum wells

We consider SiO_2/Si/SiO_2 quantum well. Since that there are six equivalent minima in the silicon conduction band, we need to use multi-valley model. To construct the effective Hamiltonian we represent it as

$$\hat{H}=\hat{H}_{Si}+\sum_i\Delta\hat{H}_i,$$

where $\hat{H}_{Si}$ is the Hamiltonian describing the motion of electrons in the bulk silicon; $\Delta\hat{H}_i$ are corrections to the effective Hamiltonian due to substitution of atoms in i-th sublattice. Using the method of invariants for heterostructure requires both materials to have the same spatial symmetry. However, in the case of Si/SiO_2 pair it is not. At the same time, SiO_2 is known to be an insulator with a wide band gap (~9 eV). Thus, we can approximately consider that electronic states of the silicon conduction band give the main contribution to the forming of the lowest states in Si quantum well. The space group of silicon O_h^7 is non-symmorphic and it is convenient to add an infinitesimal perturbation, which destroys the inversion, and use T_d^2 space group instead. In this case, six-fold degenerate state, which transforms according to the irreducible representation ${}^{\star}X_1$ of the space group O_h^7, splits into two triply degenerate states, which transform according to the irreducible representations ${}^{\star}X_1$ and ${}^{\star}X_3$ of the space group T_d^2. Calculation of the matrix elements (5) should be made in point group symmetry of node T_d. Bloch wave functions of states transforming according to the irreducible representation ${}^{\star}X_1$ of the space group T_d^2 are non-zero in the nodes of the first sublattice and have zero value in the nodes of the second sublattice ($\sin(2\pi z/a_0)$-like, a_0 is the silicon lattice constant). On the contrary, Bloch wave functions of states transforming according to the irreducible representation ${}^{\star}X_3$ have zero value in the nodes of the first sublattice and non-zero value in the nodes of the second sublattice($\cos(2\pi z/a_0)$-like). Thus, if we choose a node of the first sublattice as a center of the point symmetry, representation ${}^{\star}X_1$ of the T_d^2 space group becomes representation Γ_{15} with the transition to the point group symmetry T_d, and representation ${}^{\star}X_3$ becomes a direct sum of representations $\Gamma_1\oplus\Gamma_{12}$. Conversely, for the nodes of the second sublattice representation ${}^{\star}X_1$ becomes $\Gamma_1\oplus\Gamma_{12}$, and ${}^{\star}X_3$ becomes Γ_{15}. Difference in the behavior of the Bloch wave functions near nodes of the first and second sublattices is illustrated schematically in figure 1.

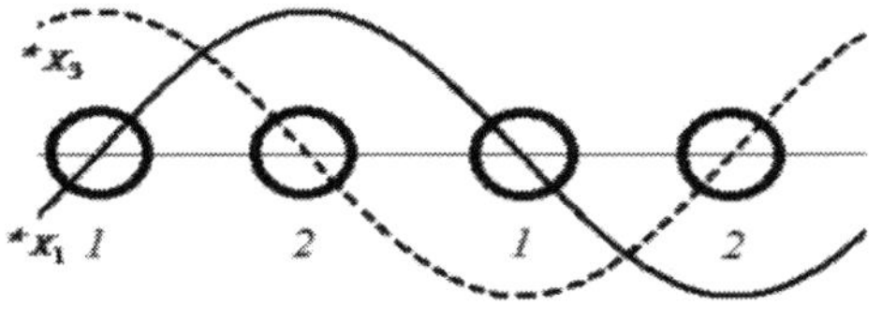

Figure 1. Behavior of the Bloch wave functions, which transform according to the irreducible representations ${}^{\star}X_1$ and ${}^{\star}X_3$, near to the nodes of crystal lattice

We introduce the following notation

$$f_i(\mathbf{k}_j^0+\mathbf{k}-\mathbf{k}_{j'}^0-\mathbf{k}')=\frac{1}{N}\sum_{\mathbf{a}}f_i(\mathbf{a})e^{-i(\mathbf{k}_j^0-\mathbf{k}_{j'}^0)\mathbf{a}}e^{-i(\mathbf{k}-\mathbf{k}')(\mathbf{a}+\chi_i)}.$$

It can be shown from definition of $f_i(\mathbf{k}-\mathbf{k}')$ that, firstly, mixing x-valley with y-valley in quantum wells and superlattices grown along [001] direction is possible only with the conservation of k_x and

k_y components. Secondly, x-z and y-z mixing is absent in approximation of small $(\mathbf{k}-\mathbf{k}')$. Thus, the task of determining the electronic spectrum in the quantum well is divided into two independent subproblems, one of which describes the *x*-*y*-mixing, and the other describes *z*-states mixing. The ground state of the quantum well is formed by *z*-minima of the conduction band of the bulk silicon. This also was noted in [2-6].

To determine the lowest electronic states of (001) Si/SiO_2 quantum well is sufficient to solve the Schrödinger equation with the following Hamiltonian

$$H(k_z,k_z')=\begin{pmatrix} H^{X_1}(k_z,k_z') & H^{X_1X_3}(k_z,k_z') \\ H^{X_3X_1}(k_z,k_z') & H^{X_3}(k_z,k_z') \end{pmatrix}+\begin{pmatrix} \Delta H^{X_1}(k_z,k_z') & \Delta H^{X_1X_3}(k_z,k_z') \\ \Delta H^{X_3X_1}(k_z,k_z') & \Delta H^{X_3}(k_z,k_z') \end{pmatrix},$$

where

$$H^{X_1}(k_z,k_z')=H^{X_3}(k_z,k_z')=\frac{\hbar^2 k_z^{\,2}}{2m_l}\delta_{k_z,k_z'},$$

$$H^{X_1X_3}(k_z,k_z')=\left[H^{X_3X_1}(k',k_z)\right]^*=iDk_z\delta_{k_z,k_z'},$$

$$\Delta H^{X_1}(k_z,k_z')=\Delta H^{X_3}(k_z,k_z')=\left[\Delta U+\frac{\hbar^2\left(k_z+k_z'\right)^2}{4}\Delta\left(\frac{1}{m_l}\right)\right]\left(f_1\left(k_z-k_z'\right)+f_2\left(k_z-k_z'\right)\right),$$

$$\Delta H^{X_1X_3}(k_z,k_z')=\left[\Delta H^{X_3X_1}(k_z',k_z)\right]^*=i\Delta D\left(k_z+k_z'\right)\left(f_1\left(k_z-k_z'\right)+f_2\left(k_z-k_z'\right)\right)$$
$$+\beta\left(k_z-k_z'\right)\left(f_1\left(k_z-k_z'\right)-f_2\left(k_z-k_z'\right)\right),$$

where m_l is longitudinal electron effective mass; D determines the position of the conduction band minimum in silicon; ΔD is a correction to D due to presence of the heterointerface; β is a constant of valley-orbit coupling. Dispersion in SiO_2 is assumed parabolic throughout the Brillouin zone with minima at the Γ-point and the effective mass the same as the free electron mass. Relationship between the constants for different sublattices is obtained by microscopic calculations.

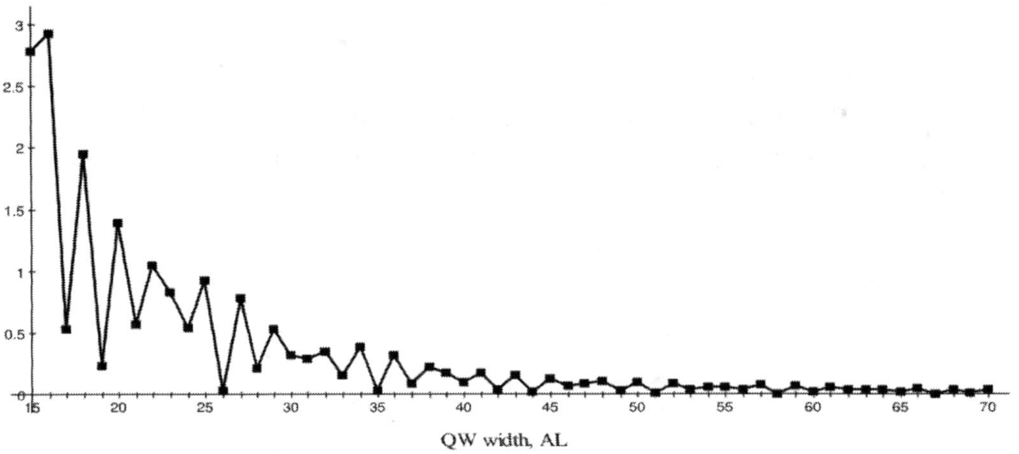

Figure 2. Valley splitting of the ground electronic state in a $SiO_2/Si/SiO_2$ finite square quantum well.

Within the proposed approach we calculated valley splitting of the ground electronic state in a $SiO_2/Si/SiO_2$ finite square quantum well as a function of the well width. Results are presented in figure 2. As can be seen from the figure 2, the dependence is nonmonotonic. The oscillations period is independent of the value of the constant of valley-orbit coupling and is determined only by the position of the valleys in the conduction band. The results are in good agreement with [4-6].

The developed approach allows us to calculate the electron wave functions in the *x*- and *k*-representations. The figure 3 shows the electron wave functions in *k*-representation without (*a*) and with (*b*) inter-valley interaction for the quantum well width of 40 atomic layers. In the first case the wave function is localized near the minimum of the conduction band. The presence of inter-valley interaction leads to the fact that the wave function is different from zero in the region between the minima.

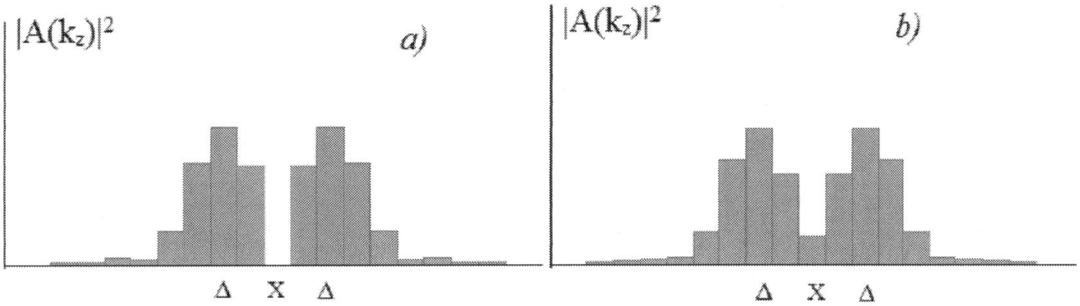

Figure 3. The square of typical wave function of the ground state in *k*-representation $|A(k_z)|^2$ around an X point of the Brillouin zone. a) Without inter-valley coupling, b) with inter-valley coupling.

The contribution of X_1 and X_3 states in forming of the envelope wave function of the ground state for well width of 20 atomic layers is shown on the left in figure 4. Each presented function does not match the symmetry of the ground state envelope wave function. However, we can restore the approximate form of the wave function as a superposition of Bloch states

$$\psi(z)=\langle z|{}^{\star}X_1,\mathbf{k}_z^0\rangle\langle{}^{\star}X_1,\mathbf{k}_z^0|\psi\rangle+\langle z|{}^{\star}X_3,\mathbf{k}_z^0\rangle\langle{}^{\star}X_3,\mathbf{k}_z^0|\psi\rangle=\sin\left(\frac{2\pi}{a_0}z\right)\psi^{X_1}(z)+\cos\left(\frac{2\pi}{a_0}z\right)\psi^{X_3}(z),$$

The square of wave function $\psi(z)$ is depicted on the right in figure 4.

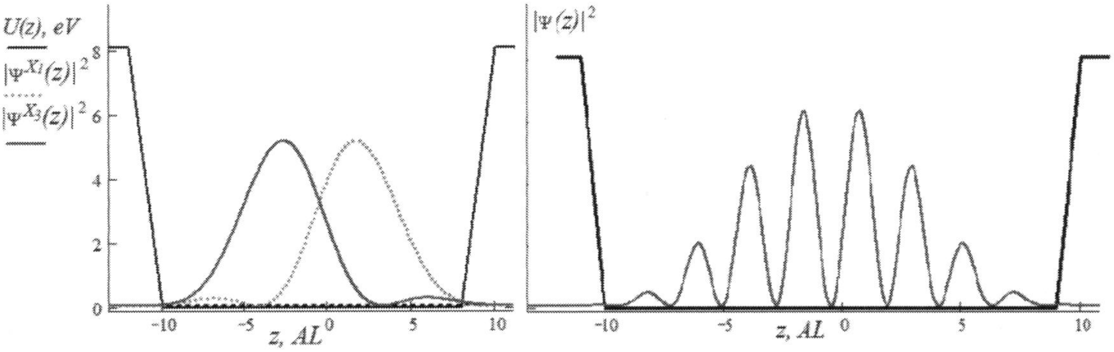

Figure 4. The left plot shows the contribution of X_1 and X_3 states in forming of the envelope wave function of the ground state for well width of 20 atomic layers. The right plot shows approximate form of the wave function as a superposition of Bloch states.

References

[1] Goswami S, Slinker K A, Friesen M, McGuire L M, Truitt J L, Tahan C, Klein L J, Chu J O, Mooney P M, van der Weide D W, Joynt R, Coppersmith S N and Eriksson M A 2007 *Nature Physics* **3**, 41
[2] Ohkawa F J and Uemura Y 1977 *J. Phys. Soc. Jpn.* **43** 907
[3] Sham L J and Nakayama M 1979 *Phys. Rev. B* **20** 734
[4] Nestoklon M O, Golub L E, Ivchenko E L 2006 *Phys. Rev. B* **73** 235334
[5] Friesen M, Chutia S, Tahan C, Coppersmith S N 2007 *Phys. Rev. B* **75** 115318
[6] Nestoklon M O, Ivchenko E L, Jancu J-M, Voisin P 2008 *Phys. Rev. B* **77** 155328
[7] Glinskii G F 2008 *Semiconductors and semiconductor nanoelectronics: symmetry and electronic states* (Saint Petersburg: Tehnolit)

Journal of Physics: Conference Series **461** (2013) 012041 doi:10.1088/1742-6596/461/1/012041

Σ^- production in pn interactions at ANKE at COSY

E Shikov, A Dzyuba, S Barsov and V Koptev

High Energy Physics Department, Petersburg Nuclear Physics Institute, RU-188350 Gatchina, Russia

E-mail: shikovegor@mail.ru

Abstract. Study of light hyperon production in nucleon-nucleon interactions is a large part of scientific program at the ANKE-spectrometer which is an internal-beam experiment at the Cooler Synchrotron (COSY-Juelich). The Σ^- production in pn collisions was studied at two proton beam momenta of 2915 MeV/c and 3015 MeV/c. The deuterium target was exploited as an effective neutron target due to measurement of "spectator" proton momentum to determine the energy of reaction. The Fermi motion allowed to cover a wide range of excess energy Q = 30 – 130 MeV where data was absent. The Σ^- production was identified by the missing mass technique. Preliminary result for the energy dependence of the $pn \to pK^+\Sigma^-$ total cross section is presented.

1. Introduction

Investigation of meson and hyperon production in nucleon-nucleon (NN) interaction at intermediate energies is necessary for solution of the fundamental problem — theoretical description of the strong interaction at low and intermediate energies. Here quark's degrees of freedom vanish, and one should use effective exchange models. Despite there are a significant database and numerous dedicated theoretical investigations, the mechanism of these processes still is far from being understood. Strangeness production in NN collisions are interesting for various reasons. A creation of an s-quark flavour, which can occur out of the vacuum as well

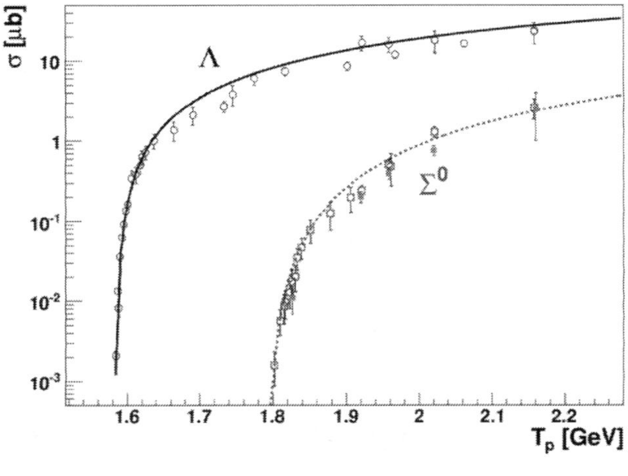

Figure 1. The energy dependence of the total cross section for $pp \to pK^+\Lambda$ (black circles) [2]–[11], $pp \to pK^+\Sigma^0$ (blue circles) [4, 5, 9, 10], $pp \to nK^+\Sigma^+$ (red dots) [9, 10]. The parametrisation of the three body $pK^+\Lambda$ phase space corrected for the $p\Lambda$ FSI is shown by the solid black line, the parametrisation of the three body $pK^+\Sigma$ phase space with a constant matrix element is shown by the blue dashed line [12].

as from the quark-antiquark sea in nucleon, allows one to investigate the internal structure of baryons through the dedicated studies of the $NN \to NKY$ reactions ($Y = \Lambda$, Σ). There are also some indications that several excited nucleon states decay of into the ΛN channel. In addition such reactions give a possibility to study nucleon-hyperon interaction which is still poorly known, but it is essential for questions related to the validity of the SU(3) flavour symmetry. A close-to-threshold regime is very suitable for such investigations because it implies a suppression of high partial waves. It simplifies the theoretical description. Moreover, final state interaction (FSI) effects are much more pronounced at such conditions.

For example, the information about the hyperon-nucleon interaction is very important for theoretical study of neutron stars radii and masses. Gravitation compression in a core of neutron star is generally considered to generate hyperon condensation, but calculations of such processes depend on properties of the *YN* interaction [1].

The energy dependence of total cross section for the $pp \to pK^+\Lambda$ reaction near threshold has been investigated in several experiments carried out at the COSY accelerator [2]– [11]. Its behavior can be described in terms of the three-body phase space modified by a strong Λ-p FSI. However, the analysis of Dalitz plot and angular distributions does not give definite answer on a possible excitation of nucleonic resonances in intermediate state [12].

On other hand, the $pp \to pK^+\Sigma^0$ reaction does not demonstrate any influence of the $\Sigma^0 - p$ FSI on the energy dependence of total cross section which is found to behave according to the three-body phase space [4, 5, 9, 11]. At the same value of excess energy the Σ^0 production total cross section is about one order of magnitude less than one for Λ production (Fig. 1). This probably reflects difference in the underlying reaction mechanisms as well as between the two FSIs involved.

Investigations of the $pp \to nK^+\Sigma^+$ reaction show that the energy dependence of total cross sections does not differ much from the three body phase space but there is an indication on some small repulsive Σ^+-n FSI [9, 10, 13]. It should be also mentioned that the ratio of total cross sections $\sigma(pp \to nK^+\Sigma^+)/\sigma(pp \to pK^+\Sigma^0) = 0.7 \pm 0.1$.

In this report we present preliminary results on the relative energy dependence of total cross section for the $pn \to pK^+\Sigma^-$ reaction near threshold which has not been measured so far.

2. Experimental setup

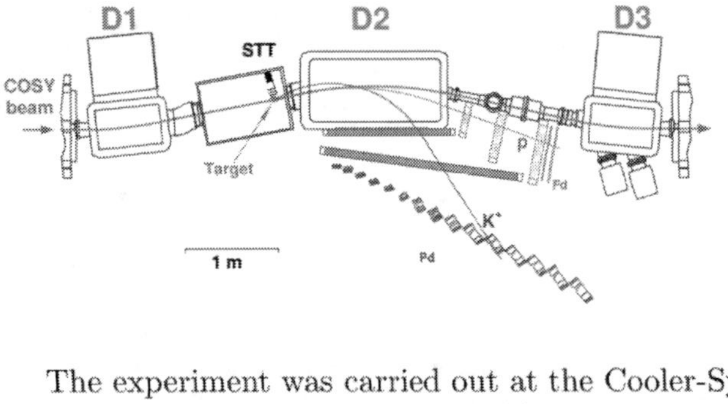

Figure 2. Sketch of the relevant parts of the ANKE detector system, showing the positions of the two bending magnets *D1* and *D3* and the target placed before the analyzing magnet *D2*. Information obtained from the forward (Fd), positive side (Pd) detectors, and the Silicon Tracking Telescope (STT) was used in this experiment. Typical trajectories, leading to the measurement of protons and K^+, are also shown.

The experiment was carried out at the Cooler-Synchrotron COSY-Juelich using unpolarized proton beam of 2915 and 3015 MeV/c momenta interacting with a deuterium cluster-jet target. Reaction products were detected in the ANKE magnetic spectrometer [16] located inside the storage ring of the COSY. The windowless cluster deuterium target was used as an effective neutron target due to the detection of low-energy protons emitted from the target. Thus, events of the $pn \to pK^+\Sigma^-$ reaction could be reconstructed in framework of the spectator model approximation.

The layout of ANKE detector system used in this measurement is sketched in Fig. 2.

3. Reaction identification

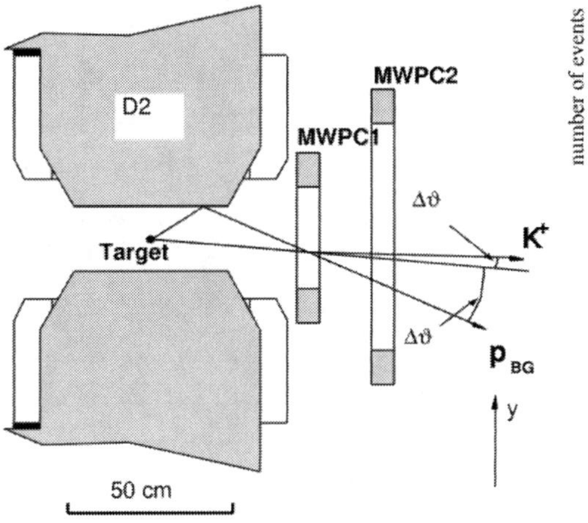

Figure 3. Principle of vertical angle cut. Ejectiles from the target (K^+) and scattering background (P_{BG}) in general have different vertical angles θ in the spectrometer.

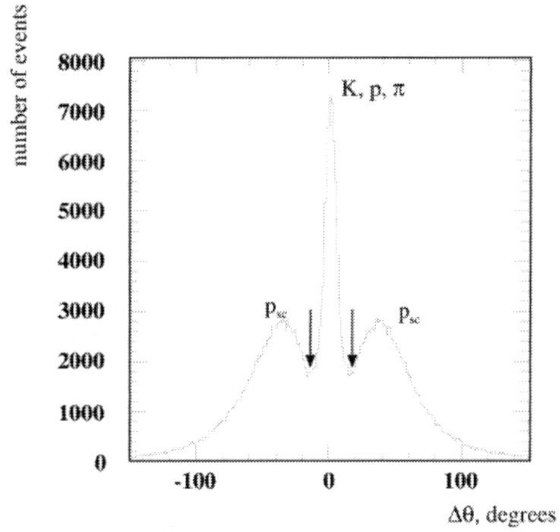

Figure 4. Vertical angle distribution summed over all the stop counters.

Kaons were identified by the Positive (Pd) detector, high momentum protons – by the Forward (Fd) detector and low energy spectator protons – by the Spectator detector. The Pd detector consists of start and stop counters used for the time-of-flight(TOF) measurement and of two multiwire proportional chambers providing momentum reconstruction and background suppression. The Fd detector system consists of three multiwire proportional chambers and of the hodoscope of scintillation counters. The Spectator detector includes two telescopes (STT). Each of them consists of three layers of position-sensitive silicon detectors with thickness of 60μ, 300μ and 5.1 mm.

So, the goal of further analysis is to select the K^+-meson in the Pd in coincidence with proton in the Fd and low-energy proton in the STT.

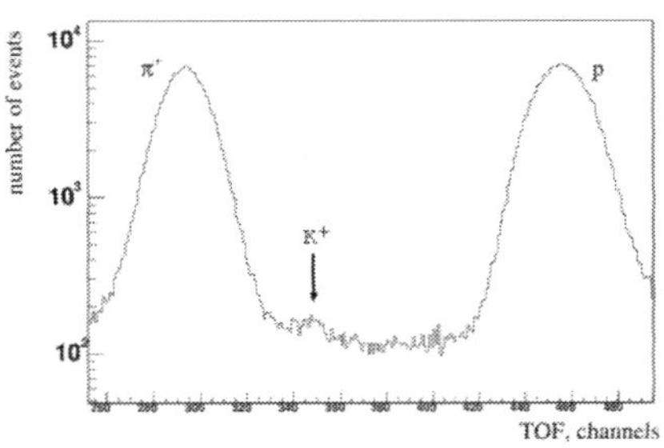

Figure 5. The total Sa-So TOF spectra summarized over all start-stop combinations.

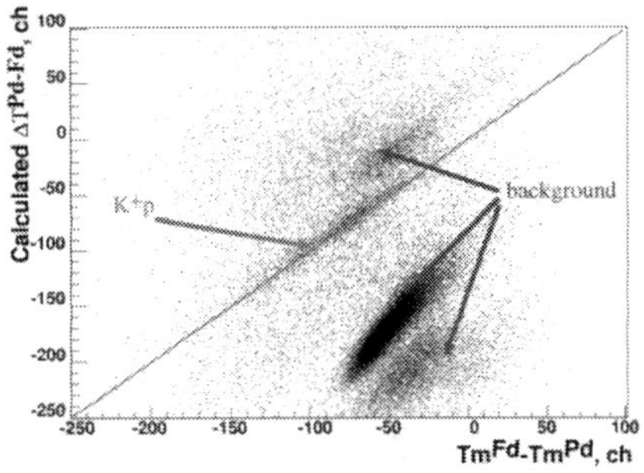

Figure 6. Example of Pd-Fd TOF from calculations vs TOF by TDC's.

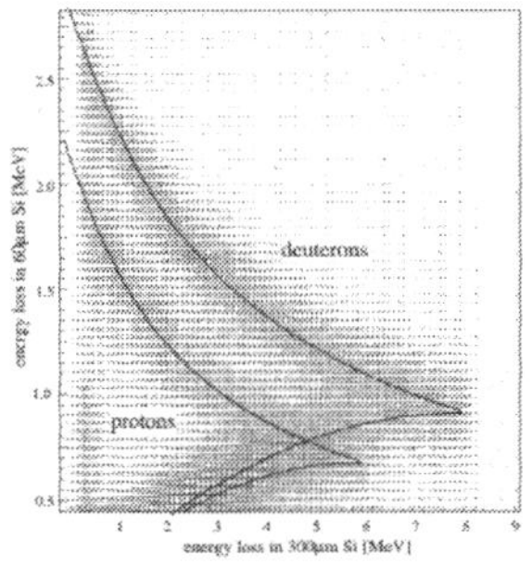

Figure 7. Energy loss in the 60 μ detector versus energy loss in the 300 μ silicon detector.

To get rid of the particles rescattered in magnet poles, the vertical angle criterion was applied. The principle of the vertical angle cut is illustrated in Fig. 3. Particles coming from the target within the angular acceptance pass through the first MWPC at some vertical coordinate y from

Journal of Physics: Conference Series **461** (2013) 012041 doi:10.1088/1742-6596/461/1/012041

which one can expect some certain vertical angle to be measured by the second MWPC. In contrast to ejectiles coming directly from the target, particles scattered in the magnet have in general much larger difference between the expected angle and the one measured in the second chamber. The experimental vertical angle distribution is shown in Fig. 4. Here, the central prominent peak is formed by ejectiles from the target while the wide side peaks are related to rescattered particles.

The main criterion for particle identification in the Pd is the time-of-flight between start and stop counters. This technique provides clear separation of pions and protons in each of start-stop combinations. The TOF distribution for the 11-th stop counter summed over the all start counters is presented in Fig. 5. The left and right peaks correspond to pions and protons coming from the target. The position of K^+ peak is marked by the arrow but it can be hardly seen because of the large background from tails of of pion and proton distributions. The further suppression of the background near K^+ peak was achieved due to coincidence with high momentum protons detected in the Fd and low-energy protons detected in the STT.

Particles detected in the Fd are dominantly protons with a very small admixture of deuterons. So, the kinematic cut which reject protons with momenta higher than it is allowed by conservation lows for the $pd \to p_{sp}pK^+\Sigma^-$ reaction ($|\vec{p}_{Fd}| < 1.8$ GeV/c) can be made.

The additional criterion was applied on the correlation between K^+ in the Pd and protons in the Fd as shown in Fig. 6. The time-of-flight difference between corresponding particles detected in Fd and Pd counters was calculated and compared with experimentaly calibrated TOF data. Points related to valid events are grouped along the diagonal of the presented histogram.

The momentum of "spectator" proton was reconstructed using the position and energy loss information from STT. The separation of protons from deuterons by the $\Delta E - E$ method is presented in Fig. 7.

4. Energy dependence

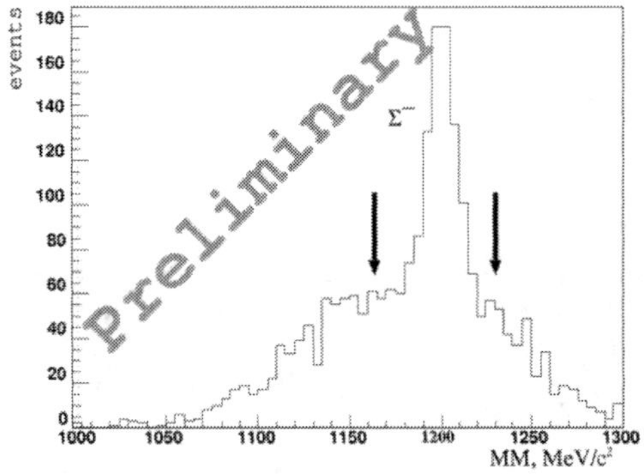

Figure 8. Missing mass distriution for $p_{beam} =$ 2915 MeV/c. The Σ^- peak is highlighted with the arrows.

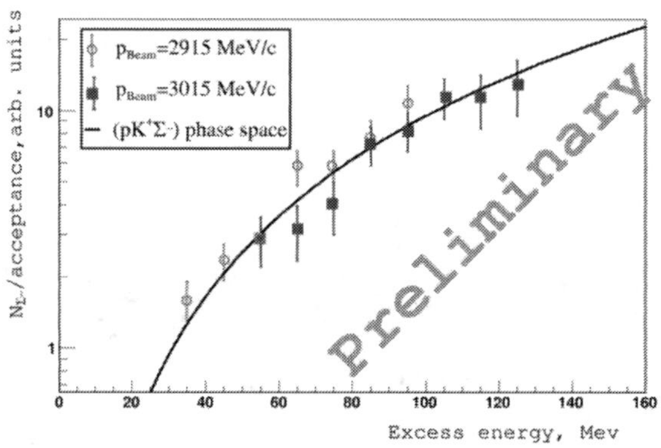

Figure 9. Energy dependence of the total cross section in arbitrary units. The red points correspond to 2915 MeV/c, the blue points correspond to 3015 MeV/c. The phase space dependency is shown with the solid line.

The final identification of reaction under investigation was done by the missing mass method:

$$M^2 = (P^{cosy} + P^{target} - P^{Pd} - P^{Fd} - P^{STT})^2$$

where P^{cosy} is the 4-momentum of the proton beam, P^{target} is the 4-momentum of the deuterium target and P^{Pd}, P^{Fd} and P^{STT} are 4-momenta of corresponding particles selected in the Pd,

Journal of Physics: Conference Series **461** (2013) 012041 doi:10.1088/1742-6596/461/1/012041

Fd and STT detectors as described above. The missing mass distribution obtained at the 2915 MeV/c beam momentum is presented in Fig. 8. It demonstrates a clear peak at 1.2 GeV/c^2 which can be interpreted as a signal from the $pd \to p_{sp}pK^+\Sigma^-$ reaction. The total number of Σ^- events under the peak is about 500 events at each of beam momenta.

Similar missing mass distributions were plotted for each excess energy bin, and the number of Σ^- events was extracted. Monte-Carlo simulations of the $pd \to p_{sp}pK^+\Sigma^-$ reaction were performed to calculate the acceptance value [14]. The preliminary energy dependence of total cross section presented in Fig. 9 follows the 3-body phase space behaviour quite well without involving any stong FSI effects.

The further analysis of the absolute normalisation and uncertainties is in progress.

The reported study was partially supported by RFBR, research project No. 12-02-31762.

References

[1] Potekhin A Yu 2010 *Uspekhi Fizicheskikh Nauk* **180** 1279
[2] Balewski J T et al. 1998 *Eur. Phys. J.* A **2** 99
[3] Bilger R *et al.* 1998 *Phys. Lett.* B **420** 217
[4] Sewerin S *et al.* 1999 *Phys. Rev. Lett.* **83** 682
[5] Kowina P *et al.* 2004 *Eur. Phys. J.* A **22** 293
[6] Abdel-Bary M *et al.* 2010 *Phys. Lett.* B **688** 142
[7] Loutitt R I *et al.* 1961 *Phys. Rev.* **123** 1465
[8] Baldini A, Flamino V, Moorhead W G, and Morison D R O 1988 *Landolt-Börnstein* New Series ed by H. Schopper Vol. 12 (Berlin: Springer-Verlag)
[9] Valdau Yu *et al.* 2007 *Phys. Lett.* B **652** 27
[10] Valdau Yu *et al.* 2010 *Phys. Rev.* C **81** 045208
[11] Abdel-Bary M *et al.* 2010 *Eur. Phys. J.* A **46** 27
[12] Sibirtsev A *et al.* 2006 *Eur. Phys. J.* A **27** 269-285
[13] Valdau Yu, Wilkin C *et al.* 2011 *Phys. Lett.* B **696** 23
[14] Shikov E 2012 *Modern problems of physics, biophysics, and information and communication technologies* (Krasnodar CSTI) pp 352-359 ISBN 978-5-91221-135-5 (in Russian)
[15] Barsov S, Dymov S *et al.* 2004 *Eur. Phys. J.* A **21** 521
[16] Barsov S *et al.* 2001 *Nucl. Inst. Meth.* A **462** 364

15th Russian Youth Conference on Physics and Astronomy (PhysicA.SPb) IOP Publishing
Journal of Physics: Conference Series **461** (2013) 012042 doi:10.1088/1742-6596/461/1/012042

Mathematical model of metal-hydride phase change applied to Yttrium

I.A. Chernov
Institute of Applied Mathematical Research, Karelian Research Centre, RAS; 185910 Pushkinskaya 11, Petrozavodsk, Russia
E-mail: IAChernov@yandex.ru

S.V. Manicheva
Karelian State Pedagogical Academy; 185035 Pushkinskaya 17, Petrozavodsk, Russia
E-mail: smanicheva@mail.ru

I.E. Gabis
Phys. Dep. of St-Petersburg State Univ., St-Petersburg, Russia
E-mail: igor.gabis@gmail.com

Abstract. We present a mathematical model for the kinetics of hydriding and dehydriding of metal powders. The single powder particle is considered. Its shape is approximated by one of the symmetric ones: sphere, long thin cylinder (wire), or flat thin plate. A few concurrent processes are considered. The model equations are derived from the mass conservation law. We consider the case of the "shrinking core" morphology, i.e. formation of the hydride skin on the surface of the particle with subsequent growth of this skin. We consider three successive stages of the phase change: skin development, skin growth, and final saturation or degassing. We apply the model to experimental data for Yttrium and show that the approximation of the experimental curves by the model ones is comparable for different cycles and different shapes for similar sets of the kinetic parameters. This also shows that shape of powder particles do not influence significantly on the kinetics of hydriding and dehydriding.

1. Introduction

Understanding the hydriding kinetics of powder particles is very important for energy storage applications [1, 2, 3]. According to the US Department of Energy objectives, loading of a hydrogen vehicle tank by 2015 must not exceed 2.5 min. This implies not only heat problems but also serious limitations on rate of hydrogen sorption by metal hydride powder. Existing articles give contradictory information about hydriding and dehydriding kinetics and its limiting factors. In [4, 5] we developed the model of hydriding of metal powders for close-to-equilibrium conditions. We considered a single powder particle. The model was verified on experimental data for uranium (prof. J. Bloch [6]) and magnesium. In [5] we studied the influence of particle's shape on the kinetics of hydriding and found out that it is insignificant.

Here we develop this model for varying temperature and pressure and also for dehydriding. Models of dehydriding are constructed in e.g. [7, 8]. Hydriding and dehydriding possess certain

Journal of Physics: Conference Series **461** (2013) 012042 doi:10.1088/1742-6596/461/1/012042

symmetry being both phase transition processes; thus it is important to describe them by a single model, so that only the parameters' values distinguish the process.

Although detailed models of hydriding and dehydriding containing almost all possible elementary reactions exist (e.g. [9, 10, 11]) and are useful in general, they are hardly suitable for evaluating the kinetis parameters of the material, because it is difficult to estimate the role of each reaction by fitting experimental data.

On the other hand, considering a single limiting process, possibly different at different times, does not reflect the nature of the complex process, thus we deny it. Our model is based on the conservation laws and takes into account the most important elementary reactions.

The model is applied to experimental data of Yttrium. Yttrium hydride is known to change transparency depending on hydrogen concentration; this is used in switchable mirrors [12] and, currently, in solar cells [13]. These hydrides have been studied, yet by other methods, in e.g. [14, 15].

The powder particle's shape is approximated by a symmetrical form; different forms provide similar quality of fitting of the experimental curves for similar sets of the kinetic parameters. This is another argument for low influence of the powder particle's shape on the kinetics of hydriding and dehydriding.

We consider a single powder particle. Its form is approximated by sphere, long thin cylinder, or thin plate. In case of cylinders and plates we neglect the butts and the smaller sides. In all cases we consider the skin of the new phase (hydride for hydriding, metal with dissolved hydrogen for dehydriding) on a core of the old phase. This core has the same form as the particle. Note that length of cylinders and face area of plates are not important for the relative quantities such as fractions of the hydride phase or of the sorbed hydrogen. By "hydride" we mean hydride phase with some dissolved hydrogen (so concentration may be above stoichiometric). We describe the following stages of hydriding or dehydriding: formation of the skin of the new phase on the particle, propagation of the phase boundary, and final saturation (degassing) of the single-phase particle.

2. The mathematical model

Due to the higher activation energy of the diffusion through the hydride compared with the overall kinetics in bulk samples (for which the penetration is through the metal) [10], one can say the diffusion through the hydride is slower compared with that through the metal. So we consider only diffusion in the new phase: in case of hydriding diffusion in metal is fast, in case of dehydriding diffusion in hydride is neglected.

The hydrogen pressure $P(t)$ and the sample's temperature $T(t)$ are considered to be known (i.e. measured in the experiment) and close to the equilibrium. The particle size is assumed to be constant during the cycle. However, it may decrease from one cycle to another since the powder grows finer during hydriding-dehydriding cycles. We assume spherical, cylindrical, or plane symmetry for the particles.

Let D be the Fick's diffusion coefficient in the new phase (i.e. metal in case of dehydriding and hydride in case of hydriding) and $c(t,r)$ and $c' = \text{const}$ be the distribution of the hydrogen concentration in the new phase and in the old phase. Here r is the radius in spherical or cylindrical coordinate system or the Cartesian coordinate on the axis orthogonal to the plane of symmetry of a plate. Due to the symmetry the concentration is independent of the other spatial variables. Let L be the size of the particle: radius in case of a sphere or cylinder and half-width in case of a plate. The size of the metal core under the hydride skin is denoted by $\rho(t)$, $\rho(0) = \rho_0$. The shape of the particle is described by the constant $\sigma = 2, 1, 0$ for sphere, cylinder, and plate respectively. In this case the Fick's equation can be written as

$$\frac{\partial c}{\partial t} = D(T)\left(\frac{\partial^2 c}{\partial r^2} + \frac{\sigma}{r}\cdot\frac{\partial c}{\partial r}\right) \qquad r \in (\rho(t), L). \tag{1}$$

At the phase boundary we have the constant concentration:

$$c(t,\rho) = \bar{c}. \tag{2}$$

The physical meaning of $\bar{c}$ is stoichiometric concentration in case of hydriding and equilibrium concentration in case of dehydriding. The local conservation law at this free boundary yields the Stefan-type equation

$$(c' - c(t,\rho))\frac{d\rho}{dt} = (1-\nu)D(T)\frac{\partial c}{\partial r}(t,\rho). \tag{3}$$

Here ν is the fraction of the diffusion flux that goes through the phase boundary from the new phase to the old one. In case of dehydriding diffusion in the old phase is very slow and is neglected; therefore $\nu = 0$. In case of hydriding we assume that $\nu = 1$ (and thus $\rho = \text{const}$) if the old (metal) phase has gas-contacting surface; this means that all diffusion flux at the phase boundary goes to the metal phase with quick diffusion and is spent on the tangential growth of the nuclei and there is no radial growth. As soon as the skin is ready we have $\nu = 0$ because redistribution is impossible due to symmetry. Models with any $\nu \in [0,1]$ were also considered.

One more boundary condition follows from the flux balance at the surface:

$$\mu s(T)P - b(T)c^2(t,L) = D(T)\frac{\partial c}{\partial r}(t,L). \tag{4}$$

In the left-hand side there are the adsorption and the desorption flux densities. All temperature-dependent parameters are assumed to obey the Arrhenius law, e.g.

$$b(T) = b_0 \exp\left(-\frac{E_b}{RT}\right),$$

where T is absolute spatially-constant temperature, R is the gas constant, b_0 and E_b are constant parameters. The desorption flux density is linear with respect to pressure, μ is the gas constant (we assumed that it is constant), s is for the adhesion coefficient (the ratio of the adsorbed hydrogen atoms to the falling ones during a unit time). Desorption flux density is proportional to the number of possible pairs of hydrogen atoms to be formed near the surface and thus c^2.

Thus the domain $[0,\rho(t))$ is occupied by the old phase, while $(\rho(t), L]$ is the new phase provided that the skin of the new phase is completed. We assume that the domain occupied by the new phase is always in the layer $\rho < L$ and is symmetrical with respect to the chosen coordinates [4]. Then the equations written above hold both in complete or incomplete skin and it can be described by one additional variable: the fraction $S(t)$ of area it covers at the surface of the particle. Initially $S(0) = 0$. The equation for S follows from the global conservation law:

$$W\dot{S}(t) = \nu D\frac{\partial c}{\partial r}(t,\rho)S + J'(1 - S(t)), \tag{5}$$

where $J' = \mu s'(T)P - b'(T)(c')^2$ is the sorption flux density for the old phase and

$$W = \int_\rho^L \left(c(t,r) - c'\right)\frac{r^\sigma}{L^\sigma}dr \tag{6}$$

is the amount of hydrogen in the skin (from ρ to L) consisting of two phases. Equation (5) shows that S hits 1 at finite time, i.e. the skin is ready at some time t^*, provided that $\nu > 0$. In case of dehydriding $\nu = 0$; then $S \to 1$ asymptotically. The skin is assumed to be ready when $1 - S$ becomes negligibly small. Indeed, differential equation (5) has a unique solution passing via any chosen point [16]; function $S \equiv 1$ is a solution in case $\nu = 0$, so no other solution can ever equal 1 (otherwise two different solution would pass via the same point). In other words, as S tends to 1, its velocity becomes lower and lower, so S never reaches the value 1. On the contrary, for $\nu > 0$ the velocity $\dot{S}$ is strictly positive, so S reaches 1 after some finite time.

Equations (1)–(6) are solved simultaneously; in [17] a numerical method is described. In case of dehydriding the process is continuous, in case of hydriding at time t^* the skin is ready and ν vanishes, so the process consists of two stages: skin formation and skin growth. One more stage of final saturation or degassing is described by the same equations except (2) with $\rho = 0$ and $j = 0$.

3. Results and discussion

We used this model for curves of hydriding and dehydriding of Y. The autoclave containing a sample with initial temperature and atomic ratio was in the chamber with the initial hydrogen pressure; the autoclave was quickly open and the pressure decrease or increase was written down. The autoclave's volume ($\approx 10^{-6}$ m^3) was little compared to the chamber's volume ($3.2 \cdot 10^{-4}$ m^3, $1.39 \cdot 10^{-3}$ m^3, $2.75 \cdot 10^{-2}$ m^3).

We had nine curves for dehydriding and four for hydriding of yttrium. The curves of the atomic ratio with respect to time were fitted by the least squares method. The fixed parameters were $\bar{c} = \hat{c} = 3 \cdot 10^{22}$ (this is the stoichiometric concentration of YH_3 calculated via yttrium's density and atomic mass) and $c' = 0$ in case of hydriding, and $\bar{c} = 0.15\hat{c}$, $c' = \hat{c}$ for dehydriding. Here and below the length, time, and amount are measured in centimeters, seconds, and atoms respectively. Fitting for three shapes for different cycles of hydriding/dehydriding is in fig. 1–2 (an atomic ratio curve for hydriding and dehydriding is shown in each figure, the incut shows a magnified part so that all four lines are visible). The fitting is equally good.

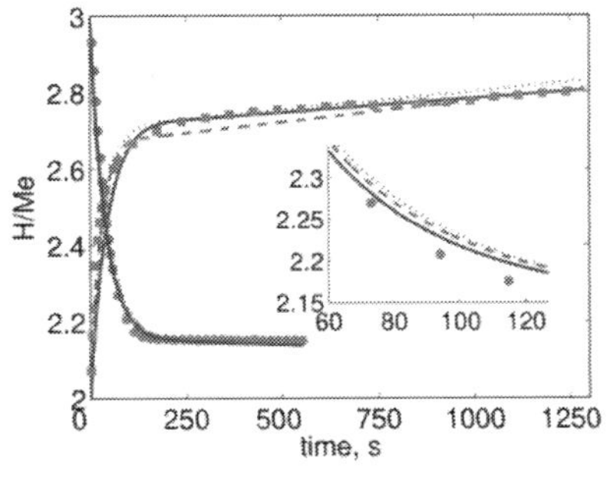

Figure 1. Curves of hydriding and dehydriding; experimental (circles), model for spheres (solid), cylinders (dashed), and flat plates (dotted).

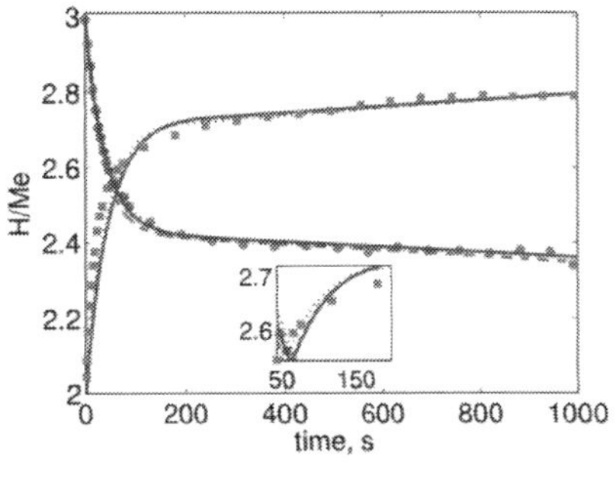

Figure 2. Curves of hydriding and dehydriding; experimental (circles), model for spheres (solid), cylinders (dashed), and flat plates (dotted).

Let us discuss the kinetic parameters. They are the same (except L and ρ_0) for different shapes. For hydriding their values are shown in table 1. Parameter b' is absent because $c' = 0$. For dehydriding the results are the same for diffusion coefficients higher than 10^{-11}; this justifies using models with quick diffusion [7, 8]. Also the sorption flux for rather low pressures (≈ 1 bar compared to ≈ 5 bar for hydriding) was much (10^5 times) lower compared to desorption flux

Table 1. Evaluations of the kinetic parameters for hydriding

$10^{12}D$	$10^9 s$	$10^7 s'$	$10^{34} b$
5.88	1.58	3.25	4.00
5.71	1.85	3.11	4.00
7.00	0.89	1.29	2.00
7.00	0.89	1.29	1.66

and thus s and s' are not evaluated (the vacuum assumption holds). The parameter b for all shapes is 10^{-31}, the average value of b' was $7.7 \cdot 10^{-31}$ and $3 \leq 10^{31} b' \leq 21$.

The difference between the desorption constant for YH_3 for hydriding (it is b) and dehydriding (it is b') is explained by different temperature: for hydriding it was ≈ 400K and for dehydriding it was ≈ 500K; this is enough to provide such difference because activation energy of desorption seems to be rather high.

Size of the particle L and the initial size of the old phase core ρ_0 differ for different shapes and cycles.

Note that according to the Buckingham π-theorem the size L can be changed in reasonable boundaries together with the parameters without influencing the results. The reason is that L fixes the length unit (other units are concentration fixed by $\bar{c}$ and time fixed by the duration of the experiment). Changing L is equivalent to changing meter to a new length unit. Thus it is impossible to evaluate L and other parameters at the same time; only the quantities ρ_0/L, DL^{-2}, b/L, s/L, etc. can be evaluated, i.e. those that do not depend on the length unit. We used this invariance to make b be the same for different cycles of dehydriding.

We see that the kinetic parameters are the same for different shapes and are similar for different experiments on hydriding and dehydriding; the approximation for different shapes is of reasonable quality. This shows that the powder particle's shape does not influence noticeably on the kinetics of hydriding and dehydriding and thus adds arguments to the main conclusion of [5].

Acknowledgments

This work was supported by the grant of St. Petersburg State University, project no. 11.38.68.2012 and by the Program of strategic development of Petrozavodsk State University, within measures on improvement of scientific research activity.

References

[1] Sakintuna B, Lamari-Darkrim F, Hirscher M and Dogan B 2007 *International Journal of Hydrogen Energy* **32** 1121–40
[2] Schlapbach L and Züttel A 2001 *Nature* **414** 353–358
[3] Satyapal S, Petrovic J, Read C, Thomas G and Ordaz G 2007 *Catalysis Today* **120** 246–256
[4] Chernov I, Bloch J and Gabis I 2008 *International Journal of Hydrogen Energy* **33** 5589–5595
[5] Chernov I, Bloch J, Voit A and Gabis I 2010 *International Journal of Hydrogen Energy* **35** 253–258
[6] Bloch J 2003 *J. Alloys Cmpd.* **361** 130–137
[7] Zaika Y, Chernov I and Gabis I 2005 *J. Alloys Cmpd.* **404–406C** 332–334
[8] Zaika Y and Rodchenkova N 2009 *Applied Mathematical Modelling* **33** 3776–3791
[9] Castro F and Meyer G 2002 *J. Alloys Cmpd.* **330–332** 59–63
[10] Bloch J and Mintz M 1997 *J. Alloys Cmpd.* **253–254** 529–541
[11] Martin M, Gommel C, Borkhart C and Fromm E 1996 *J. Alloys Cmpd.* **238** 193–201
[12] Nagengast D, Kerssemakers J, van Gogh A, Dam B and Griessen R 1999 *Applied Physical Letters* **75** 17–24

[13] Mongstad T, Platzer-Björkman C, Karazhanov S, Holt A, Maehlen J and Hauback B 2011 *J. Alloys Cmpd.* **509** S812–S816
[14] Majer G, Gottwald J, Peterson D and Barnes R 2002 *J. Alloys Cmpd.* **330–332** 438–442
[15] dos Santos A, Ambrosio R and Ticianelli E 2004 *International Journal of Hydrogen Energy* **29** 1253–1261
[16] Burkill J 1962 *The theory of ordinary differential equations* (Edinburgh-London: Oliver and Boyd)
[17] Chernov I 2009 *Electronic Transactions on Numerical Analysis* **35** 40–56

Journal of Physics: Conference Series **461** (2013) 012043 doi:10.1088/1742-6596/461/1/012043

Numerical simulation of synthesis of fullerenes by the arc discharge method

A Afanasyeva[1], S Ponyaev[2], V Krgiganovskaya[1,3]

[1] St.Petersburg State Polytechnic University, Saint-Petersburg, Russia
[2] Ioffe Physical-Technical Institute of the Russian Academy of Sciences
[3] University of Amsterdam, Amsterdam, Netherlands

mail: arina.afan@gmail.com

Abstract. A mathematical model of the fullerene formation process by carbon vapor clustering in an arc discharge camera was developed. Numerical simulation of formation of fullerene clusters up to C_{60} and C_{70} from carbon vapor produced by an arc discharge in a plasma reactor camera was carried out. A geometrical model of the reactor was built, and a numerical model of the processes occurring in the gas flow in the reactor and at clustering of carbon vapor including 82 components and 522 chemical reactions was developed.

1. Introduction

Fullerenes and their derivatives are recently discovered forms of carbon. They attract the interest of researchers due to new, unexpected properties, which are the result of an unusual molecular structure of a hollow cage. Investigations of physical and chemical properties of fullerenes show a potential of these materials for use in many fields of science and technology, such as electronics, optoelectronics, electrochemistry and chemistry. [1]

At present few applications of fullerenes in industry are commercialized, which is due to a very high cost of fullerene production. In order to make fullerene production cost efficient, a detailed study of fullerene formation in arc discharge reactors is needed.

Arc discharge is the most commonly used method of fullerene production, it provides the maximum yield (up to 12% from the raw soot) and also enables one to control the composition of raw soot (ratio C_{60}/C_{70} etc.) by varying production process parameters (such as pressure, temperature, camera configuration, etc.). Numerical modeling is a necessary stage for development and optimization of technology. The major goal is to find the conditions which provide the maximum content of fullerenes in raw soot.

2. Mathematical model

To describe a turbulent flow of a multicomponent gas mixture, Reynolds-Averaged Navier-Stokes (RANS) equations are used. As a result of solution of these equations, averaged characteristics of the flow are obtained.

The continuity equation and the balance mass of the mixture components are

$$\frac{\partial \rho}{\partial t} + \frac{\partial \rho u_j}{\partial x_j} = 0, \qquad (1)$$

$$\frac{\partial \rho Y_\alpha}{\partial t}+\frac{\partial \rho u_j Y_\alpha}{\partial x_j}=\frac{\partial}{\partial x_j}\left(\frac{\mu}{Sc}+\frac{\mu_t}{Sc_t}\right)\frac{\partial Y_\alpha}{\partial x_j}+\rho\,\dot{r}_\alpha, \qquad (2)$$

the equation of momentum transfer is

$$\frac{\partial \rho u_i}{\partial t}+\frac{\partial \rho u_j u_i}{\partial x_j}=-\frac{\partial P}{\partial x_i}+\frac{\partial \tau_{ij}}{\partial x_j}+\rho g, \qquad (3)$$

i=1,2,3,

the equation of enthalpy transfer is

$$\frac{\partial \rho h}{\partial t}+\frac{\partial \rho u_j h}{\partial x_j}=\frac{\partial}{\partial x_j}\left(\frac{\mu}{\mathrm{Pr}}+\frac{\mu_t}{\mathrm{Pr}_t}\right)\frac{\partial h}{\partial x_j}-\frac{\partial q_{rj}}{\partial x_j}, \qquad (4)$$

The equations of kinetic energy of turbulence and dissipation velocity are

$$\frac{\partial k}{\partial t}+\bar{u}_j\frac{\partial k}{\partial x_j}=\tau_{ij}\frac{\partial \bar{u}_i}{\partial x_j}-\varepsilon+\frac{\partial}{\partial x_j}\left[\left(v+v_t/\sigma_k\right)\frac{\partial k}{\partial x_j}\right], \qquad (5)$$

$$\frac{\partial \varepsilon}{\partial t}+\bar{u}_j\frac{\partial \varepsilon}{\partial x_j}=C_{\varepsilon_1}f_1\tau_{ij}\frac{\varepsilon}{k}\frac{\partial \bar{u}_i}{\partial x_j}-C_{\varepsilon_2}f_2\frac{\varepsilon^2}{k}+\frac{\partial}{\partial x_j}\left[\left(v+v_t\sigma_\varepsilon\right)\frac{\partial \varepsilon}{\partial x_j}\right]. \qquad (6)$$

These equations are formulated for the case of essentially subsonic flows at a constant pressure. Here $\dot{r}_\alpha$ are the sources of the mass of the mixture components due to chemical reactions, ρ is the density, u is the velocity, h is the enthalpy, Y_a is the mass fraction of component a, Sc is the Strouhal number, μ is the dynamic viscosity, τ_{ij} are the components of stress tensor, Pr is a Prandtl number, ξ is the velocity of dissipation, k is the kinetic energy, ν and ν_t are the kinematic viscosity and turbulent kinematic viscosity, respectively, C and f are the model coefficients, and q_{rj} are the heat sources due to chemical reactions (equal to the negative reaction enthalpy q_r=-ΔH_r)

The scheme of chemical reactions of carbon vapor condensation used in our study is based on a generalization of experimental data [2].

There are no thermodynamic data in the literature for all the clusters of carbon higher than C_{10} (excluding fullerenes), and the simplest piecewise-linear approximation is used in the model. This method of evaluation of thermodynamic data provides a smooth change in the Gibbs free energy as a function of cluster size required by the model and, apparently, is sufficient to solve a number of problems.

The thermal effect of a reaction is calculated as

$$\Delta H_r^0=\sum H_{prod}-\sum H_{reac}=\sum_i v_i(a_{1i}RT+a_{6i}R)-\sum_j v_j(a_{1j}RT+a_{6j}R), \qquad (5)$$

where v are the molar coefficients in chemical reactions, a_{ni} and a_{nj} are the coefficients in the equations of enthalpy of the resulting and initial reagents of the reaction, respectively.

The model also uses the equation of state of an ideal gas to describe the flow of helium

Journal of Physics: Conference Series **461** (2013) 012043 doi:10.1088/1742-6596/461/1/012043

$$P = R\rho T \sum_{a \neq C} \frac{Y_a}{M_a}., \quad (6)$$

where P is the gas pressure, R is the gas constant, ρ is the density, Y_a is the mass fraction of component a, and M_a is the molar mass of component a.

To calculate the temperature dependence of the rate constant of a reaction, standard Arrhenius formula is employed

$$k_i = A_i T^{\beta_i} \exp\left(-\frac{E_i}{RT}\right), \quad (7)$$

where A_i, β_i are the coefficients in Arrhenius equation, and E_i is the activation energy.

3. Results and discussion

The numerical experiment is performed in OpenFOAM (Open Source Field Operation And Manipulation CFD ToolBox) - freely available software for numerical simulation of problems in mechanics. [3] The use of OpenFOAM for the simulation of arc discharge reactors is possible because OpenFOAM is oriented at numerical simulation of flows of multicomponent gases, the chemical interaction between components in the multicomponent gas mixture being considered.

At present there is no complete physical model of the electric arc with strong evaporation of electrodes, and exact calculations of these parameters are impossible. Thus, it is useful to calculate a 2-dimensional model of a turbulent jet coming from the anode electrode, which is an idealized representation of the flow of a helium-carbon mixture in the interelectrode gap. Similar calculations are presented in [2]. In this model the following boundary conditions were established: a fixed velocity v = 29.34 m/s and fixed temperature T=3000 K were set at the jet inlet, the gas mixture consisted of 90% He and 10% C; constant pressure P=50 kPa and zero gradient for velocity and temperature were set at the outlet.

As a result of calculations of simplified tasks, concentrations of carbon clusters in the mass fractions after the establishment of an equilibrium concentration at the point of maximum output of fullerenes (at the jet center) were obtained (Fig.1). The mass fractions of most stable fullerenes C_{60} and C_{70} after establishment of the equilibrium composition of the gas mixture were calculated (the mass fraction of C_{60} was found to be 0.78‰ and mass fraction of C_{70} was found to be 0.10‰, the total content of carbon clusters in the gas mixture was 61‰, thus the total content of fullerenes C_{60} and C_{70} was 1.45%). Fig.2 shows the temperature distribution at the jet center.

Figure 1. Concentrations of fullerenes C_{60} and C_{70} along the jet central line (X axis).

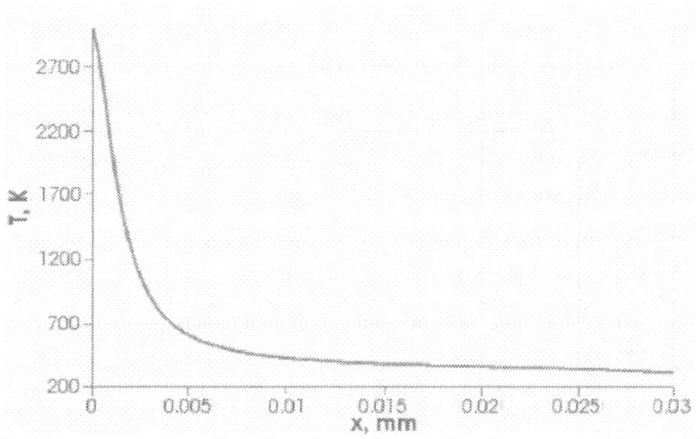

Figure 2. Temperature distribution along the jet central line (X axis).

The calculation of the tasks in the real geometry was conducted in several steps. First the helium flow was calculated, the chemical interaction being excluded. The calculations were performed by solver sonicFoam (for calculations of the flow of a viscous compressible gas) to establish the velocity field. The flow model was turbulent and the k-epsilon model of turbulence was selected.

Fig. 3 shows the fields of temperature and concentration of fullerene C_{60} in the cross section of the arc discharge camera. The maximum content of fullerenes is in a fairly narrow temperature area. In the high temperature field (near the interelectrode gap) the content of the fullerenes is negligible.

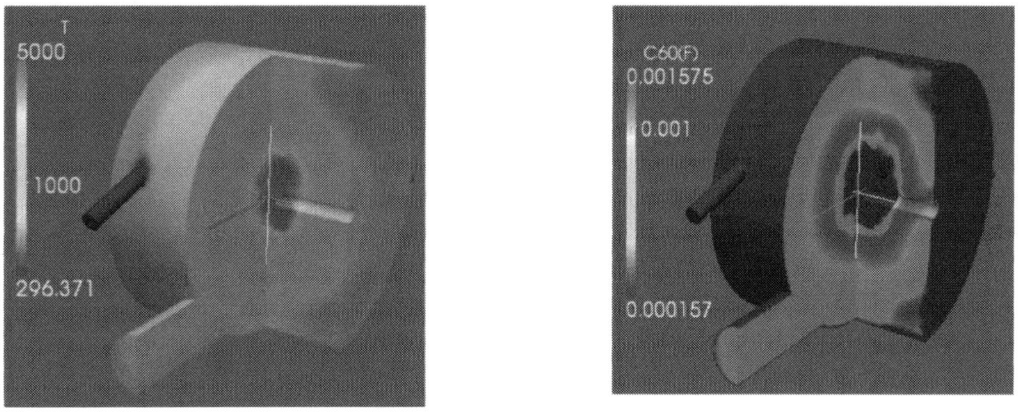

Figure 3. a) temperature and b) C_{60} concentration fields in the camera cross section.

4. Conclusions

The mathematical model of fullerene formation from a carbon vapor in an inert environment (helium in our case) has been developed. The numerical simulation of this process includes 82 components of the gas mixture and 522 chemical reactions. A strong dependence of the rate of carbon cluster formation on temperature has been revealed. The problem of a carbon vapor jet flow in the two-dimensional geometry has been solved. The distribution of the concentrations of the components has been obtained, the formation of fullerenes C_{60}, C_{70} has been shown and the temperature zones in which the speed of fullerene formation is maximal have been defined.

Modeling of the flow in a three-dimensional reactor using the real geometry of the setup built at Ioffe Physico-Technical Institute has been performed. Distributions of velocity and temperature in the reactor chamber have been found and the distribution of fullerene C_{60} in the camera has been obtained. It is important to note that in order to improve the model it is necessary to sort out the chemical reactions that make major contributions into fullerene formation and exclude insignificant reactions.

This is a complicated task because the formation of fullerenes of carbon vapor occurs sequentially as the cluster size increases.

To clarify the kinetics of the process it is necessary to include the coefficients of the dependence of the reaction rate constants on temperature and pressure into the Arrhenius equation, to find E_a values in the Arrhenius equation (in our model it is zero for most reactions), and also to include the reactions with participation of carbon ions.

The model can also be improved by introducing electromagnetic interactions.

Acknowledgments

The work was supported by contract No. 02.740.11.0201 of the federal target program "Scientific and Scientific-Pedagogical Personnel of the Innovative Russia", the topic "Mechanics of Impulse Processes in Gases and Plasma".

References

[1] 2010. *Handbook of nanophysics. Clusters and fullerenes* / ed Klaus D Sattler. CRC Press pp 536-671

[2] Krestinin A V, A P Moravskii and P A Tesner. 1998. A kinetic model of formation of fullerenes C_{60} and C_{70} in condensation of carbon vapor. *Chemical Physics Reports* **17(9)**: 1678-1707.

[3] Official website OpenFOAM: http://www.openfoam.com/.

Journal of Physics: Conference Series **461** (2013) 012044 doi:10.1088/1742-6596/461/1/012044

Vibrations in amorphous solids beyond the Ioffe-Regel criterion

Y M Beltukov[1], **V I Kozub**[1] **and D A Parshin**[2]

[1]A. F. Ioffe Physical-Technical Institute, 194021 Saint Petersburg, Russia
[2]Saint Petersburg State Polytechnical University, 195251 Saint Petersburg, Russia

Abstract. We consider vibrations in *3d* random harmonic lattices with translational invariance as a model of amorphous solid. Above some frequency $\omega_{\rm IR}$, corresponding to the Ioffe-Regel crossover, notion of phonons becomes ill defined. They cannot propagate through the lattice and transfer energy. Nevertheless most of the vibrations in this range are not localized. Changing strength of disorder we can vary $\omega_{\rm IR}$ from zero value (when rigidity is zero and there are no phonons in the lattice) up to a typical frequency in the system. Above $\omega_{\rm IR}$ a displacement structure factor $S(\mathbf{q},\omega)$ coincides well with a structure factor of random walk on the lattice. As a result the vibrational line width $\Gamma(q) = D_u q^2$ where D_u is a diffusion coefficient of particle displacements. Our findings may have important consequence for the interpretation of experimental data on inelastic x-ray scattering in glasses.

1. Introduction

Propagation of vibrational excitations in disordered systems is one of the advanced problems in condensed matter physics. In perfect crystals these excitations are well defined plane waves (phonons) having frequency ω and wave vector $\mathbf{q}$. However in amorphous solids the notion of phonons can be lost due to strong scattering on disorder. Then the natural question arises about the mechanism of energy and momentum transfer in this case.

Some time ago, delocalized vibrations in glasses of a new type, different from phonons, were introduced. They were called *diffusons* [1, 2, 3]. These are vibrations spreading through the system not ballistically, as phonons (on distances of the order of mean free path) but by means of diffusion. It is an important class of excitations which occupy in glasses the dominant part of the spectrum [3]. In these papers it was put forward the hypothesis that the boundary between phonons and diffusons is determined by the Ioffe-Regel criterion for phonons.

Therefore, as we believe, it is important to study properties of diffusons systematically in systems where they exist. They bring a new physics to our understanding of vibrational properties in strongly disordered systems and energy/heat transfer in glasses. To study these important properties we should have a model, sufficiently simple, to describe all of them.

Recently, using a random matrix approach, we developed a simple scalar model of a *3d* disordered harmonic lattice with translational invariance [4, 5] which, as we believe, can represent typical properties of vibrations in disordered systems. It was shown that for certain type of disorder the rigidity of the system is zero and usual phonons cannot propagate through the lattice. However almost all vibrations in the lattice are delocalized and therefore can participate in the heat transfer. To start with, for the sake of clarity of further consideration, we outline below the main properties of our model.

2. A random matrix approach

The vibrational properties of a mechanical system of N particles are determined by a dynamical matrix $M_{ij} = \Phi_{ij}/\sqrt{m_i m_j}$, where Φ_{ij} is a force constant matrix and m_i are particle masses. The matrices M and Φ are real, symmetric and *positive definite* matrices $N \times N$ (we consider for simplicity a scalar model). The last condition is important. It ensures mechanical stability of the system. For a free mechanical system it is necessary to satisfy also *translation invariance* conditions

$$\sum_i M_{ij} = \sum_j M_{ij} = 0 \tag{1}$$

(for simplicity we consider below all masses $m_i = 1$). It ensures that the potential energy of the system depends only on the differences of particle displacements.

We take the dynamical matrix in the form [5]

$$M = AA^T + \mu M_0. \tag{2}$$

Here A is a random matrix $N \times N$ built on a simple cubic lattice with N particles and interaction between nearest neighbors only. The only non zero non-diagonal matrix elements A_{ij} between the nearest neighbors are taken as independent random numbers from Gaussian distribution with zero mean $\langle A_{ij} \rangle = 0$ and unit variance $\langle A_{ij}^2 \rangle = 1$. The variance controls the degree of disorder in the lattice. To ensure the translational invariance the diagonal elements are calculated as a minus sum of non-diagonal elements $A_{ii} = -\sum_{j \neq i} A_{ji}$. M_0 is a simple crystalline dynamical matrix with unit springs between the nearest neighbors.

If the first term AA^T in Eq. (2) is responsible for the disorder in the system, the second term μM_0 describes the ordered part of the Hamiltonian. The parameter μ controls the relative amplitude of this part and the rigidity of the lattice. It can vary in the interval $0 \leq \mu < \infty$, changing the rigidity and relative amount of disorder. In this paper we have mainly considered the case of strong disorder when $0 \leq \mu \leq 1$ and fluctuating part of the dynamical matrix is bigger then the ordered part. Such form of the dynamical matrix guaranties the mechanical stability of the system for any positive value of μ.

3. Phonons and the Ioffe-Regel criterion

To find the phonon dispersion curve $\omega(\mathbf{q})$ and phonon mean free path $l(\omega)$ we should calculate space and time Fourier transform of the particle displacement field $u(\mathbf{r}, t)$. For that we ascribed to all the particles at the initial moment $t = 0$ random displacements (with Gaussian distribution with zero mean and unit variance) and zero velocities. Then, numerically solving Newton equations, we analyzed the particle dynamics at $t \neq 0$.

Let $u(\mathbf{r}_i, t)$ be the i-th particle displacement as a function of particle coordinate $\mathbf{r}_i$ and time t. We define the displacement structure factor (DSF) of the displacement field as follows

$$S(\mathbf{q}, \omega) = \frac{2}{NT} \left| \sum_{i=1}^{N} e^{-i\mathbf{q}\mathbf{r}_i} \int_0^T u(\mathbf{r}_i, t) e^{i\omega t} dt \right|^2. \tag{3}$$

For better frequency resolution the upper time limit T was taken sufficiently large. Since vectors $\mathbf{r}_i$ in a cubic lattice are discrete, the wave vectors $\mathbf{q} \equiv \mathbf{q}_n$ are also discrete and are defined on the corresponding reciprocal lattice.

To analyze phonon excitations, we fit the structure factor with the Lorentz distribution

$$S(\mathbf{q}, \omega) \propto \frac{1}{(\omega - \omega_{\mathbf{q}})^2 + (\Delta\omega)^2}. \tag{4}$$

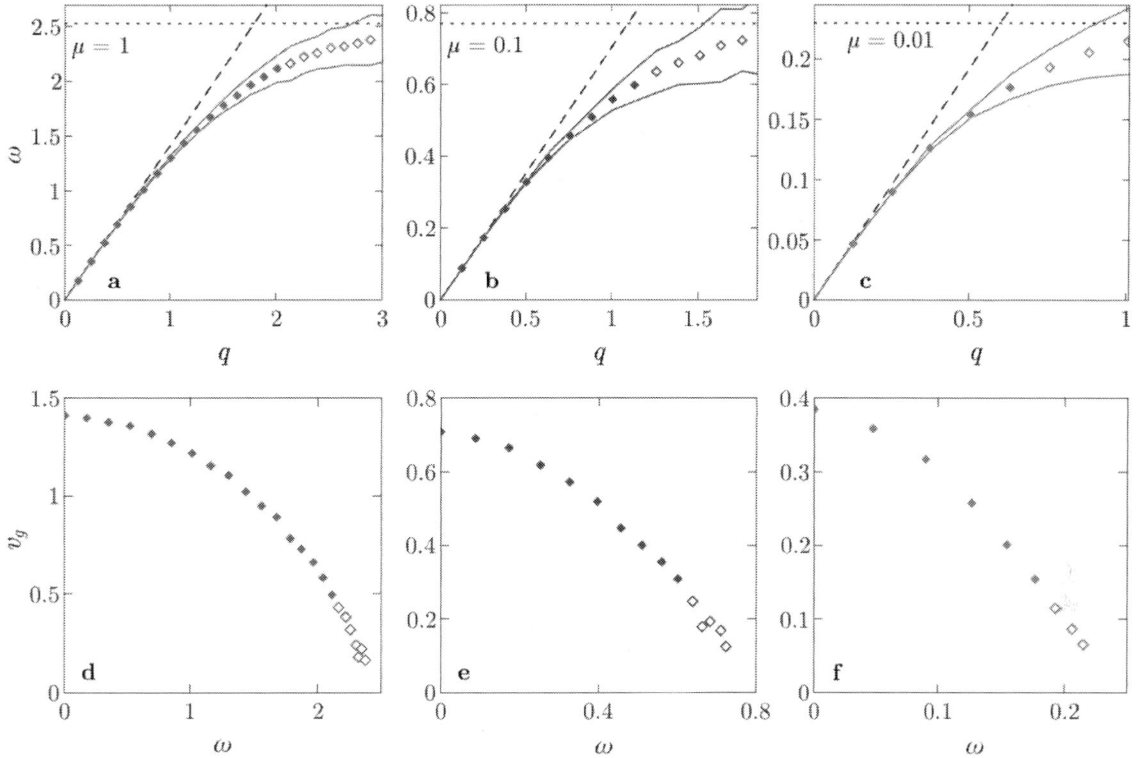

Figure 1. (a–c) The dependence $\omega_{\mathbf{q}}$ on q for $\mathbf{q} \parallel \langle 100 \rangle$ direction for various μ (1, 0.1, 0.01) in a cubic sample with $N = 50^3$ (one realization). Filled and open diamonds are the maximums of $S(\mathbf{q}, \omega)$ as a function of ω for each discrete value of q_n for frequencies below and above the Ioffe-Regel crossover correspondingly (see text below for details). Solid lines correspond to halves of the maximums. Dashed lines show $\omega = vq$ linear dependence with sound velocity v. (d–e) The group velocity $v_g = d\omega/dq$ as a function of ω.

From this fit we can find both the phonon frequency $\omega_{\mathbf{q}}$ and the phonon line width $\Delta\omega$. The results for $\omega_{\mathbf{q}}$ are shown on Fig. 1 for three values of μ and $\mathbf{q} \parallel \langle 100 \rangle$. For sufficiently small values of wave vector q we see a nice linear dispersion curve $\omega_q = vq$.

The phonons are well defined excitations if their mean free path $l = v_g/2\Delta\omega$ exceeds the phonon half wave length $l > \lambda/2 = \pi/q$ (Ioffe-Regel criterium for phonons). We will call the corresponding crossover frequency as ω_{IR}. Fig. 2 shows the ratio l/λ as a function of ω for several values of μ and different directions of the wave vector $\mathbf{q}$. The boundary between filled and open symbols (the full horizontal line) corresponds to frequency ω_{IR}. Thus filled and open symbols on Figs. 1 and 2 belong to phonons with frequencies below and above the Ioffe-Regel crossover frequency correspondingly.

4. Diffusons

Above ω_{IR} phonons cease to exist as well defined excitations. They are smoothly transformed to diffusons. Since the position of center of inertia is conserved $\sum_i u_i(t) = 0$ (all masses are equal). Therefore we can consider a diffusion of displacements like a diffusion of "particles" in a lattice where the total number of particles is conserved.

Consider first the case of $\mu = 0$ when phonons are absent and only diffusons are present in the lattice. Fig. 3 shows the structure factor $S(\mathbf{q}, \omega)$ as a function of wave vector q for three

Journal of Physics: Conference Series **461** (2013) 012044 doi:10.1088/1742-6596/461/1/012044

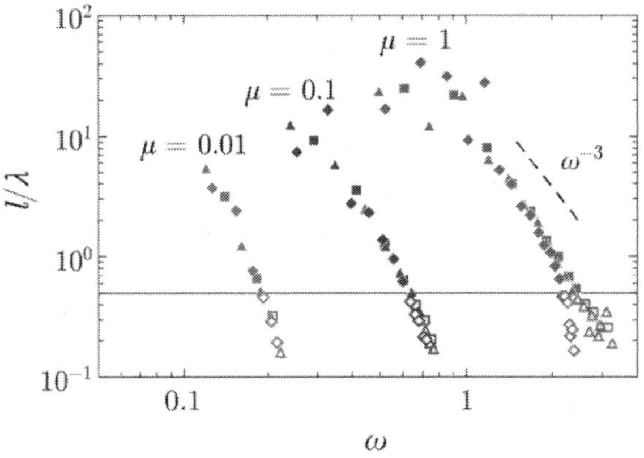

Figure 2. The ratio $l(\omega)/\lambda$ as a function of ω for different μ. Different symbols correspond to different $\mathbf{q}$ directions. ◇ for $\mathbf{q} \parallel \langle 100 \rangle$, △ for $\mathbf{q} \parallel \langle 110 \rangle$, □ for $\mathbf{q} \parallel \langle 111 \rangle$. The full horizontal line (separating filled and open symbols) corresponds to Ioffe-Regel crossover $l(\omega) = \lambda/2$.

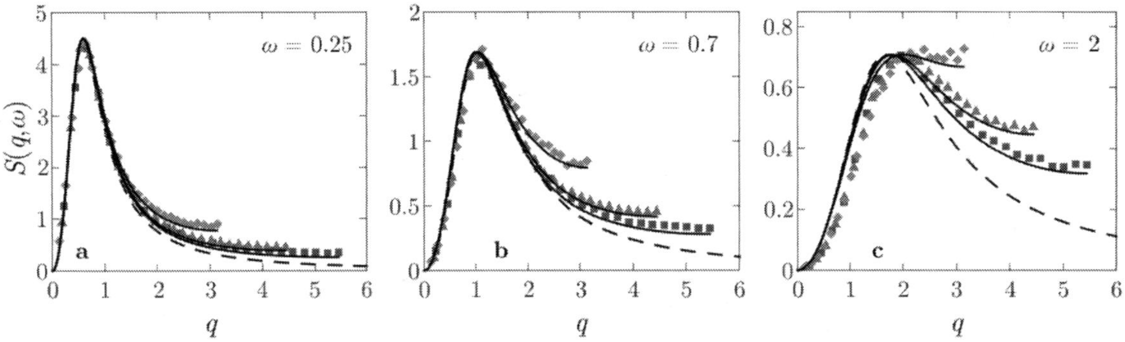

Figure 3. The displacement structure factor $S(\mathbf{q}, \omega)$, Eq. (3) (symbols) for $\mu = 0$ and for three different frequencies. The sample size is $N = 50^3$. The averaging is performed over 300 realizations. Different symbols correspond to different $\mathbf{q}$ directions as explained on Fig. 2. Full lines correspond to the structure factor $S_{\rm rw}(\mathbf{q}, \omega)$ of the random walk on the lattice given by Eq. (5) with $D_{\rm rw} = 0.7$. Dashed line corresponds to the limit $q \ll 1$ (see Eq. (7)).

different directions in $\mathbf{q}$ space (symbols) and for three different frequencies ω. Let us compare this displacement structure factor with structure factor of the random walk $S_{\rm rw}(\mathbf{q}, \omega)$ on the lattice. As was shown in [6] for the case of the random walk on a lattice, $S_{\rm rw}(\mathbf{q}, \omega)$ is given by expression

$$S_{\rm rw}(\mathbf{q}, \omega) = \frac{2\Gamma(\mathbf{q})}{\omega^2 + \Gamma^2(\mathbf{q})}, \quad \Gamma(\mathbf{q}) = 4D_{\rm rw}\left(\sin^2\frac{q_x}{2} + \sin^2\frac{q_y}{2} + \sin^2\frac{q_z}{2}\right) \tag{5}$$

where $D_{\rm rw}$ is a diffusion constant of the random walk. For small values of $q \ll 1$

$$\Gamma(\mathbf{q}) = D_{\rm rw} q^2 \tag{6}$$

and in the continuum limit we have the well known result for the diffusion structure factor

$$S_{\rm rw}(\mathbf{q}, \omega) = \frac{2D_{\rm rw} q^2}{D_{\rm rw}^2 q^4 + \omega^2}. \tag{7}$$

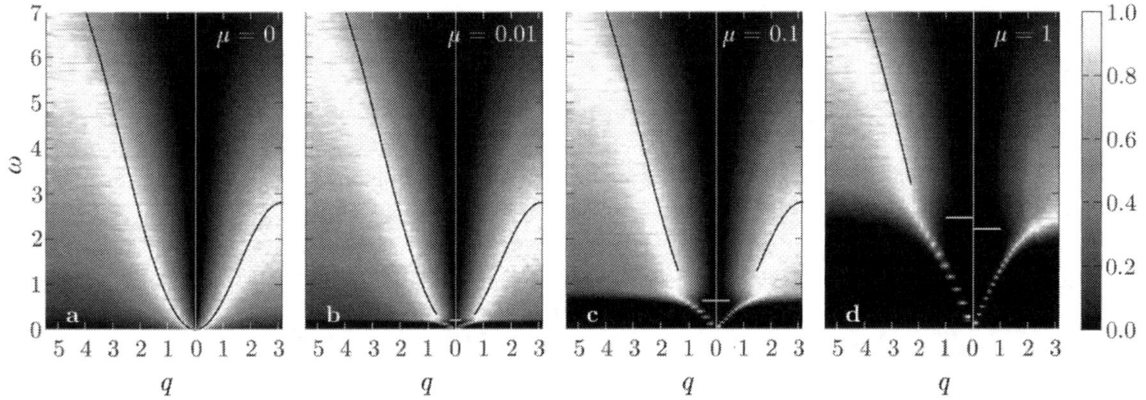

Figure 4. The normalized structure factor $S_n(\mathbf{q}, \omega)$ as a function of q for some direction in $\mathbf{q}$ space and for each frequency ω for various values of μ (0, 0.01, 0.1, 1). The sample size is $N = 50^3$. The averaging is performed over 100 realizations. Left sides of all plots are for $\mathbf{q} \parallel \langle 111 \rangle$, right sides are for $\mathbf{q} \parallel \langle 100 \rangle$. White horizontal dashes show the Ioffe-Regel crossover frequency ω_{IR}. For $\mu = 1$ the frequency ω_{IR} is slightly different for different $\mathbf{q}$ directions.

A comparison of the displacement structure factor $S(\mathbf{q}, \omega)$, (3), and the structure factor of the random walk $S_{\text{rw}}(\mathbf{q}, \omega)$, (5), is shown on Fig. 3. One fitting parameter was the diffusion coefficient D_{rw}. From comparison of these data we obtained $D_{\text{rw}} \approx 0.7$. It means that the diffusion coefficient of particle displacements $D_u \approx 0.7$. Another fitting parameter was a height $h(\omega)$ of the random walk structure factor in the maximum. According to Eq. (5), in the maximum $\Gamma(\mathbf{q}) = \omega$ and $h(\omega) = 1/\omega$, but to fit the data points on Fig. 3 we used slightly higher values of $h(\omega)$. The small difference between $h(\omega)$ and $1/\omega$ can be explained by different frequency dependencies of the density of states for vibrations and for the random walk.

Now let us analyze the displacement structure factor $S(\mathbf{q}, \omega)$ for $\mu \neq 0$. For better visual effect we will show a map of the function $S(\mathbf{q}, \omega)$ on the plane (ω, q) for different directions in $\mathbf{q}$ space. To do that, for each frequency ω we have found the maximum $S(\mathbf{q}, \omega)$ as a function of q along some directions in $\mathbf{q}$ space. Then we normalized function $S(\mathbf{q}, \omega)$ along this line $\omega = const$ to the magnitude of this maximum.

The results are shown on Fig. 4 for four different values of μ and two directions in $\mathbf{q}$ space. The white color corresponds to the maximum when normalized structure factor $S_n(\mathbf{q}, \omega) = 1$ while the black color to the case where $S_n(\mathbf{q}, \omega) = 0$. For $\mu \neq 0$ we can see clearly two types of excitations in the lattice. At low enough frequencies, below ω_{IR}, we see phonons with well defined dispersion law $\omega_{\mathbf{q}}$, the same as in the previous Section. At the Ioffe-Regel crossover frequency ω_{IR}, the structure factor strongly broadens and phonon dispersion line disappears. Above ω_{IR} the displacement structure factor coincides well with the structure factor for $\mu = 0$ case shown on Fig. 4a, which corresponds to diffusons. The maximum of the normalized structure factor $S_n(\mathbf{q}, \omega)$ (white regions) agrees well with the maximum of the random walk structure factor $S_{\text{rw}}(\mathbf{q}, \omega)$ (black line). Deviations from $S_{\text{rw}}(\mathbf{q}, \omega)$ take place at high frequencies near the localization threshold.

The vibrational line width above ω_{IR} is $\Gamma(q) = D_u q^2$ (Eq. 6). Such quadratic dependence of $\Gamma(q)$ was found in many glasses in the experiments on inelastic x-ray scattering, see for example [7, 8] and references therein. It was also found in molecular dynamic simulation of amorphous silicon [9]. However in these and other papers this line width was attributed to phonons without discussion of its physical origin. We guess that the observed q^2 dependence of $\Gamma(q)$ has nothing to do with phonons and is in fact related to diffusons.

5. Conclusion

We have found that the delocalized vibrational excitations in this disordered lattice are of two types. At low frequencies below the Ioffe-Regel crossover, $\omega < \omega_{\rm IR}$, they are the usual phonons (plane waves) which can be characterized by frequency ω and wave vector $\mathbf{q}$ and line width $\Delta\omega$. However, with increasing of ω, due to the disorder-induced scattering, the phonon line width $\Delta\omega$ increases rapidly and at some frequency $\omega \approx \omega_{\rm IR}$ the phonon mean free path l becomes of the order of the wave length λ.

As a result, at higher frequencies the original notion of phonons is lost and delocalized vibrational modes have a diffusive nature. They are similar to *diffusons* introduced by Allen and Feldman, et al. [3]. The diffusons again can be characterized by frequency ω, but have no well defined wave vector $\mathbf{q}$. Above $\omega \approx \omega_{\rm IR}$ the structure factor of particle displacements $S(\mathbf{q},\omega)$ becomes very similar to the structure factor $S_{\rm rw}(\mathbf{q},\omega)$ of a random walk on the lattice. The former has a broad maximum as a function of q at $q = \sqrt{\omega/D_u}$, where D_u is a diffusion coefficient of the particle displacements.

Presumably diffusons are also accounted for the mysterious q^2 dependence of the vibrational line width $\Gamma(q)$ observed in many experiments on inelastic x-ray scattering in glasses. Therefore we think that it is necessary to take them into account in interpretation of experimental data.

We are very grateful to V. L. Gurevich and Anne Tanguy for many stimulating discussions and gratefully acknowledge interesting discussions with B. Rufflé and E. Courtens as well. One of the authors (DAP) thanks the University Lyon 1 for hospitality. This work was supported by St. Petersburg Government (diploma project no. 2.4/29-06/143C), Dynasty Foundation, RF President Grant "Leading Scientific Schools" NSh-5442.2012.2 and Russian Ministry of Education and Science (contract N 14.740.11.0892).

References

[1] P. B. Allen, J. L. Feldman, Phys. Rev. B **48**, 12581 (1993).
[2] J. L. Feldman, M. D. Kluge, P. B. Allen, F. Wooten, Phys. Rev. B **48**, 12589 (1993).
[3] P. B. Allen, J. L. Feldman, J. Fabian, F. Wooten, Phil. Mag. B **79**, 1715 (1999).
[4] Y.M. Beltukov and D.A. Parshin, Physics of the Solid State **53**, 151 (2011) (Fizika Tverdogo Tela, **53**, 142 (2011)).
[5] Y.M. Beltukov and D.A. Parshin, JETP Letters **93**, 598 (2011) (Pis'ma v ZhETF **93**, 660 (2011)).
[6] Diffusion in Condensed Matter. Methods, Materials, Models, ed. Paul Heitjans, Jörg Kärger, Springer Berlin Heidelberg New York, 2005, p.745.
[7] F. Sette, M. H. Krisch, C. Masciovecchio, G. Ruocco, G. Monaco, Science **280**, 1550 (1998).
[8] G. Ruocco and F. Sette, J. Phys.: Condens. Matter **13**, 9141 (2001).
[9] J. K. Christie, S. N. Taraskin, S. R. Elliott, J. Non-Cryst.Sol. **353**, 2272 (2007).

Journal of Physics: Conference Series **461** (2013) 012045 doi:10.1088/1742-6596/461/1/012045

Spatial Evolution of a Strong Field of Few-cycle Light Beam in Dielectric Media with Induced Plasma Nonlinearity.

S.A. Stumpf, A.A. Korolev, S.A. Kozlov

National Research University of Information Technology, Mechanics and Optics, 49 Kronverksky pr., Saint-Petersburg, Russia 197101.

stumpf@rain.ifmo.ru, kozlov@mail.ifmo.ru

Abstract. The paper reports results of computer simulation of strong light beam propagation in dielectric media in case of plasma generation. We investigate an extra-broadening of radiation spectrum to a 'violet' wing of visible range. We show that the resulting pulse spectrum is represented by sequence of well-separated maximums, broadening as propagation distance increases. Experimental data are compared with simulation results, showing a good mutual correspondence of spectral representations.

1. Introduction.

A propagation of high-intense femtosecond pulses through dielectric media without its optical breakdown allows observing of nonlinear optical phenomena more shaped and illustrative than for longer pulses. Nowadays spectral supercontinuum generation is investigated deeply and complexly for self-action of intense single pulse in dielectric media. So a significant attention is paid to examination of simultaneous propagation of two or more intense femtosecond pulses with different spectra, because spectral extra-broadening of such complex beam demonstrates new peculiarities and gives a notably richer output spectrum.

A number of papers [1-5] investigate a propagation of intense two-color beam through air in conditions when inharmonic oscillations of valence electrons convert to a plasma nonlinearity. Effective generation of wide-range IR-radiation (wavelength up to 1 mm) is experimentally demonstrated. This method allows acquiring the most intense THz-radiation [2], widely applicable [3-5]. So an actual problem is optimal conditions for such generation. This problem requires a theory of two femtosecond pulse interaction propagating in optical media upon plasma excitation, which was derived by us in paper [6]. This theory gives an excellent correlation between computed data and experimental results of observation of quazi-periodic dependence of generated THz radiation power on interacting pulses spatial shift.

Recent years studies examine experimentally an amplification of spectral short-wave wing caused by plasma nonlinearity and free electron motion [7-10]. The main source of a generation of light quants with higher energy (and, though, shorter wavelength) is assumed to be a conversion of kinetic energy accumulated by free electron in an oscillating electrical field to an electromagnetic wave at its recombination with an originating ion. Our dynamic model contains all necessary components to describe such an interaction: a wave equation describing self-action of light wave; a description of plasma excitation by a strong light beam and opposite process – recombination as the wave passes by, an equation of ponderomotive dynamics of free electron in an oscillating electric field.

Journal of Physics: Conference Series **461** (2013) 012045 doi:10.1088/1742-6596/461/1/012045

The present paper reports main results of application of the model to an investigation of spectral short-wave wing amplification of high-intense femtosecond pulse propagating through dielectric media in case of with plasma excitation. We show the resulting spectrum, present its spatial dynamics and compare it with experimental data obtained in [11]

2. A strong few-cycle pulse evolution equation for isotropic dielectric media with plasma nonlinearity.

We have deduced an equation for strong few-cycle pulse dynamics in transparent isotropic dielectric media in [6], on a basis of density matrix formalism in three-band optical media approach. The higher conductivity subband is assumed to be a free electron motion state for a describing of plasma nonlinearity. We assumed that a field of radiation is essentially smaller than electric field of unitary ionized atom on a distance equal to Bohr radius $E_g = 10^9\ V/m$, i.e. we are considering a radiation with intensity $I << 10^{17}\ W/cm^2$. This allows use of habitual energetic model of a dielectric with quasi-discrete energy state structure. We consider a ESP interaction with dielectric medium, assuming that the spectrum of pulse, though it is very wide, is bounded by a material transparency range.This assumption is correct up to a several femtoseconds duration of optical pulse. In circumstances of non-resonant (in single-photon approximation) interaction we assume that linear part of dielectric polarization principally exceeds the nonlinear part, depending on a higher field orders. This assumption, for example, in fused silica, is met for pulse intensities up to $I \sim 5\cdot 10^{14} W / cm^2$. So this makes possible to limit ourselves to analyzing of third-order nonlinearity of media polarization. Thus we analyze only electronic nonlinearity, neglecting other mechanisms in few-cycle field because of their inertia characteristic times cardinally exceed pulse length [12, 13].

The result has a form of system of waveform evolution equation and dynamic material ones, considering inertial part of field-cube electron nonlinearity as well as inertial plasma nonlinearity induced in pulse strong field. Neglecting a self-reflected wave, as grounded in [14, 15], we combine this system to united evolution equation (1):

$$\begin{cases} \dfrac{\partial E}{\partial z} - a\dfrac{\partial^3 E}{\partial \tau^3} + g\dfrac{\partial E^3}{\partial \tau} + \dfrac{\partial}{\partial \tau}\left(g^{(1)} E \left(\dfrac{\partial E}{\partial \tau}\right)^2 + g^{(2)} E^2 \dfrac{\partial^2 E}{\partial \tau^2}\right) + \dfrac{2\pi}{cn_0} j = 0 \\ \dfrac{\partial \rho}{\partial \tau} + \dfrac{\rho}{\tau_p} = \alpha E^2 \\ \dfrac{\partial}{\partial \tau} j + \dfrac{j}{\tau_c} = \beta \rho E^3, \end{cases} \tag{1}$$

where E is pulse field electrical component, z is a spatial coordinate collinear to pulse propagation direction, $\tau = t - \frac{n_0}{c} z$, c – vacuum light speed, n_0 and a characterize a dependence of linear refraction index n on a frequency ω

$$n(\omega) = n_0 + ca\omega^2, \tag{2}$$

which can be easily refined, when necessary, with higher order addictives, by expanding (3) with higher order field time derivatives [15]; j stands for density of ionized electrons current, ρ – for their density; factors g, $g^{(1)}$, $g^{(2)}$ describe nonlinearity refraction index of dielectric media

$$n_2(\omega) = n_2^0 + A\omega^2, \tag{3}$$

where $n_2^0 = \frac{3}{2}cg + \frac{1}{2}\tau_{12}cg^{(1)}$, $A = \frac{1}{2}c(g^{(1)} - 3g^{(2)})$, $\alpha = \frac{1}{2n_0} \frac{|p_{23}|^2}{T_{21}(\hbar\omega_{32})^2}$, $\beta = \frac{e^2}{m_e^* c} \cdot \frac{n_0^2 - 1}{\hbar\omega_{21}}$ characterizes inertial plasma nonlinearity of dielectric media (see [6]), p_{ij} are scalar values of dipole moments of interstate transitions $i \to j$, ω_{ij} – frequencies according to these transition energy, T_{ij} are interstate relaxation times and τ_{ij} are state excitation relaxation times, m_e^* is an effective electron mass in conductivity band, $\sigma = 2\tau_c^{-1}$ is defined by average collision relaxation time τ_c of free electrons, which has an order of tens femtoseconds for most dielectrics.

3. A broadening of high-intense few-cycle pulse spectrum to 'violet' wing while propagating in SF6.

Pictures 1-3 illustrate results of numerical simulations of few-cycle pulse represented as

$$E(t) = E_0 \exp(-(tt_u^{-1})^2)\sin(\omega_0 t), \quad (4)$$

where E_0 – electric field amplitude, t_u – input pulse duration, ω_0 – central wavelength, propagating in dielectric media. Modeled media is chosen to be sulfur hexafluoride (SF_6) having an electrical resistance index several times higher than atmosphere air. We use the following values for media properties: a linear refraction index $n_0 = 1.0021$ ([16]), nonlinear refraction index $n_2 = 1.6 \cdot 10^{-19} cm^2/W$ ([17, 18]).

Figure 1 shows the extrabroadening of pulse spectrum in a wavelength range of $350 - 950\,nm$, at pulse properties: duration $t_u = 30\,fs$, central wavelength $\lambda_0 = 2\pi c/\omega_0 = 800nm$, input peak intensity $I = 1 \cdot 10^{14} W/cm^2$, propagation distance $z = 6.4\,mm$ (additional picture gives an overview of spectrum in wider range $350nm - 8\mu m$). For better illustration, this spectrum is compared with the one obtained by modeling the same circumstances and propagation regime, except plasma nonlinearity, that is supposed to be absent in some model media (dashed line). It is notable that plasma nonlinearity causes a drastic broadening of pulse spectrum into a short-wave range, having a form of separated spectral peak sequence. And when considering only inertia-less nonlinearity, we obtain some generation of longer wave components.

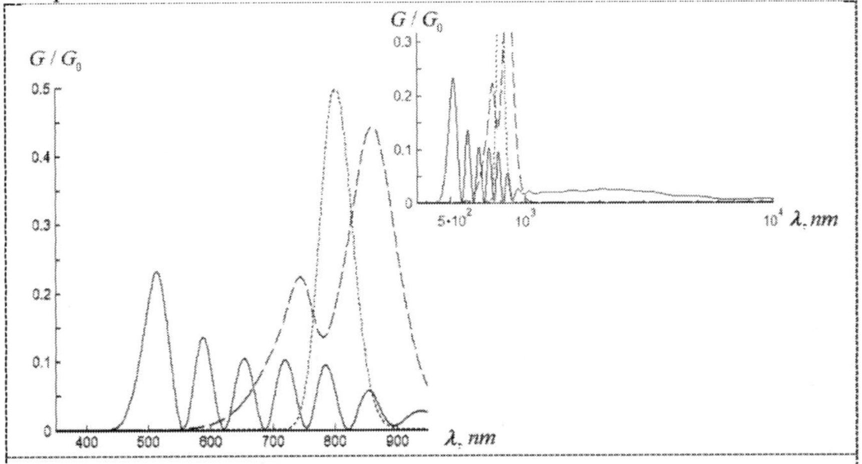

Figure 1. Extrabroadening of pulse spectrum caused by plasma nonlinearity (solid line), compared with the one in case of plasma neglecting (dashed) and initial spectrum (dotted).

Figure 2 illustrates this effect in dynamics, as it increases and evolves as the pulse propagates for a longer distance. This leads us to a conclusion that a propagation of a separated femtosecond pulse through a dielectric media with notable plasma nonlinearity may be accompanied with generation of high-frequency spectral components, dominating frequency 'green shift', in this case – to approximately $500\,nm$. It is obvious, that this phenomena differs from well-known harmonic generation, caused by inertia-less nonlinearity of third and higher orders.

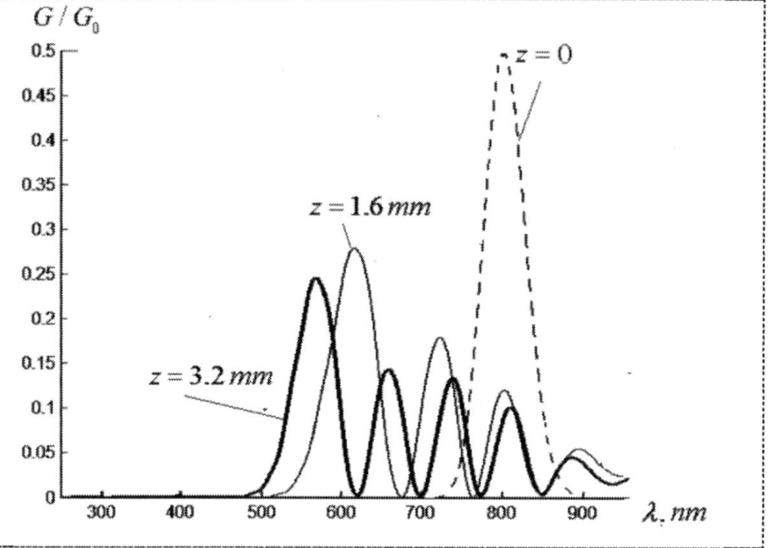

Figure 2. Spatial dynamic of broadening pulse spectrum, amplifying itself as distance increases.

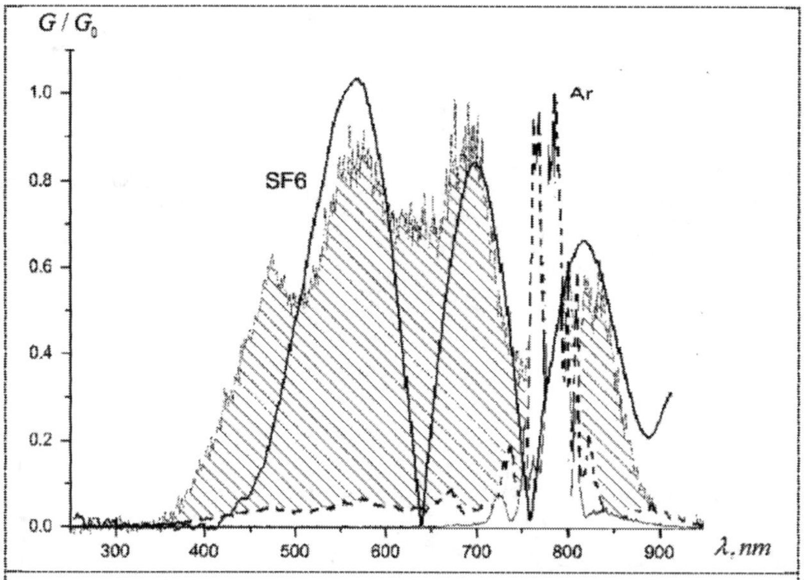

Figure 3. A combination of modelled (solid line) and experimental (shaded area) data on examined spectral broadening in sulphur hexafluoride.

Figure 3 combines a spectrum of modeled output pulse and experimental data on femtosecond pulse propagation in sulfur hexafluoride, published in [11]. Experimentally detected spectrum broadening, forming several high-frequency maximums (shaded area) correlates well with numerically simulated effect, reflecting, as well, appearance of several contrast maximums (solid line). We see the good accordance between quantity and positions of spectral peaks for experimental and simulated data. Higher contrast for modeled picture may be grounded by ideality of input pulse spectra and spatial profile, neglecting of diffraction and boundary effects. This allows to conclude that the approach derived in the present report is practically consistent and may be applied for investigations of interaction between high-intense laser pulses with dielectric media, considering media ionization and pre-breakdown phenomena.

4. Conclusion.

The study reports an application of dynamic interaction model to a simulation of high-intense femtosecond pulse propagation through dielectric media, in case pulse properties cause the initiation of optical breakdown, generation of free electrons. It is shown that the model demonstrates a generation of several isolated spectral maximums between main and second harmonics of initiative pulse, and besides the quantity and intensity of these maximums is dependent on propagation distance. Simulation results are compared with experimental data of investigation of femtosecond pulse propagation in sulfur hexafluoride, showing a notably high level of data correspondence. That enables concluding the definite practical value of derived approach and prospective application of the model to planning experiments, processing experimental data and explaining a physical nature of observed phenomena.

References

[1] Karpowicz N., Lu X., Zhang X.-C. J. Modern Opt. 2009. V.56, N.10. P. 1137–1150.
[2] Dai J., Karpowicz N., Zhang X.-C. Phys. Rev. Lett. 2009. V.103, 023001 (4 p.)
[3] Garnov S.V. Proc. Int. Symp. "Topical problems of biophotonics – 2009", P. 275.
[4] Achmetzhanov R.A., Ilyakov I.E. et al. JETP 2009. V.136. №3(9). P. 431-441.
[5] Andreev A.A., Bespalov V.G., Gorodetsky A.A. et. al. Opt. Spectr. 2009 V.107. P.569-576.
[6] Stumpf S.A., Korolev A.A., Kozlov S.A. Izv. RAN, Ser. Phys., 2007. T.71, №2, P.158-161.
[7] Couairon A., Bergé L., Phys. Rev. Lett., 2002, v.88, N 13, P.135003(4).
[8] Kosareva O.G., Panov N.A., Kandidov V.P., Chin S.L. et al. Appl. Phys. B, 2009, v. 91, N3, P.35-43.
[9] Akozbek N., Trushin S.A., et al. New Journal of Physics, 2006, v.8, N 9, P.177-188.
[10] Trushin S.A., Kosma K., Fuß W., Schmid W.E. Opt. Lett., 2007, v. 32, N16, P.2432-2434.
[11] Zhavoronkov N. Opt. Lett. 2011. V.36. N.4. P. 529-531.
[12] Berkovsky A.N., Kozlov S.A., Shpolyansky Yu.A. Topics in coherent and nonlinear optics, 170-188, SPbSU ITMO, Saint-Petersburg, 2004.
[13] Husakou A.V., Herrmann J. Phys. Rev. Lett. 87, 203901 (1-4), 2001.
[14] Kozlov S.A., Sazonov S.V. JETP, 111, 404–418, 1997.
[15] Bespalov V.G., Kozlov S.A., Shpolyansky Yu.A.,Walmsley I.A. Phys. Rev. A. 66, 0138111 (1-4), 2002.
[16] Leonard, P. J. Atomic Data and Nuclear Data Tables. 1974. V.14. P.21-37.
[17] Vuong L. T., Lopez-Martens R. B., Hauri C., Gaeta A. L. Opt. Express. 2008. V.16. P. 390-401.
[18] Nibbering E. T. J., Grillon G., Franco M., Prade B., Mysyrowicz A. J. Opt. Soc. Am. B. 1997. V.14. P.62-65.

Journal of Physics: Conference Series **461** (2013) 012046 doi:10.1088/1742-6596/461/1/012046

Estimated probability of the partial coverage of QSOs by intervening H_2-clouds at formation of QSO absorption-line spectra

D D Ofengeim[1], A V Ivanchik[1,2], A D Kaminker[1,2], S A Balashev[2]

[1] St.-Petersburg State Polytechnical University, Politekhnicheskaya 29, 195251 St.-Petersburg, Russia

[2] Ioffe Physical-Technical Institute, Politekhnicheskaya 26, 194021 St.-Petersburg, Russia

E-mail: ddofengeim@gmail.com

Abstract. We have estimated the probability of a partial coverage of QSO broad-line regions by intervening H_2-clouds. This effect has been revealed by an analysis of H_2 absorption systems in QSO spectra [1, 2]. Accounting of the effect may change significantly physical parameters of interstellar clouds derived from the spectral analysis [2].

We show that the probability of incomplete coverage turns out to be not lower than about 8%. Actually, a frequency of the effect revealing may occur essentially higher than this quantity, encouraging us in further systematic searches of the incomplete coverage in the H_2 absorption QSO spectra.

1. Introduction

It was widely accepted until recently that intervening H_2 absorption clouds fully cover cosmologically remote quasars (QSOs) usually treated as point-like powerful sources of radiation. The interstellar clouds in remote galaxies situated on lines of sight between QSOs and an observer imprint a set of absorption lines into initial QSO spectra which consist of a smooth continuum and broad emission lines. The latter are formed within a broad-line region (BLR) in a wider vicinity of a central QSO machine.

The partial coverage of a BLR by an intervening absorption cloud was firstly reported in [1] and investigated in detail by [2]. It was demonstrated that the H_2-bearing cloud covers the QSO 1232+082 ($z_{em} = 2.57$) intrinsic continuum source completely but only a part of the BLR. The details of this unique effect and different alternative interpretations were discussed in the paper by [2].

Recently another H_2 absorption system with partial coverage of the appropriate QSO has been revealed in Q 0528-250 spectrum [3]. Since taking into account of the partial coverage leads to serious changes of the results of QSO spectra analysis (e.g. column densities of absorbers), it is important to investigate possibility of the partial coverage effect in other QSO H_2 absorption systems. Following [2], one can introduce *noncoverage factor* f as the ratio of a light flux passing by the cloud (i.e. came to the detector without absorption) to a flux which would be detected in absence of the cloud. It is useful to calculate a distribution function of f, i.e. a probability to reveal $f > f_0$ (f_0 is an assigned value) for an arbitrary absorption system.

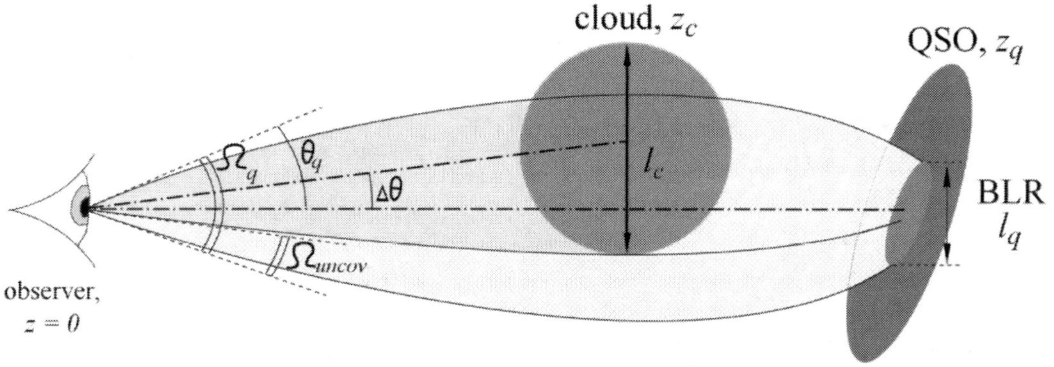

Figure 1. Schematic illustration of an absorption cloud with a transverse size l_c (grey circle) situated between the observer and QSO broad-line region (BLR) with a transverse size l_q (red ellipse); θ_q is an angular size of BLR, $\Delta\theta$ is an angle between the line-of-sights from the observer to the centres of QSO and cloud. Ω_q is a solid angle (light cone) of the whole BLR radiation flux, Ω_{uncov} is a solid angle filled by non-absorbed radiation. Radiation from the QSO without H_2-absorption systems is shown as yellow area of the light cone and radiation containing the absorption systems – as a dimly coloured area. Note that the light cone is curved due to the expansion of the Universe.

Note that there is an important reason to search the partial coverage phenomenon for molecular clouds in contrast with H I clouds (mostly forming Lyα forest) which is likely to have transverse sizes well exceeding the probable sizes of QSO BLRs (e.g. [4]).

In the present paper, we build the probability function of noncoverage factor f for QSO BLRs with $z_{em} = z_q$ (QSOs) by absorption systems with $z_{abs} = z_c$ (clouds). We consider an arbitrary angular distance between line-of-sights to the centres of the cloud and corresponding BLR, as well as an arbitrary ratio κ of the cloud to BLR transverse sizes. In Section 2 we introduce noncoverage factor f and consider the main parameters determinative f. In Section 3 we calculate distribution of noncoverage factor for certain distributions of physical parameters. In Section 4 we discuss obtained results and their relation to observations.

2. Geometry of partial coverage

A sketch of partial coverage is represented on Fig.1. Light from the QSO propagates through the Universe and it is registered by an observer. The observer detects a light cone within an angle θ_q or solid angle Ω_q. Some part of light passes by the cloud and comes to the observer without formation of an absorption-line system in a spectrum. This part is comprised within a solid angle Ω_{uncov}. The rest light is partly screened by the cloud in such a way that absorption of molecules in the cloud imprints a set of absorption lines in the initial spectrum. In result the observer detects a complex radiation from QSO with spectrum integrated over angles. A screened (*covered*) flux of radiation may be approximately estimated as a flux comprised by a solid angle $\Omega_{cov} = \Omega_q - \Omega_{uncov}$.

Noncoverage factor f is a ratio of the flux which goes by the cloud without absorption to the flux of radiation which would come to the observer if there was no cloud on its path. Let us assume roughly that both the screened and unscreened fluxes are uniformly distributed within their solid angles. Thus f may be estimated as a ratio of the solid angle Ω_{uncov} to the whole solid angle Ω_q of BLR observations:

$$f = \frac{\Omega_{uncov}}{\Omega_q}, \qquad (1)$$

Journal of Physics: Conference Series **461** (2013) 012046 doi:10.1088/1742-6596/461/1/012046

where at $\theta_q \ll 1$ one can write $\Omega_q = \pi\theta_q^2$.

Using the well-known definition of the *angular size distance*, $D_A(z)$, for cosmologically distant objects with a proper transverse size l at a cosmological redshift z, one can determine the angular size [5] $\theta = l/D_A(z)$, where $D_A = c d_A(z)/H_0$ is the angular size distance, H_0 is the *Hubble constant* at present, c is the speed of light and the dimensionless value $d_A(z)$ in the standard ΛCDM cosmological model is:

$$d_A(z) = \frac{1}{1+z}\int_0^z \frac{dz'}{\sqrt{\mathbf{\Omega}_m(1+z')^3 + \mathbf{\Omega}_\Lambda}}. \qquad (2)$$

Here $\mathbf{\Omega}_m$ is the dimensionless cold matter density parameter and $\mathbf{\Omega}_\Lambda = \Lambda c^2/(8\pi G)$ is determined by the cosmological constant Λ, G – gravitational constant. According to Fig. 1 one can set $\theta_q = \theta/2$ and employ the equation $\theta_q = lH_0/(2cd_A(z))$. Then using this relation we obtain the final equation for Ω_q. Estimations of the values Ω_{uncov} and f will be outlined below.

Let us demonstrate that the phenomenon of incomplete covering is amplified by cosmological properties of the space-time. Fig. 2 displays two dependencies of an angular size of a cosmologically distant object (with proper linear size l and redshift z) on the distance $L = ct(z)$, where $t(z)$ is the standard expression for the cosmological time of the light-signal propagation from the object to observer. Thus the green curve in Fig. 2 demonstrates the dependence of $\theta(z)$ on the light-propagation distance $L = ct(z)$ in expanding ΛCDM model. In the Euclidean (stationary) cosmological model it is a hyperbolic dependence $\theta = l/L$ on the distance L to the object.

One can see that an angular size in the expanding Universe is larger than it would be expected in the stationary model, and it grows with L at $L > 3$ Gpc. This is well known phenomenon in the Friedmann cosmological models (e.g. [6]). Specific light trajectories in such models relatively to the observer reference frame are schematically drawn in Fig. 1.

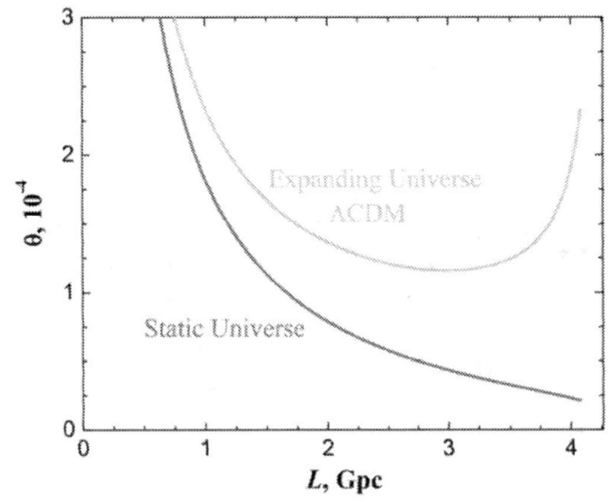

Figure 2. Angular size (in arcsec) $\theta(z)$ of an object with a proper transverse size $l = 1$ pc and cosmological redshift z versus distance $L = ct$; light passes from the object to the observer for cosmological time $t(z)$ in expanding (green curve) and stationary (Euclidean; red curve) Universes.

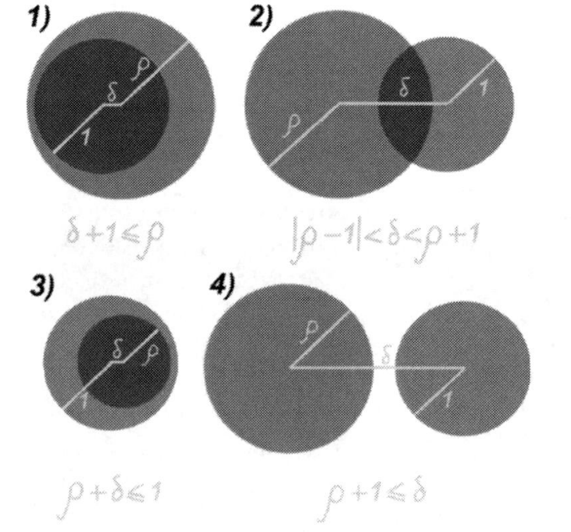

Figure 3. Four types of coverage. Types of mutual relations between cloud (grey circles) and QSO BLR (red circles) in the observer reference frame: 1) is full coverage (widely known case), 2) – crescent-like coverage, 3) – an "annular" coverage, 4) – full lack of coverage.

Thus, if BLR transverse size l_q is comparable with the cloud size l_c, then the phenomenon of partial coverage may occur for essentially different z_q and z_c. The condition $L > 3$ Gpc corresponds to $z > z_* \approx 1.64$. For detected H_2 absorption systems in QSO spectra both values $z_{em} = z_q$ and $z_{abs} = z_c$ typically exceed the value z_*, so it is necessary to take into account the cosmological expansion.

Assuming that BLRs and clouds are spherical one can determine Ω_{uncov} (see Eq. (1)) using an estimation of two partly overlapping circle areas. A circle conventionally representing the BLR may be characterized by the radius θ_q and a circle representing the cloud – by the radius θ_c. Then the square of the BRL circle is $\pi\theta_q^2 = \Omega_q$. Let us introduce also an overlapping square S (in units of rad^2) of the BLR and cloud circles. Then one can define $f = 1 - S/\Omega_q$, and additionally $\rho = \theta_c/\theta_q$, $\kappa = l_c/l_q$. In these terms we have $\rho(\kappa, z_q, z_c) = \kappa\, d_A(z_q)/d_A(z_c)$, $f = 1 - s/\pi$, where $s = S/\theta_q^2$. Note that ρ may be either $>$ or $< \kappa$. It is useful to introduce also a relative angular deviation $\delta = \Delta\theta/\theta_q$, where $\Delta\theta$ is an angular distance between line-of-sights to the centres of BRL and cloud. In our schematic approach δ characterizes a distance between circle centres. One can compose the resulting expression for $f = f(\kappa, \delta, z_q, z_c)$ as function of four parameters κ, δ, z_q and z_c.

Fig. 3 displays all possible relative positions of two conventional circles referred to the BLR and cloud; it is suitable to represent all values in units θ_q, then the radius of the BLR is 1.

1) *Full coverage.* All flux from the QSO goes through the cloud, $f = 0$. The condition is: $1 + \delta \le \rho$.
2) *Crescent coverage.* The condition for this case is: $|\rho - 1| < \delta < \rho + 1$. After some geometric calculations with $f(s) = 1 - s/\pi$ we obtain

$$f = 1 - \frac{\rho^2}{\pi}\arccos\frac{\rho^2 + \delta^2 - 1}{2\rho\delta} - \frac{1}{\pi}\arccos\frac{1 + \delta^2 - \rho^2}{2\delta} + \frac{1}{2\pi}\sqrt{[\rho^2 - (\delta - 1)^2]\,[(\delta + 1)^2 - \rho^2]}. \tag{3}$$

3) *Annular coverage.* The angular size of the cloud should be less then the BLR size. The condition is $\rho + \delta \le 1$; $\rho \le 1$, $\delta \le 1$. In this case we get $f = 1 - \rho^2$.
4) *Full noncoverage.* Relative angular deviation is large and all flux from the QSO goes by the cloud, i.e. the cloud remains unobservable in the QSO spectrum, i.e. $f = 1$. The condition of this case is $1 + \rho \le \delta$.

3. Distribution of noncoverage factor

Let us determine a distribution of noncoverage factor f, i.e. probability $P(f > f_0)$ to detect the absorption systems in QSO spectra with noncoverage factor exceeding a value f_0. We fix z_q and z_c and treat the parameters κ and δ as arbitrary ones. Firstly note that the very fact of absorption-lines registration excludes the case (4) with $f = 1$, i.e the probability of full noncoverage $P(f = 1) = 0$. According to the conditions of Section 2 it is true when $\delta < 1 + \rho(\kappa, z_q, z_c)$. We assume the uniform distribution of the deviation parameter δ and introduce the probability P and probability density function Φ_δ according to the standard definition: $P(\delta' < \delta < \delta' + d\delta') = \Phi_\delta(\delta')\, d\delta'$. In such a way we use:

$$\Phi_\delta(\delta') = \frac{\Theta(1 + \rho(\kappa, z_q, z_c) - \delta')}{1 + \rho(\kappa, z_q, z_c)}, \tag{4}$$

where $\Theta(x)$ is the Heaviside function.

Distribution of κ is determined by distributions of l_c and l_q. Hereafter we assume that $l_q = 0.1$ pc and consider only dependence of the distribution on l_c, which is assumed [7] to be $l_c \sim 0.1 \div 10$ pc. The probability density function Φ_κ for κ-distribution $P(\kappa' < \kappa < \kappa' + d\kappa') =$

$\Phi_\kappa(\kappa')\,d\kappa'$ may be represented as $\Phi_\kappa(\kappa') = \psi(\kappa')\Theta(\kappa'-\kappa_{min})\Theta(\kappa_{max}-\kappa')$, where $\psi(\kappa')$ describes a distribution function over cloud sizes l_c; below we consider the function $\psi(\kappa)$ as a power-law distribution in a general form: $\psi(\kappa) = (1-\beta)\kappa^{-\beta}\,(\kappa_{max}^{1-\beta} - \kappa_{min}^{1-\beta})^{-1}$, uniform distribution over κ corresponding to $\beta = 0$.

We assume that the parameter δ is fixed and calculate the reciprocal function $\kappa(f, \delta, z_q, z_c)$. Actually, the reciprocal function $\kappa(f, \delta)$ at certain z_q and z_c can be represented analytically in five regions completely covering the whole domain of permissible values f and δ. Our calculations show that $\kappa(f, \delta)$ in all cases is a decreasing function of f, i.e. $\kappa < \kappa_0 = \kappa(f_0)$ when $f > f_0$, and $f = 1$ at $\kappa = 0$. On the contrary, the greater δ, the greater $f \leq 1$ can be obtained. Summing up all points discussed above one can write

$$P(f > f_0) = \int_0^\infty \int_0^{\kappa(f_0,\delta,z_q,z_c)} \Phi_\kappa(\kappa')\;\Phi_\delta(\delta')d\kappa' d\delta'. \tag{5}$$

Resulting dependencies $P(f > f_0)$ are given in Figs. 4, 5, where $P(f > 0) = 1 - \mathcal{P}(f = 0) \neq 1$, and $\mathcal{P}(f = 0)$ is a probability of full coverage.

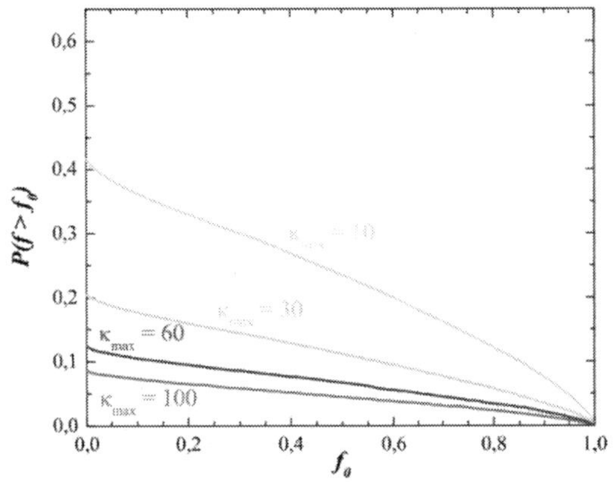

Figure 4. $P(f > f_0)$ **for uniform distribution of cloud sizes at** $z_q = 2.57$, $z_c = 2.33$. **Coloured curves correspond to different** κ_{max}.

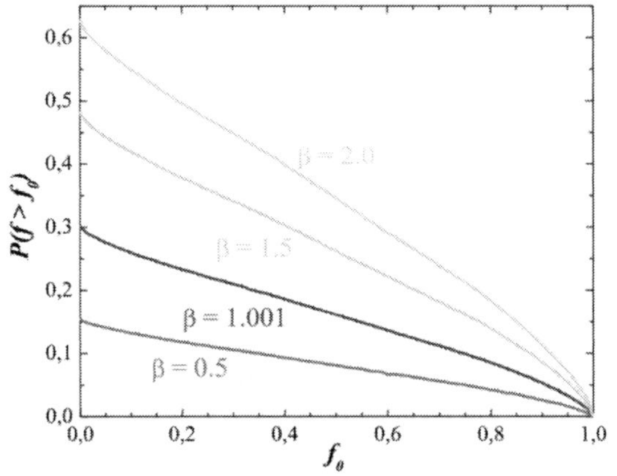

Figure 5. $P(f > f_0)$ **for power-low distribution of cloud sizes at** $z_q = 2.57$, $z_c = 2.33$. **Coloured curves correspond to different** β.

4. Results and conclusions

In the present paper we investigate the probability of the partial coverage of a QSO BLR by an intervening H_2-cloud. Partial coverage is estimated by *noncoverage factor*, i.e. the ratio of a solid angle comprising the whole BLR emission to a solid angle of an uncovered part. A *distribution of noncoverage factor* was calculated for fixed redshifts of the QSO and cloud in the cases of uniform and power-law distributions of cloud sizes.

Using the obtained value, $P(f > f_0)$, one can estimate the probability to reveal the incomplete coverage at a chosen level $f \geq 0.02$ which might affect the absorption lines analysis and obtained physical conditions (see [2]). It is shown that in the most unfavourable case of uniform κ-distribution and sufficiently large $\kappa_{max} = 100$ *the probability to reveal the lowest noncoverage factor is* $P(f > 0.02) \geq 0.08$. In more realistic cases the probability can reach values exceeding 0.3. It means that at the analysis of QSO spectra one should expect the possibility of the incomplete coverage phenomenon.

Two molecular clouds with actually estimated linear sizes has $l_c \sim 0.1 \div 0.2$ pc. It means that appropriate $\kappa \sim 1 \div 2$ at $l_q \sim 0.1$. It is not likely that all ~ 20 registered molecular H_2-clouds at high redshifts correspond to a small sizes. So, it is very important to accumulate some additional information on the distribution of the clouds, as well as the QSO BLRs, over their sizes. In the case of the power-law distribution it would increase minimal $P(f > 0.02)$ estimation. It is good reason to investigate QSO absorption-line spectra with systematic search for the partial coverage. Being revealed the partial coverage phenomenon may lead to essential revision of interstellar medium parameters at high redshift $z \sim 2 - 4$.

4.1. acknowledgments

This work was supported by Ministry of Education and Science of Russian Federation (Agreement No. 8409), by the State Program "Leading Scientific Schools of Russian Federation (grant NSh 4035.2012.2), and by RFBR (grant No. 11-02-01018-a), also SB thanks RF Presedent Programme (grant MK-4861.2013.2).

References

[1] Ivanchik A V, Petitjean P, Balashev S A, Srianand R, Varshalovich D A, Ledoux C and Noeterdaeme P, 2010, *MNRAS*, **404**, 1583
[2] Balashev S A, Petitjean P, Ivanchik A V, Ledoux C, Srianand R, Noterdaeme P and Varshalovich D A, 2011, *MNRAS*, **418**, 357
[3] Balashev S A, Klimenko V V, Ivanchik A V, Ledoux C, Noterdaeme P, Petitjean P, Varshalovich D A, 2013, MNRAS (in preparation)
[4] Dave R, Oppenheimer B D, Katz N, Kollomeier J A and Weinberg D H, 2010, *MNRAS*, **408**, 2051
[5] Kayser R, Helbig P, Schramm T, 1997, *A& A*, **318**, 680
[6] Zel'dovich Ya B, Novikov I D, 1983, *The Structure and Evolution of the Universe*, Univ. Chicago Press, Chicago.
[7] Rauch M, 1998, *ARA& A*, **36**, 267

Journal of Physics: Conference Series **461** (2013) 012047 doi:10.1088/1742-6596/461/1/012047

Stabilization of atoms in strong non-classical electromagnetic fields

A V Bogatskaya[1] and **A M Popov**[1,2]
[1] Department of Physics, Moscow State University, 119991, Moscow, Russian Federation
[2] Institute of Nuclear Physics Moscow State University, 119234, Moscow, Russian Federation

E-mail: annabogatskaya@gmail.com

Abstract. The dynamics of an atom in a strong single mode non-classical electromagnetic cavity field is studied. The reconstruction of the energy spectrum of atom and the effect of the ionization suppression in the presence of non-classical field are discussed. It is found that Kramers – Henneberger mechanism of the ionization suppression is realized in strong nonclassical fields also. For squeezed vacuum states the threshold of the stabilization is approximately three times lower than for classical (coherent) state of the same intensity.

1. Introduction

Recent progress in laser physics makes it possible to obtain so-called squeezed states of electromagnetic wave field [1-4]. Such states with the large amount of energy are an example of macroscopic quantum effect, and hence they are of significant interest from both theoretical and practical points of view. Among different problems of interaction of non-classical electromagnetic fields with matter the case of high-intensity fields seems to be attractive and promising as high-order multiphoton processes should be taken into account [5]. The interest for such fields is based on the fact that the dispersion of the number of photons for strong squeezed states is much larger than the dispersion of photons for the classical electromagnetic field with the same intensity. Hence probabilities or cross sections for different multiphoton processes can be extremely large in such squeezed states of the field. This idea was confirmed for a number of processes, among them are the nonlinear polarization response and high-order nonlinear susceptibilities [6,7], strong-field atomic dynamics and multiphoton ionization [8,9], laser-induced free-free transitions for atomic and negative ion systems [10].

The effect of the ionization suppression (or stabilization) is one of the most intriguing nonlinear effects arising in the interaction of the strong electromagnetic fields with atoms or molecules. This effect was firstly predicted in theoretical papers [11-13] and later it was observed in experiments [14-16]. Two different mechanisms of the stabilization phenomenon can be distinguished. One of them is the interference stabilization proposed in [11] for Rydberg atomic states. Another one is the adiabatic or Kramers – Henneberger (KH) stabilization [12, 13]. This mechanism is predicted to exist for both ground and excited states of a quantum system. The detail description of the stabilization phenomenon can be found in [17, 18]. For both cases the stabilization is the result of the reconstruction of atomic spectra in the presence of the strong external electromagnetic field and formation of the atom

Journal of Physics: Conference Series **461** (2013) 012047 doi:10.1088/1742-6596/461/1/012047

"dressed" by the field and resistant against ionization. In the strong field limit the energy spectrum structure of this "dressed" atom can be differ significantly from the spectrum of the field-free atom.

Till nowadays the problem of the strong-field atomic stabilization was studied only in the classical field limit. The only exception is the recent paper [19], where the dynamics of highly excited (Rydberg) states in the intense non-classical electromagnetic field was analyzed. In particular, it was found that such states are significantly more resistant against ionization if the field mode is in the squeezed vacuum state.

The interaction of the quantum system being initially in the ground state with the strong single mode non-classical electromagnetic field is studied in this paper for the case of the single photon transition from the ground state to the continuum. For the optical frequency band such situation is realized dominantly for negative atomic ions which have the only bound state in the discrete spectrum[1]. The reconstruction of the atomic spectrum in the presence of the non-classical field and the formation of the "dressed" atom are analyzed. The stabilization phenomenon similar to the KH stabilization in the classical field is found to exist for the case of non-classical electromagnetic field states.

2. Quantized electromagnetic field and its interaction with an atomic quantum system

In this paper we will consider only the single mode of the electromagnetic field of frequency ω. The interaction of such a field with an atomic quantum system can be realized experimentally in a cavity or a resonator of micron size. Each radiative field mode can be considered as an oscillator with the Hamiltonian H_f. In the "pseudocoordinate" representation the expression for H_f can be written as

$$H_f = \frac{\hbar\omega}{2}\left(-\frac{\partial^2}{\partial\varepsilon^2}+\varepsilon^2\right). \tag{1}$$

Here ε is the dimensionless electric field strength, $\varepsilon_0 = \sqrt{4\pi\hbar\omega/L^3}$ is the normalization constant and L is the spatial size of the cavity.

From the quantum-mechanical point of view the classical electromagnetic field is a coherent state of the field mode with a large average number of photons [20]. This state can be decomposed over the stationary (Fock) states

$$\phi_c(\varepsilon) = \sum_k \alpha_k |k\rangle, \tag{2}$$

with the amplitudes α_k given by a Poisson statistics.

Among different non-classical states of the field mode the most interesting is the squeezed vacuum state. This state is given by the function

$$\phi_{sq}(\varepsilon) = \frac{1}{\sqrt{\beta\sqrt{\pi}}}\exp\left(-\varepsilon^2/(2\beta^2)\right), \tag{3}$$

where β is the parameter of squeezing. Though average electric field strength in a squeezed vacuum state is equal to zero for any instant of time, the average field energy for the case $\beta << 1$ (or $\beta >> 1$) is high and given by the expression:

$$\langle E\rangle = \hbar\omega\left(\langle k\rangle + 1/2\right). \tag{4}$$

Here $\langle k\rangle = (\beta - 1/\beta)^2/4$ is the average number of photons in a squeezed vacuum state. The decomposition of the squeezed vacuum state over the Fock states

$$\psi_{sq}(\varepsilon) = \sum_k \alpha_{2k}|2k\rangle, \tag{5}$$

[1] One of the examples of such a system is a negative Hydrogen ion. Its ionization potential is equal to 0.75 eV, and the spatial size is of several atomic units.

gives the non-Poisson statistics of photon with

$$|\alpha_{2k}|^2 = \frac{1}{\sqrt{1+\langle k\rangle}} \cdot \frac{(2k)!}{2^{2k}(k!)^2}\left(\frac{\langle k\rangle}{1+\langle k\rangle}\right)^k, \tag{6}$$

and all odd coefficients equal to zero. The comparative analysis of the photon statistics for coherent and squeezed vacuum state is presented in [8, 9]. As the dispersion of the number of photons for coherent and squeezed vacuum states are $D^{coh} = \langle k\rangle$ и $D^{sq} = 2(\langle k\rangle^2 + \langle k\rangle)$, one obtains that for $\langle k\rangle >> 1$ the probability to find the number of photons significantly different from the average value $\langle k\rangle$ is extremely high in the squeezed state.

The interaction of atomic system with a single-mode quantized electromagnetic field is considered in the dipole approximation. The total Hamiltonian of the system "atom + electromagnetic field" is given by

$$H(\vec{r},\varepsilon) = H_0(\vec{r}) + H_f - \varepsilon_0(\vec{d}\vec{e}_\lambda)\varepsilon, \tag{7}$$

where H_0 is the atomic Hamiltonian with the only stationary state in discrete spectrum $|g\rangle$ and the energy E_g, and the continuum states $|E\rangle$; $\vec{r}$ is electronic coordinate; $\vec{d}$ is the dipole moment and $\vec{e}_\lambda$ is the polarization of the field. Further, we will suppose that $\hbar\omega > I$ ($I = -E_g$ is the ionization potential).

The total time-dependent wave-function of the system $|\Psi\rangle$ we will expand in terms of interaction-free eigenstates of the atomic and light field Hamiltonians:

$$\Psi(\vec{r},\varepsilon,t) = \sum_k C_{g,k}(t)|g\rangle|k\rangle \exp\left(-\frac{i}{\hbar}(E_g + E_k)t\right) + \sum_k \int dE \cdot C_{E,k}(t)\cdot|E\rangle|k\rangle \exp\left(-\frac{i}{\hbar}(E + E_k)t\right), \tag{8}$$

Here $E_k = \hbar\omega(k + 1/2)$ is the energy of the field subsystem, and coefficients $C_{g,k}$ ($C_{E,k}$) are the amplitudes to find the atom in a state $|g\rangle$ ($|E\rangle$), and the field in a state $|k\rangle$. Substituting (8) to the nonstationary Schroedinger equation, one obtains:

$$\begin{aligned} i\hbar\frac{dC_{E,k}}{dt} &= -\frac{\varepsilon_0}{\sqrt{2}} d_{Eg} \exp\left(\frac{i}{\hbar}(E - E_g)t\right)\left\{C_{g,k+1}\sqrt{k+1}\exp(-i\omega t) + C_{g,k-1}\sqrt{k}\exp(i\omega t)\right\} \\ &- \frac{\varepsilon_0}{\sqrt{2}}\int dE' d_{EE'} \exp\left(\frac{i}{\hbar}(E - E')t\right)\left\{C_{E',k+1}\sqrt{k+1}\exp(-i\omega t) + C_{E',k-1}\sqrt{k}\exp(i\omega t)\right\} \\ i\hbar\frac{dC_{g,k}}{dt} &= -\frac{\varepsilon_0}{\sqrt{2}}\int dE' d_{gE'} \exp\left(\frac{i}{\hbar}(E_g - E')t\right)\left\{C_{E',k+1}\sqrt{k+1}\exp(-i\omega t) + C_{E',k-1}\sqrt{k}\exp(i\omega t)\right\} \end{aligned} \tag{9}$$

Here d_{gE}, $d_{EE'}$ are the atomic dipole matrix elements.

3. AC Stark shift in the non-classical electromagnetic field

Let us start from the study of the reconstruction of the energy spectrum and the atomic potential by the non-classical light field. We suppose that for initial instant of time the atomic state is $\varphi_g(\vec{r}) = |g\rangle$, and the state of the field mode is $\phi^{field}(\varepsilon) = \sum_k \alpha_k|k\rangle$ with definite values of α_k. In the second order of perturbation theory for low enough time of consideration one derives from the set of equations (9)

$$C_{gk}^{(2)} = -\frac{i}{\hbar}\frac{\varepsilon_0^2}{2}\alpha_k \int \frac{|d_{Eg}|^2}{\hbar}\left\{\frac{2\omega_{Eg}}{\omega_{Eg}^2 - \omega^2}k + \frac{1}{\omega_{Eg} + \omega}\right\}dE \cdot t,$$

and hence the function of the system "atom + electromagnetic field" can be written in a form:

$$\Psi(\vec{r},\varepsilon,t) \approx \sum_k \alpha_k \left(1 - \frac{i}{\hbar}\frac{\varepsilon_0^2}{2}\chi_k t\right)|g\rangle|k\rangle \exp\left(-\frac{i}{\hbar}(E_g + E_k)t\right) + continuum, \tag{10}$$

where $\chi_k = \int \frac{|d_{Eg}|^2}{\hbar}\left\{\frac{2\omega_{Eg}}{\omega_{Eg}^2 - \omega^2}k + \frac{1}{\omega_{Eg} + \omega}\right\}dE$, and «*continuum*» implies the superposition of states to find electron in continuum. From (10) one obtains the following expression for the atomic wave function of the bound state

$$\psi(\vec{r},t) \approx \sum_k |\alpha_k|^2 \left(1 - \frac{i}{\hbar}\frac{\varepsilon_0^2}{2}\chi_k t\right)\varphi_g(\vec{r})\exp\left(-\frac{i}{\hbar}E_g t\right) \approx \varphi_g(\vec{r})\exp\left(-\frac{i}{\hbar}(E_g + \delta E_g)t\right),$$

where

$$\delta E_g = -\frac{\varepsilon_0^2}{2}\int \frac{|d_{Eg}|^2}{\hbar}\left\{\frac{2\omega_{Eg}}{\omega_{Eg}^2 - \omega^2}\langle k\rangle + \frac{1}{\omega_{Eg} + \omega}\right\}dE$$

is the shift of the energy of the bound state $\varphi_g(\vec{r})$. It is important to notice, that the shift obtained is defined only by the average number of quanta $\langle k\rangle$ and do not depend on the photon statistics. The shift appears to exist also for the case of $\langle k\rangle = 0$. It is well known Lamb shift, that appears from interaction of the atomic system with the electromagnetic vacuum. For intense fields ($\langle k\rangle >> 1$) this effect can be neglected. Then, taking into account the expression $\langle \varepsilon^2\rangle = 2\langle k\rangle\varepsilon_0^2$ [8, 9], we derive:

$$\delta E_g = -\chi\langle\varepsilon^2\rangle/4, \quad \chi = \int \frac{|d_{Eg}|^2}{\hbar}\frac{2\omega_{Eg}}{\omega_{Eg}^2 - \omega^2}dE, \tag{11}$$

where χ is the atomic susceptibility. The expression (11) corresponds to the semiclassical limit of the problem [17], when the atomic subsystem is considered as a quantum one, while the field is supposed to be classical. Indeed, if the light mode is in the coherent state with $\langle k\rangle >> 1$, then $\langle\varepsilon^2\rangle \cong \langle\varepsilon\rangle^2$ and expression (11) can be written in a typical form

$$\delta E_g = -\chi\langle\varepsilon\rangle^2/4, \tag{12}$$

where $\langle\varepsilon\rangle$ is the averaged over quantum state amplitude value of the electric field strength.

4. Ionization suppression in a strong non-classical field

In the classical field limit the stabilization of atoms is a result of the reconstruction of the atomic spectrum by a strong electromagnetic field and the formation of the "dressed" atom resistant against ionization. For the case of single photon transition to the continuum this "dressed" atom exists in a form of KH atom [17,18]. The emergence of the KH potential is typically associated with the nearly free oscillations of the atomic electron in a strong laser field with the amplitude $a_e = e\varepsilon/m\omega^2$. It was found [21], that in relatively weak fields corresponding to the quiver motion amplitude $a_e \le a_0$ (a_0 is the typical size of the quantum system) the structure of the energy spectrum of the KH atom corresponds to the field-free atomic spectrum with AC Stark shift taken into account. It was demonstrated in section 3, that the AC Stark shift in the quantized field does not feels the photon statistics and depends only on the average number of photons. It means that at least in such intensity limit the reconstruction of the atomic energy spectrum in a quantized field can be interpreted in terms of appearance of the KH atom. As the intense non-classical field can be found in the state with zero value of the average electric field strength, we conclude that the emergence of the KH potential do not result from the quiver motion of the atomic electron.

It was found in [21, 22] that for the classical electromagnetic field limit the KH stabilization can be interpreted in terms of coherent repopulation of the atomic continuum states by free-free transitions and the interference of the different multiphoton order transitions from the initial bound state $|g\rangle$ to the final continuum state $|E\rangle$. Similar approach can be developed for non-classical field states also. In this paper we restrict ourselves to the consideration of the interference of the direct single-photon transition to the continuum and the three-photon transition via the intermediate continuum state (figure 1). In this case the total amplitude of the single- and three-photon transition to the continuum state $|E\rangle$ and the number of quanta in the field mode equal to k can be written in a form:

$$C_{E,k} = C^{(1)}_{E,k} + C^{(3)}_{E,k} = -\frac{i}{\hbar}\frac{\varepsilon_0}{\sqrt{2}}\alpha_{k+1}\sqrt{k+1}\cdot \left\{ d_{E,i} + \frac{\varepsilon_0^2}{2} k \int dE'\, dE'' \frac{d_{EE''}\, d_{E'',E'}\, d_{E',i}}{\hbar^2} \frac{1}{(\omega_{E',i}-\omega)(\omega_{E'',i}-2\omega)} \right\} \int \exp(i(\omega_{E,i}-\omega)t)\,dt. \quad (13)$$

The probability of ionization can be derived for (13) as a result of summation over photon number states and integration over energy spectrum of photoelectrons $W_i = \sum_k \int |C_{E,k}|^2 dE$.

Taking into account that $\sum_k |\alpha_k|^2 k^2 = \langle k^2 \rangle$ and applying the pole approximation [17] for calculation of the integral in (13) we obtain the expression for the ionization rate of quantum system

$$\dot{W} = \frac{2\pi}{\hbar}|d_{E,i}|^2 \frac{\varepsilon_0^2}{2}\left[\langle k\rangle - \pi^2\varepsilon_0^2 |\langle E|d|E+\hbar\omega\rangle|^2 (\langle k^2\rangle - \langle k\rangle)\right]. \quad (14)$$

Here $E = \hbar\omega - I$ and I is the ionization potential.

The result obtained is the generalization of the expression from [22] for the arbitrary state of the field. If the state of the field is the coherent one, the expression for the ionization rate takes the form

$$\dot{W} = \frac{2\pi}{\hbar}|d_{E,i}|^2 \frac{\varepsilon_0^2}{2}\langle k\rangle\left[1 - \pi^2\varepsilon_0^2 |\langle E|d|E+\hbar\omega\rangle|^2 \langle k\rangle\right], \quad (15)$$

which can be easily obtained in the frames of the semiclassical theory [22]. The appearance of the stabilization phenomenon follows from (15) and the threshold of the stabilization can be estimated as

$$\langle k\rangle^{(threshold)} \approx \frac{L^3}{4\pi^3\hbar\omega|\langle E|d|E+\hbar\omega\rangle|^2}. \quad (16)$$

Estimating the value of the free-free matrix element transition as $\langle E|d|E+\hbar\omega\rangle \sim ea_0/\hbar\omega$, where a_0 is the typical size of the atomic system, we find that

$$\langle k\rangle^{(threshold)} \approx \frac{1}{4\pi^3}\frac{\hbar\omega}{Ry}(L/a_0)^3, \quad (17)$$

where $Ry \approx 13.6$ eV. For example, for the cavity of the micron size and $\hbar\omega = 2$ eV we have $\langle k\rangle^{(threshold)} \sim 10^{10}$, that is approximately corresponds to the intensity $\sim 10^{14}$ W/cm^2.

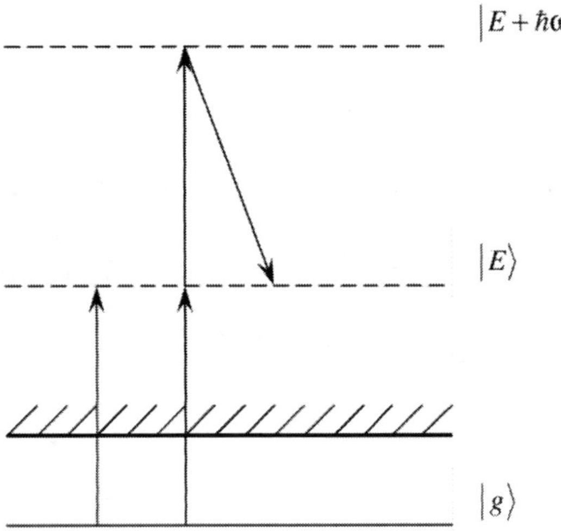

Figure 1. A scheme of bound-continuum and continuum-continuum transitions taken into account in the model.

Different situation arises for the non-classical field. For example, for the squeezed vacuum state we obtain from (15)

$$\dot{W} = \frac{2\pi}{\hbar}\left|d_{E,i}\right|^2 \frac{\varepsilon_0^2}{2}\langle k\rangle\left[1-\pi^2\varepsilon_0^2\left|\langle E|d|E+\hbar\omega\rangle\right|^2(3\langle k\rangle+1)\right]. \quad (18)$$

It means that for the squeezed vacuum state the stabilization threshold is approximately three times lower in comparison with the coherent state of the same intensity.

5. Conclusions

Tthe phenomenon of the field reconstruction and formation of the “dressed” atom (KH atom) in the single mode cavity field is studied in the paper. It is found that for single-photon bound of the discrete spectrum and continuum this reconstruction depends on only the average number of quanta but not the photon statistic of the field mode. It is demonstrated that the KH stabilization appears to exist in the non-classical light also as a result of nonlinearity of the matrix element transition to the continuum. For squeezed field states the threshold of the stabilization can be several times lower than for classical field.

It should be emphasized also that decreasing of microcavity size leads to the decreasing of the number of photons necessary for stabilization. In principle, it is possible to suppose the situation when the stabilization will result only from few-photon interaction with an atom. It follows from (17) that in such a case the size of the cavity should be close to the size of the quantum system. For visible or IR frequency band it means that the size of the quantum system should be significantly greater than the atomic size. Probably, such situation can be realized in the modern nanostructure objects (quantum dots), where the Wannier-Mott exciton can serve as an object stable to ionization [24,25].

Acknowledgments

This work was supported by Russian Foundation for Basic Research (grant 12-02-00064) by the Dynasty Foundation (program of support for students).

References

[1] Slusher R E, Holleberg L W, Yurke B, Mertz J C 1985 *Phys. Rev. Lett.* **55** 2409
[2] Ling-An Wu, Kimble H J, Hall J H and Huifa Wu 1986 *Phys. Rev. Lett.* **57** 2520
[3] Shangqing L and Yansong C 1995 *J. Opt. Soc. Am. B* **12** 829
[4] Brattke S, Varcoe B T H and Walther H 2001 *Phys. Rev. Lett.* **86** .3534
[5] Bykov V P 1991 *Intense Laser Phenomena and Related Subjects* (Singapore: World-Scientific)
[6] Janszky J and Yushin Y 1989 *Phys. Lett. A* **137** 451
[7] Masalov A V 1991 *Opt. Spectrosc.* (USSR) **70** (3) 382
[8] Popov A M and Tikhonova O V 2002 *JETP* **95** 844
[9] Popov A M and Tikhonova O V 2006 *Laser Phys. Lett.* **3** 195
[10] Burenkov I A and Tikhonova O V 2010 *J. Phys. B* **43** 235401
[11] Fedorov M V and Movsesian A M 1988 *J. Phys. B* **21** L155
[12] Gavrila M and Kaminski J Z 1984 *Phys. Rev. Lett.* **52** 614
[13] Pont M and Gavrila M 1990 *Phys. Rev. Lett.* **65** 2362
[14] Jones R R, Schumacher D and Bucksbaum P 1993 *Phys. Rev. A* **47** R49
[15] Hoogenraad J, Vrijen R and Noordam L 1994 *Phys. Rev. A* **50** 4133
[16] van Druten N J, Constantinescu R C, Schins J M et al 1997 *Phys. Rev. A* **55** 622
[17] Fedorov M V 1998 *Atomic and free electrons in a strong light field* (Singapore: World-Scientific)
[18] Popov A M, Tikhonova O V and Volkova E A 2003 *J. Phys. B* **36** R125
[19] Sharapova P R and Tikhonova O V 2012 *Quantum Electronics* **42** 197
[20] Glauber R J 1963 *Phys.Rev.* **131** 2766
[21] Popov A M, Tikhonova O V and Volkova E A 2000 *Laser Phys.* **10** 188
[22] Tikhonova O V, Popov A M and Fedorov M V 2002 *Phys. Rev. A* **65** 053404
[24] Agranovich V, Benisty H and Weisbuch C 1997 *Solid State Commun.* **102** 631
[25] Agranovich V, La Rocca G and Bassani F 1997 *JETP Lett.* **66** 748

AUTHOR INDEX

AUTHOR INDEX